The Continental Crust:
its Composition and Evolution

GEOSCIENCE TEXTS

SERIES EDITOR
A. HALLAM
Lapworth Professor of Geology
University of Birmingham

Principles of Mineral Behaviour
A. PUTNIS and J. D. C. MCCONNELL

An Introduction to Ore Geology
A. M. EVANS

Sedimentary Petrology: an Introduction
MAURICE E. TUCKER

An Introduction to Geophysical Exploration
P. KEAREY and M. BROOKS

GEOSCIENCE TEXTS

The Continental Crust: its Composition and Evolution

An Examination of the Geochemical Record Preserved in Sedimentary Rocks

STUART ROSS TAYLOR
MA, DSc(Oxon), MSc(NZ), PhD(Indiana), FAA

SCOTT M. McLENNAN
MSc (Western Ontario), PhD(ANU)

Research School of Earth Sciences
The Australian National University
Canberra, Australia

BLACKWELL SCIENTIFIC PUBLICATIONS

OXFORD LONDON EDINBURGH
BOSTON PALO ALTO MELBOURNE

© 1985 by
Blackwell Scientific Publications
Editorial offices:
Osney Mead, Oxford, OX2 0EL
8 John Street, London, WC1N 2ES
23 Ainslie Place, Edinburgh, EH3 6AJ
52 Beacon Street, Boston
  Massachusetts 02108, USA
744 Cowper Street, Palo Alto
  California 94301, USA
107 Barry Street, Carlton
  Victoria 3053, Australia

All rights reserved. No part of this publication
may be reproduced, stored in a retrieval
system, or transmitted, in any form or by any
means, electronic, mechanical, photocopying,
recording or otherwise without the prior
permission of the copyright owner

First published 1985

Set, printed and bound
in Great Britain by
Adlard & Son Ltd
Dorking, Surrey

DISTRIBUTORS

USA and Canada
  Blackwell Scientific Publications Inc
  P O Box 50009, Palo Alto
  California 94303

Australia
  Blackwell Scientific Book Distributors
  31 Advantage Road, Highett
  Victoria 3190

British Library
Cataloguing in Publication Data

Taylor, Stuart Ross
  The continental crust.—(Geoscience texts)
  1. Earth—Crust
  I. Title II. McLennan, Scott M.
  III. Series
  551.3'3   QE511

ISBN 0-632-01148-3

Dedicated to
Noël,
Susanna, Judith and Helen
S.R.T.

and to
Fiona
S.M.M.

Contents

Preface, xi

Acknowledgements, xv

1 Some Perspectives, 1
   1.1 Basic data, 1
   1.2 COCORP and crustal structure, 5
      Notes and references, 8

2 The Present Upper Crust, 9
   2.1 The unique upper crust, 9
   2.2 Methods for determining composition, 9
   2.3 Some geochemical influences on sedimentary rock composition, 12
      2.3.1 Weathering, 13
          Water-rock partition coefficients, 14
      2.3.2 Erosion, 16
      2.3.3 Physical transport and deposition, 17
      2.3.4 Ocean water chemistry, 22
          Residence times, 22
      2.3.5 Diagenesis and metamorphism, 23
      2.3.6 Sedimentary rocks as crustal samples, 24
   2.4 Rare earth elements in sedimentary rocks, 29
      2.4.1 The post-Archean record, 29
      2.4.2 Mineralogical controls, 35
      2.4.3 Effects of weathering, 38
      2.4.4 Effects of diagenesis and metamorphism, 41
      2.4.5 Origin and significance of Eu-anomalies in sedimentary rocks, 42
   2.5 Other upper crustal samples, 42
      2.5.1 Loess and crustal composition, 42
      2.5.2 Tektites, 45
   2.6 The chemical composition of the upper crust, 45
      2.6.1 The rare earth elements, 45
      2.6.2 The heat producing elements (K, Th, U), 46
      2.6.3 Rubidium and strontium, 48
      2.6.4 Other elements, 48
      2.6.5 Heat flow and the thickness of the upper crust, 49
      2.6.6 Mineralogy of the upper crust, 49
   2.7 Summary, 50
      Notes and references, 52

3 Models of Total Crustal Composition, 57
   3.1 Boundary conditions from the upper crustal compositions, 57
   3.2 Heat flow and radioactive element abundances, 58
   3.3 The andesite model, 59
      3.3.1 The Ni and Cr problem, 62
   3.4 Other primary mantle sources, 64
   3.5 The Archean 'bimodal' basic-felsic igneous model, 65
   3.6 A proposed bulk crustal composition, 66
      3.6.1 Some comparisons, 69
   3.7 Summary, 70
      Notes and references, 71

4 The Lower Crust, 73
   4.1 Sampling problems, 73
   4.2 The granulite facies terrains, 75
        4.2.1 The Lewisian gneiss complex, 77
        4.2.2 Geochemical aspects of other granulite terrains, 80
   4.3 Xenoliths, 81
   4.4 Basic or acidic lower crusts?, 84
   4.5 Nd and Sr isotopic studies, 86
   4.6 Electrical conductivity, 88
   4.7 Proposed compositions for the lower crust, 90
   4.8 Summary, 93
        Notes and references, 93

5 Uniformity of Crustal Composition with Time, 96
   5.1 Measurement of secular variations, 96
   5.2 The sedimentary record, 96
        5.2.1 Secular trends, 96
        5.2.2 Major elements, 97
        5.2.3 Trace elements, 99
   5.3 Role of sedimentary recycling, 105
   5.4 Nd model ages of sediments, 110
   5.5 Heat flow provinces and crustal uniformity, 112
   5.6 Summary, 114
        Notes and references, 114

6 Greywackes: Provenance and Tectonic Significance, 117
   6.1 Greywackes and the upper crust, 117
   6.2 Deep-sea sands, 117
        6.2.1 Tectonic settings, 118
        6.2.2 Petrography, 118
        6.2.3 Major element chemistry, 119
        6.2.4 Trace element data, 120
   6.3 Phanerozoic greywackes, 121
        6.3.1 Tectonic settings, 122
        6.3.2 Petrography, 123
        6.3.3 Major element chemistry, 124
        6.3.4 Trace element data, 127
   6.4 Archean greywackes, 129
        6.4.1 Petrography, 131
        6.4.2 Major element chemistry, 131
        6.4.3 Trace element data, 133
        6.4.4 Provenance, 133
   6.5 Comparisons, 133
        6.5.1 Petrography, 133
        6.5.2 Major element chemistry, 136
        6.5.3 Trace element data, 138
        6.5.4 Early and late Archean greywackes, 138
        6.5.5 Archean turbidites as upper crustal samples, 139
   6.6 Summary, 139
        Notes and references, 140

7 The Archean Crust, 143
   7.1 The problems, 143
   7.2 The Archean record, 143
   7.3 Low-grade terrains, 143
        7.3.1 Volcanic rocks, 144
        7.3.2 Sr and Nd isotopes in Archean volcanics, 150
        7.3.3 Sedimentary rocks, 151
        7.3.4 The basement of greenstone belts, 152

7.4 High-grade terrains, 153
    7.4.1 Relationships between high-grade and low-grade terrains, 154
7.5 Cratonic terrains, 154
7.6 Tectonic models, 155
7.7 Crustal thickness, 156
7.8 Regional geochemistry, 157
    7.8.1 Early Archean greenstone belts, 162
        (a) Pilbara Block, Western Australia, 164
        (b) Barberton Mountain Land, South Africa, 164
    7.8.2 Late Archean greenstone belts, 154
        (a) Yilgarn Block, Western Australia, 165
        (b) Yellowknife Supergroup, Slave Province, Canada, 167
        (c) South Pass Greenstone Belt, Wind River Range, USA, 170
        (d) Other areas, 171
    7.8.3 High-grade terrains, 171
        (a) Godthâb–Isukasia Region, West Greenland, 171
        (b) Other areas, 173
7.9 Trace element systematics, 173
    7.9.1 Rare earth elements (REE), 174
    7.9.2 La–Th–Sc systematics, 174
    7.9.3 Ferromagnesian trace elements, 175
    7.9.4 A mixing model, 176
    7.9.5 Comparisons between volcanic and sedimentary sequences, 179
    7.9.6 Comparisons between early and late Archean greenstone belts, 179
7.10 Composition of the Archean crust, 180
    7.10.1 The upper crust, 180
    7.10.2 Total crustal compositions, 182
    7.10.3 Differentiation of the Archean crust, 184
7.11 Summary, 185
    Notes and references, 186

8 The Archean–Proterozoic Boundary, 191

8.1 The Archean–Proterozoic boundary defined, 191
8.2 Geochemistry of early Proterozoic sedimentary rocks, 192
    8.2.1 Huronian Supergroup, 192
    8.2.2 Wopmay Orogen, 196
    8.2.3 Pine Creek Geosyncline, 196
    8.2.4 Hamersley Basin, 200
8.3 Late Archean igneous activity and crustal evolution, 203
    8.3.1 Recycling and the sedimentary mass, 203
8.4 Late Archean of Southern Africa, 204
8.5 Early Proterozoic greenstone belts, 205
8.6 Summary, 206
    Notes and references, 206

9 Models for the Origin of the Continental Crust, 209

9.1 Introduction, 209
9.2 The Archean crust, 209
9.3 The Archean–Proterozoic transition, 213
9.4 The post-Archean crust, 216
9.5 The formation of granites, 217
    9.5.1 Early Archean granites, 218
    9.5.2 Late Archean potassium granites, 218
    9.5.3 Proterozoic granites, 220
    9.5.4 Phanerozoic granites, 221
9.6 Anorthosites and crustal growth, 224
    9.6.1 Archean anorthosites, 225
    9.6.2 Proterozoic anorthosites, 226
9.7 The development of the present-day plate-tectonic regime, 229
9.8 Summary, 229
    Notes and references, 230

10 The Growth Rate of the Crust, 233
- 10.1 The nature of the problem, 233
- 10.2 Models of crustal growth, 233
- 10.3 Geological constraints, 235
  - 10.3.1 The freeboard argument, 235
  - 10.3.2 Sedimentation rates and crustal recycling, 239
  - 10.3.3 Measured and inferred growth rates, 243
- 10.4 Istopic constraints, 244
  - 10.4.1 Early crustal growth (>3.2 Ae), 244
  - 10.4.2 Late Archean crustal growth (3.2–2.5 Ae), 246
  - 10.4.3 Post-Archean crustal growth (<2.5 Ae), 248
- 10.5 Summary: a preferred model of crustal growth, 249
  - Notes and references, 251

11 Crust–Mantle Relationships, 256
- 11.1 The origin of the earth, 256
- 11.2 The primitive terrestrial mantle, 260
  - 11.2.1 Upper and lower mantle compositions, 261
  - 11.2.2 Primitive mantle composition, 263
- 11.3 Effect of crustal extraction, 266
  - 11.3.1 Extent of terrestrial differentiation, 267
- 11.4 Crustal recycling: evidence from island-arc volcanics, 268
  - 11.4.1 Geochemical evidence, 268
  - 11.4.2 Pb, Sr, Nd and Hf isotopes, 269
  - 11.4.3 ^{10}Be evidence, 271
  - 11.4.4 Th/U ratios, 272
- 11.5 The oceanic crust, 272
- 11.6 Summary, 275
  - Notes and references, 275

12 Early Planetary Crusts, 278
- 12.1 The 800 million year gap, 278
- 12.2 Primary and secondary crusts, 278
- 12.3 The lunar crust, 279
- 12.4 The Mercurian crust, 282
- 12.5 The Martian crust, 284
- 12.6 Venus, 286
- 12.7 The Galilean and Saturnian satellites, 287
- 12.8 Planetary constraints on expanding Earth hypotheses, 290
- 12.9 Meteorite flux and the cratering record, 291
- 12.10 The early Earth, 293
- 12.11 Summary, 294
  - Notes and references, 295

Appendices, 297
1. Reference abbreviations, 297
2. Chondritic rare earth element normalizing factors, 298
3. Ionic radii for cations in Angstrom units, 299
4. Isotopic notation, 300

Author Index, 302

Subject Index, 307

Preface

Many trace elements are highly concentrated in the continental crust of the Earth, relative to estimates of whole earth composition. The crust thus possesses a geochemical importance out of proportion to its mass, and its origin and composition are important in geochemistry and cosmochemistry. Because of this, geochemists need to produce estimates of crustal composition. Although the task is not easy, such workers should not be dismayed by 'The subtilty of nature, the secret recesses of truth, the obscurity of things, the difficulty of experiment, the implication of causes and the infirmity of man's discerning power' as Sir Francis Bacon remarked [1].

The geological processes responsible for generating the continental crust operate on a scale which has involved much of the mantle. Taking a broad overview is always difficult in geology, where, for practical reasons, the minutae of the subject are mostly investigated. We should recall, in this context, the wise advice of Sir William Hamilton who commented in 1773, after studying the Italian volcanoes, that 'We are apt to judge of the great operations of Nature on too confined a plan' [2].

In 1964, one of the authors was led by the apparent uniformity of the rare earth element patterns in sedimentary rocks to use this as a guide to their average abundances in the upper crust [3]. This concept formed the basis for constructing an abundance table based on values for common igneous rocks which the sediments were sampling. Such information then led to a consideration of processes which might generate the crust. Present-day crustal growth is most readily accounted for by volcanism in orogenic zones and island arcs. Thus the 'Andesite Model' arose, and survived for a number of years as a testable model [4].

The question of secular variations in crustal composition could also be addressed using the REE abundances in sedimentary rocks and work began on this problem in 1969 (although somewhat delayed by the return of lunar samples and their subsequent investigation). Meanwhile, increasing knowledge of Archean terrains [5] led to the realization that the Archean crust was different in composition, was probably generated by different processes and may comprise a significant, possibly major fraction of the material making up the continental crust. The recognition of a major break in upper crustal composition between the Archean and post-Archean record became apparent from the REE studies of sedimentary rocks [6]. These investigations have now evolved to the point where a general synthesis appears to be useful, and this forms one rationale for the present book.

We have not attempted to compose a review in the customary usage of the term. The amount of information about the crust, following 200 years of geological investigation, probably exceeds the ability of individuals to synthesize. Instead we have tried to resolve some of the problems of composition and evolution of the crust by leaning heavily on our own research work. This is biased toward studies involving trace element geochemistry, particularly of sedimentary rocks. In a book which deals with many topics we have endeavoured to resist the temptations of digression. For example, we have not discussed in detail such interesting topics as the origin of orogenic andesites, for which adequate recent treatments exist. Problems of crustal structure lie outside our area of expertise and are treated only in passing. In addition, a very large amount of new structural information is being accumulated at present (see Section 1.2).

Although the crust is complex in detail, as every geologist engaged in field work is acutely aware, the processes of erosion and subsequent deposition of sediments derived from the continents provide a sampling of the exposed crust. A primary theme in this book is that some of the chemical elements (notably the rare-earth elements, but also including thorium and scandium) are transferred virtually quantitatively from the crust into clastic sediments during erosion of the upper crust and as a group are not fractionated significantly during the processes of sedimentation and diagenesis. Efficient mixing from source rocks of diverse composition appears to occur during these processes. Accordingly, the abundances of these elements can provide an index of upper crustal composition, and so may be used to trace the evolution of the upper continental crust.

Although there are many ways in which the topics addressed in this book might have been organized, we begin with an assessment of the composition of the presently exposed upper crust. This leads us in turn to consider possible models for the bulk crustal composition, which must contain lower concentrations of the heat producing elements. Such estimates enable us to infer possible compositions for the lower crust, which forms, together with the lower mantle, one of the two least understood zones of the Earth.

The next question to be addressed concerns the uniformity of crustal composition with time. Has there been any secular change? It is concluded that the present composition has persisted for about 2500 million years, but that the Archean crust was different in composition. Because of the importance of this conclusion, an extensive discussion is given of the Archean sedimentary record. In particular, the question whether Archean sedimentary rocks are providing an overall average sample of the crust at that time is addressed at length. The nature of the Archean–Proterozoic boundary, a fundamental discontinuity in the history preserved in sedimentary rocks is next examined, followed by a discussion of models for the derivation of Archean and Post-Archean crusts.

A fundamental question concerns the rate of growth of the crust. Our assessment is that crustal growth was neither early nor uniform, but proceeded in an episodic fashion, with much of the growth occurring between about 3000 and 2500 million years ago. Recycling of the crust through the mantle appears to have been minor, from a wide variety of observations. One conclusion that emerges from this study is that the rates of secular change and of crust building processes proceed on time scales of billions (10^9) of years. Thus the author of Psalm 90 who commented that 'a thousand years... are but as yesterday when it is past, and as a watch in the night' [7] was underestimating time-scales for the evolution of the Earth by several orders of magnitude!

Since the crust is derived ultimately from the mantle of the Earth, it is relevant to consider both the composition of the primitive mantle and the effects of crustal generation on mantle composition. At least 30% of the mantle has been affected by this process. Finally, it is of interest to consider the question of early (pre 3.8 aeon) crusts on the Earth, and also to compare the continental and oceanic crusts of the Earth with the surface of the Moon, Mars, Mercury, Venus and the Galilean satellites of Jupiter.

The contemplation of sediments lying on the surface of the Earth thus leads by rather short steps to the consideration of quite fundamental questions, reinforcing the old adage that, in science, nothing occurs in isolation, and that all phenomena are linked together. Indeed, H. C. Sorby perceived this in 1908 when he wrote that 'Possibly many may think that the deposition and consolidation of fine-grained mud must be a very simple matter, and the results of little interest. However... it is soon found to be so complex a question... that one might feel inclined to abandon the enquiry, were it not that so much of the history of our rocks appears to be written in this language' [8].

Because of the wide number of topics addressed, we have endeavoured to provide an adequate reference list, essential in a work of this kind. Without proper references to the literature, the reader has to accept unsupported statements as dogma, or revealed truth. Accordingly, we have attempted to list appropriate references, so that the interested reader can pursue topics in depth. In general we have referred to the latest literature, partly to save space, and because references to earlier works can be found therein. We have adopted a numbering system for citations in the text, placing the references at the end of each chapter, listing authors, year, title, journal reference, volume number and initial or cited page. Papers involving more than two authors are cited by the initial author followed by '*et al.*' A listing of the more obscure abbreviations for journal titles is given in Appendix 1. Our literature survey has been carried through to January 1984 with some additions to March 1984. We have also attempted to evaluate the literature and to provide our interpretation on controversial issues. It is no service to the reader to merely record the existence of controversies without comment since these commonly deal with important issues.

Books frequently are not subject to the benefits of peer review. In order to avoid what we regard as a serious deficiency in many works, we have prevailed on a number of our colleagues to read the material in this book when it was in draft form. Their invaluable comments saved us from various errors and we hope have made the book a more reliable source of information (see

Acknowledgments). A majority of the trace element data reported in this book for sedimentary rocks come from our own laboratory. This is not intentional, but is forced upon us by various factors. These include the rather recent development of interest in these rocks by geochemists, so that few other appropriate analytical data have yet appeared in the literature. We have chosen to report all major element data as volatile-free ($-H_2O$, CO_2, etc.) in order to facilitate comparative studies. All iron values are reported as FeO, to overcome the uncertainties associated with oxidation states in surficial environments. In the various tables, a dash indicates no analysis is available and an asterisk indicates that the element concentration was below detection limits.

Our interest in the problem of the composition and evolution of the continental crust arose partly from a desire to test whether secular variations existed. The results of our studies have led to the conclusion that there is a major dichotomy between the Archean and later times both in the composition and mechanism of continental growth. Although we have long been adherents to a Lyellian approach to geological problems, it seems that the presently observable processes for continental growth extend back only to the base of the Proterozoic. Just as it has proven risky to infer lunar petrogenesis from terrestrial analogues, so it also seems hazardous to extrapolate from presently observable processes into that 'dark backward and abysm of time' [9] which we refer to as the Archean.

Notes and references

1 This translation from Latin of aphorism 92 from '*Novum organum*' by Francis Bacon (1620) is given by Peter Medawar (1979) in '*Advice to a Young Scientist*'. Pan Books. p. 6.
2 Hamilton, W. (1773) *Observations on Mt. Vesuvius, Mt Etna and other Volcanoes*. T. Cadell, London. p. 161. (Better known to history as the husband of Emma, Lady Hamilton, this distinguished naturalist and Fellow of the Royal Society wrote one of the first modern works on volcanology.)
3 Taylor, S.R. (1964) Abundance of chemical elements in the continental crust: A new table. *GCA*, **28**, 1273.
4 Wilson, J.T. (1952) Orogenesis as the fundamental geological process. *Trans. AGU*, **33**, 444; Taylor, S.R. (1967) The origin and growth of continents. *Tectonophys.*, **4**, 17; Taylor, S.R. (1977) Island arc models and the composition of the continental crust. *AGU Ewing Series* **I**, 325.
5 Terrain is used throughout this book in preference to the obsolescent term terrane. Bates, R.L. & Jackson, J.A. (1980) *Glossary of Geology*. (2nd Ed.) AGI, Falls Church, Virginia. p. 645.
6 Taylor, S.R. & McLennan, S.M. (1981) The composition and evolution of the continental crust: Rare earth element evidence from sedimentary rocks. *Phil. Trans. Roy. Soc.* (Lond.), **A301**, 381.
7 Psalm 90, verse 4, The Bible, King James Version, Oxford University Press.
8 Sorby, H.C. (1908) On the application of quantitative methods to the structure and history of rocks. *QJGS*, **64**, 190. We are indebted to Paul Potter, Barry Maynard and Wayne Pryor (*Sedimentology of Shale*. Springer-Verlag, 1980) for drawing attention to this quotation. (Sorby is famous for his comment that meteoritic chondrules had formed as molten drops in a 'fiery rain', an observation, like that of Darwin's 'warm little pond' as a site for the origin of life, that has not been improved upon in 100 years.)
9 Shakespeare, W. (1606) Macbeth, Act 1 Scene III.

Acknowledgements

In writing a book covering a broad field, it is difficult to be expert in the many topics which are addressed. For this reason, and to ensure a measure of peer review, we prevailed upon a number of our colleagues to read the various chapters in draft form. We are indebted to the following for performing this arduous task:

Lew Ashwal	Chapter 4
Ian Campbell	Chapter 9
Malcolm McCulloch	Chapters 4, 11
Ken Eriksson	Chapter 7
Barry Maynard	Chapter 6
Denis Shaw	Chapters 2, 3
Jan Veizer	Chapters 5, 10
Grant Young	Chapter 8

The work reported here on the geochemistry of sedimentary rocks began in March, 1969, and has involved a number of people. Lyn Oates assisted in the initial sample selection in a very cold core shed. Maureen Kaye provided a large body of XRF and XRD data which we still mine with profit. Weldon Nance and Pat Oswald-Sealy contributed much analytical data. Our ideas and interpretations have benefited from cooperative work with Petr Jakěs (who we thank for great insights), Weldon Nance, Owan Bavinton, Ken Eriksson, Vic McGregor, Alfred Kroner, Grant Young, Brian Fryer, George Jenner, Roberta Rudnick and Mukul Bhatia. Stephen Moorbath, Alec Trendall, Bruce Chappell and John Ferguson contributed samples and constructive comments at various stages. The planetary photographs in Chapter 12 are courtesy of NASA.

Production of a book is critically dependent on typing skills and we are grateful to Karen Buckley, Patricia Primmer and Gail Stewart, whose heroic efforts overcame long-standing administrative deficiencies. J. M. G. Shelley made a major contribution by transforming our laboratory computer into a word processor.

Finally, we acknowledge the encouragement of Robert Campbell and the staff of Blackwell Scientific Publications for their assistance.

1

Some Perspectives

The continental crust of the planet Earth provided a platform on which the later stages of the evolution of life occurred. These led, on one of the three available continental masses, to the appearance of *Homo sapiens* [1]. Thus the processes by which the crust itself arose are of direct interest, for if the crust did not exist, life, restricted to oceans and volcanic islands, would have taken a different course [2]. Since the current oceanic crustal record does not persist beyond about 200 million years, except for small fragments, the continental crust is the repository of the geological record in more remote epochs. Thus the memory of early events will be found only in the continental crust. Although the mass of the crust, relative to that of the planet, is small (about half of one per cent), it contains substantial concentrations of many elements relative to their abundance in the entire Earth. Furthermore, the crust may be unique to the Earth, as are the oceans. Nothing similar appears on the surface of the Moon, Mercury or Mars. The two small elevated regions on Venus (Ishtar Terra and Aphrodite Terra) seem unlikely to resemble the terrestrial continents in composition or origin, emphasizing the uniqueness of the surface of our planet in the solar system [3].

Many questions arise. What is the composition and structure of the crust? Does it vary in composition with depth? Was it formed early in Earth history, or late? Was it formed uniformly through time or in a series of episodic events? Is it the remnant of an early sialic crust, possibly produced by accretion from space? If it is produced from the mantle, is it accreted by vertical or lateral processes? What is the mechanism for deriving it from the mantle? What role does oceanic or island arc volcanism play? Is underplating a viable concept? What is the role of the oceanic crust? Is continental crust recycled through the mantle, or is it relatively stable, once produced, and thus impervious to subduction? What is the rate of erosion and destruction of the continental crust? What is the significance of the high electrical conductivity measured deep within the crust? Has the composition of the crust changed or has it been the same throughout time? In this book, we make some attempt to answer these questions and to discuss the composition, origin and evolution of the continental crust of the Earth, primarily using the record of these events preserved in the sedimentary rocks.

1.1 Basic data

A basic observation, unique to the Earth, is that the mean elevation of the continental masses is 126 m above mean sea-level [5], while the mean depth of the oceans is 3.8 km below that datum. This dichotomy is apparent from a study of the hypsometric curve (Fig. 1.1) which shows the percentage of land at various elevations. It has long been realized that this reflects a fundamental difference in density and hence in composition, between the continental and oceanic crusts. The curve for Venus is shown for comparison to reinforce the unique nature of the surface of the Earth.

Only 29% of the land area is higher than 1 km and only 13% lies above 2 km. The land area of the various continents covers 29% of the surface of the Earth.

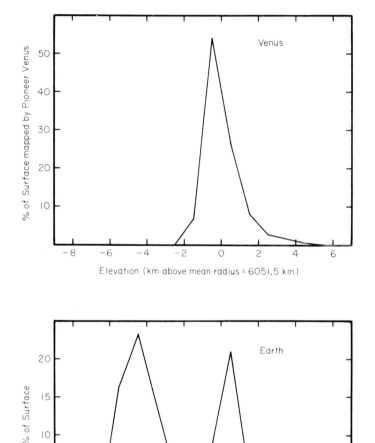

Fig. 1.1. Hypsometric curves from the Earth (lower) and Venus (upper), showing the percentage of global topography versus elevation. The bimodal curve for the Earth is mainly a result of crustal differentiation, which forms both light continental crust and denser oceanic crust. No such difference apparently exists on Venus. Data interval is 1 km (from [4, 5]).

Table 1.1. Continental areal extent, mean height and % flooding (adapted from [5]).

	Area (10^6 km^2)	Mean height (m)	% below sea-level
Africa	35.722	244	14.5
Agulhas	0.134	−3222	100.0
Antarctica	18.984	−344	78.4
Arabia	4.595	470	10.6
Australia	14.163	−244	34.9
Central America	1.344	−579	52.8
Eurasia	62.160	361	25.8
India	4.648	236	14.9
Jan Mayen	0.055	−1600	99.4
New Zealand	3.980	−1568	90.3
North America	39.793	96	30.6
Rockall	0.459	−1702	100.0
Seychelles	0.386	−1706	99.9
South America	23.982	149	19.0
Total	210.405	126	30.6

When submerged continental areas are included, this figure rises to 41% [5] (Table 1.1). A two-fold division into continental shields covering 106×10^6 km^2 and younger folded belts (42×10^6 km^2) was made by Poldervaart [6]. A slightly different approach lists the areas of the exposed basement provinces of differing ages (Table 1.2). They total 25% of the surface area of the globe and 29.5% are of rocks younger in age than 450 m.y. [7]. The abundances of plutonic rocks in the upper crust (Table 1.3) show that granitic rocks dominate. The total mass of sediments is 2.7×10^{24} g of which 1.88×10^{24} g represents continental sediments, the rest being pelagic and shelf sediments [9]. Table 1.4 provides some basic information on the abundances of the common sedimentary rock types.

We assume a mean thickness of 40 km for the continental crust (Fig. 1.2 [5, 11]) but note that in many regions, the base of the crust is not well established. The Mohorovičić discontinuity appears to be joining a long list of scientific parameters, of which the Hubble Constant is the most famous, which

Table 1.2. Area of basement provinces [7].

Age (m.y.)	Area (10^6 km^2)	Percentage
0–450	38.2	29.5
450–900	41.1	31.8
900–1350	14.6	11.3
1350–1800	8.7	6.7
1800–2250	19.4	15.0
2250–2700	6.2	4.8
2700–3150	1.1	0.8
	129.4	99.9

Table 1.3. Abundance of plutonic rocks in the upper crust [8].

	Volume percentage
Granite, granodiorite	77
Quartz diorite	8
Diorite	1
Gabbro	13
Syenites, anorthosites and peridotites	1
	100

Table 1.4. Sedimentary rock data [10].

Present continental sedimentary rock volume	5.8×10^8 km^3
Geosynclinal areas	75%
Platforms	25%
Shales	72%
Carbonates	15%
Sandstones	11%
Evaporites	2%

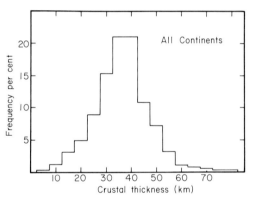

Fig. 1.2. Frequency histogram of crustal thickness [5]. The difficulties of arriving at an average thickness are obvious and compounded by the current uncertainty about the location of the Moho. We adopt a conventional value of 40 km.

have proven elusive. Apart from local variations due to tectonic events (e.g. Himalayas, Andes) there does not appear to be any good evidence for variations or trends in crustal thickness with time [12]. Neither seismic data nor depths derived from metamorphic studies reveal any real differences for the Archean, Proterozoic or Phanerozoic segments.

If we assume a mean density of 2.8 g cm^{-3} and a thickness of 40 km, then the mass of the crust, including the continental shelf areas is 23.6×10^{24} g. This is 0.57% of the mass of the mantle. This estimate exceeds that based on Cogley [5] by about 15%, but we consider that both are within error limits on account of the uncertainty regarding thickness. The average seismic P-wave velocity for the crust is about 6.3 km s^{-1} but ranges from 5.8 to over 7.0 km s^{-1} [13].

Upper crustal velocities, excluding the very near surface sedimentary layers, are typically about 6.1 km s^{-1} while lower crustal P-wave velocities are slightly higher, typically 6.5–6.9 km s^{-1} [13] although values as high as 7.6 km s^{-1} have been reported [14]. The surface heat flow shows interesting variations with age [15]. For Mesozoic and younger sites, the average is about 72 mW m^{-2}. Palaeozoic and Proterozoic sites are fairly uniform at about 51–54 mW m^{-2}, while Archean sites have significantly lower values. The near-surface geothermal gradient for crustal sections ranges from 10–50°C km^{-1} averaging about 30°C km^{-1}. The gradient becomes less steep with depth, so that typical temperatures at the base of the crust are about 500–600°C. In Fig. 1.3, we show a range of oceanic and continental geotherms from Chapman & Pollack [16]. These relate heat flow (mW m^{-2}), depth and temperature. The curvature near the surface is due to crustal heat production. The solidus temperatures for volatile-free and hydrous mantle peridotite are also shown. The most realistic situation for mantle melting will be intermediate.

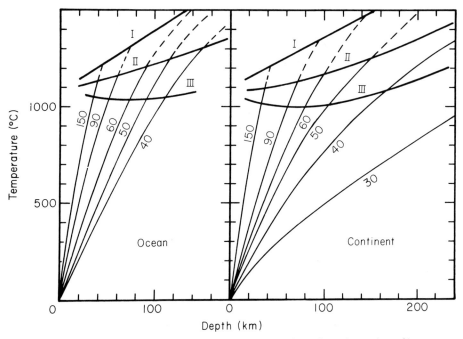

Fig. 1.3. Families of geotherms for oceanic and continental regions, for various values of heat flow, expressed in mW m^{-2}. The geotherms show near-surface curvature due to the production of radioactive heat in crustal rocks. I is the mantle solidus for dry upper mantle rocks, III is the solidus for hydrous upper mantle, and II is an intermediate solidus allowing for the presence of other volatiles such as CO_2. (Adapted from [16].)

1.2 COCORP and crustal structure

In this book, we are concerned principally with the geochemical record in sedimentary rocks as a tracer of crustal evolution. We do not discuss the structure of the continental crust, except in passing, for several reasons.

Firstly, the huge amount of information in the literature refers principally to the upper crust. Next, the subject is in a rapid state of change due to the extensive body of new results being generated by seismic studies such as being carried out by the Consortium for Continental Reflection Profiling (COCORP) [17]. These data are rapidly changing our ideas about many features including crustal structure, layering in the deep crust, the heterogeneity of the crust and the nature of the Mohorovičić discontinuity. It is premature in a work dealing primarily with geochemistry to do more than to comment briefly upon these aspects.

A summary of current progress in this field usefully forms a framework for the geochemical discussions, particularly since the new results provide an emerging interpretation of continental crustal structure in the context of plate tectonics. The principal effect of the COCORP data is that they shed light on the structure of the deep crust. Although perhaps not totally unexpected, the limited COCORP data available indicate extreme heterogeneity both within and between all the present sites. This heterogeneity often appears to be on scales of a few kilometres or less. An unfortunate aspect is that this makes it difficult to reconcile the broader sampling obtainable from seismic velocity data with specific rock types which have little lateral or vertical continuity. The observed rapid lateral and vertical changes in properties observed at each site, and the unique nature of each site indicate that extensive mixing and deformation of the deep crust appears to be common. This may be understood in terms of plate tectonic models.

A principal conclusion is that no simple layered model of overall crustal structure is realistic. However, much localized layering has been revealed. Some deep layers may be buried slices of sedimentary rocks, others perhaps represent buried oceanic crust. Curiously, such layering appears to be less common in the upper crust, a fact consistent with the notion that it is dominated by the presence of massive granitic intrusions, as well as highly deformed sediments.

At middle crustal levels, subparallel but discontinuous reflectors are more common. Inversions in seismic velocity also frequently appear and simple models in which seismic velocities increase steadily with depth do not seem to fit the new data. Many layers appear to be present, but there is not a unique low velocity layer within the crust. The layering is also discontinuous.

The deeper crust, like the intermediate levels discussed above, contains many flat or low-dipping discontinuous reflectors which sometimes increase as the Moho is approached. In some cases, these effects may be due to buried oceanic crust. The crust–mantle boundary often appears to be laminated [14]. As remarked earlier, a consequence of the heterogeneity of the deep crust is that is is difficult to relate seismic velocity data to rock type. The velocity data samples on a much larger volume than the scale of the observed heterogeneities. Accordingly, this casts some doubts on previous attempts [18] to relate deep crustal rock types to the seismic velocity data.

The surest approach to identify the nature of the reflecting layers may have to rely upon those examples which can be identified near surface where they can be accessible to drilling. From the examples in the Appalachians, it seems clear that thin sheets can be thrust for hundreds of kilometres laterally, deeply buried, and so emplaced in the deep crust. There are many indications of extreme structural complexity due to such tectonic processes, which seem to have mixed the crust on a large scale.

The deep crust thus appears to be a 'jumbled mix of components.... Fragments from the sea-floor perhaps 10 km or a few tens of kilometres thick are piled up and slid onto the edge of the continent, sometimes travelling hundreds of kilometres over sediments of the margin as sea-floor is destroyed' [19]. Such structures find a ready explanation in terms of the Wilson cycles of opening and closing ocean basins. During the stages of closure, much material including seamounts, island arcs and so forth may be jumbled together and accreted to the continents, forming various examples of accreted and suspect terrains. Most examples of suspect terrains appear to have been volcanic or sedimentary portions of island-arcs or oceanic crust, rather than fragments of pre-existing continents. Wrangellia is the type locality [20]. The present western margin of the Pacific, and Indonesia, form current examples of this process in operation.

The results from the COCORP programme also shed fresh light on the nature of the Moho. Just as it is not longer tenable to think of simple layered crustal models, so the concept of the Moho as a simple boundary marking the transition from crustal seismic velocities of 6.5–7.0 km s^{-1} to around 8.1 km s^{-1} is no longer viable. Higher crustal velocities up to 7.6 km s^{-1}, much evidence of lamination extending over many kilometres near the base of the crust and a gradual transition to mantle velocities as high as 8.6 km s^{-1}, all have to be fitted into our new understanding of the physical relationship between the crust and mantle.

The continental crust also appears to be linked to the underlying mantle in a manner not yet well understood. In oceanic regions, the lithosphere extends for about 100 km to the top of the low velocity zone (LVZ) in the mantle. This region, where S- and to a lesser extent P-waves show a drop of a few per cent in velocity, is generally interpreted to be the location of small (1%) amounts of melt. The bottom of the oceanic lithosphere coincides with the lower range for the stability field of amphibole, whose breakdown, and dehydration, may release water and hence induce melting. The situation beneath the continents is less clear. Particularly in shield areas, the LVZ appears to be absent and the depth of the lithosphere is deeper, extending to over 200 km. It has been suggested that the crust and mantle are linked to this depth. One possible explanation is that this represents refractory peridotite from which a basaltic melt has been extracted [21].

It is clear that many of the questions about the structure of the continental crust, although exceedingly complex in detail, may be understood within the general context of plate tectonic theory, with cycles of ocean basin opening and closing. The present state of our knowledge may be likened to our knowledge of terrestrial geography during the period of the great voyages of exploration. New discoveries are constantly arriving, but enough information is at hand to construct a usable global picture, frequently in error in detail, occasionally with some erroneous concept such as 'terra australis incognita', but broadly correct. Accordingly, we note the cautionary tales inherent in the new structural information being currently revealed by COCORP and others. With these in mind, we now proceed, in subsequent chapters, to report on our findings on continental origin, growth and evolution as revealed principally by the geochemical record preserved in the sedimentary rocks.

Notes and references

1 Pollard, W.F. (1979) The prevalence of Earth-like planets. *Am. Sci.*, **67**, 654. Pollard makes the interesting comment that continental drift produced three separate major regions, America, Africa-Asia and Australia, on which evolution could proceed independently. Primates appeared on two of these platforms, but the genus *Homo* arose on only one. Thus even on a planet favourably endowed for life, the chances of evolution proceeding in a given direction appear small.
2 Monod, J. (1974) *Chance and Necessity*. Collins, Glasgow.
3 Campbell, I.H. & Taylor, S.R. (1983) No water, no granites—no oceans, no continents. *GRL*, **10**, 1061.
4 Head, J. *et al.* (1981) Topography of Venus and the Earth. *Am. Sci.*, **69**, 614; Masursky, H. *et al.* (1980) Pioneer Venus radar results: Geology from images and altimetry. *JGR*, **85**, 8232.
5 Cogley, J.G. (1984) Continental margins and the extent and number of continents. *RGSP*, **22**, 101. The data in this paper supersede those in the often-quoted work by Kossinna, E. (1933) Die Erdoberflache. *In:* Gutenberg, B. (ed.), *Handbuch der Geophysik*, **2**, 869.
6 Poldervaart, A. (1955) Chemistry of the Earth's crust. *GSA Spec. Pap.*, **62**, 119.
7 Hurley, P.M. & Rand, J.R. (1969) Pre-drift continental nuclei, *Science*, **164**, 1229.
8 Wedepohl, K.H. (1969) *Handbook of Geochemistry*. Springer-Verlag, Berlin. I, 244.
9 Veizer, J., pers. comm.
10 Wedepohl, K.H. (1969) *Handbook of Geochemistry*. Springer-Verlag, Berlin. I, 267; Garrels, R.M. & Mackenzie, F.T. (1971) *Evolution of Sedimentary Rocks*. Norton.
11 Soller, D.R. *et al.* (1982) A new global thickness map. *Tectonics*, **1**, 125.
12 Condie, K.C. (1973) Archean magmatism and crustal thickening. *GSA Bull.*, **84**, 2981.
13 Smithson, S.B. *et al.* (1981) Mean crustal velocity. *EPSL*, **53**, 323.
14 Finlayson, D.M. *et al.* (1979) Explosion seismic profiles, and implications for crustal evolution in southeastern Australia. *BMRJ*, **4**, 243.
15 Morgan, P. (1984) The thermal structure and thermal evolution of the continental lithosphere. *PCE*, **15**, Chap. 2.
16 Chapman, D.S. & Pollack, H.N. (1977) Regional geotherms and lithosphere thickness. *Geol.*, **5**, 265; Pollack, H.N. & Chapman, D.S. (1977) On the regional variation of heat flow, geotherms and lithospheric thickness. *Tectonophysics*, **38**, 279.
17 Smithson, S.B. (1979) Aspects of continental crustal structure and growth: targets for scientific deep drilling. *Contrib. Geol. Univ. Wyoming*, **17**, 65; Oliver, J. (1982) Probing the structure of the deep continental crust. *Science*, **216**, 689; Oliver, J. *et al.* (1983) COCORP and the continental crust. *JGR*, **88**, 3329.
18 Christensen, N.I. & Fountain, D.M. (1975) Constitution of the lower continental crust based on experimental studies of seismic velocities in granulite. *GSA Bull.*, **86**, 227.
19 Oliver, J. *et al.* (1983) COCORP and the continental crust. *JGR*, **88**, 3340.
20 A special issue of *JGR* (1982), **87**, 3631, contains several articles dealing with accretion tetonics and provides much insight into the nature of suspect terrains.
21 Jordan, T.H. (1981) Continents as a chemical boundary layer. *Phil. Trans. Roy. Soc.*, **A301**, 359; (1978) Composition and development of the continental tectosphere. *Nature*, **274**, 544.

2
The Present Upper Crust

2.1 The unique upper crust

The chemical composition of the exposed upper continental crust is a critical parameter in geochemical calculations. This portion of the Earth is most accessible to direct observation and sampling and can be examined with a minor amount of effort in comparison with the oceanic crust, the deep continental crust, the mantle and core of the Earth, and the surfaces of other planets. This ease of access is offset by the general complexity of surficial geology, obvious in geological maps of any scale, so that adequate sampling becomes the major task in deciphering the composition. In this chapter, we address these problems and discuss some solutions which have been attempted. In the final sections, we provide our assessment of the chemical composition of the upper crust.

2.2 Methods for determining composition

There have been three basic approaches in establishing the composition of the exposed upper continental crust. The first has been to estimate areal or volumetric proportions of the principal rock types which are exposed and, using typical compositions, calculate an average composition from these values [1, 2]. The second, and closely related method is to carry out wide-scale sampling with analysis of individual or composite samples which are representative of very large areas [3–8]. These approaches are tedious and may be biased during the averaging process. Thus if there is a secular or episodic variation in composition it may be concealed during data compilation through combining rocks of unlike composition or of differing ages. Nevertheless, such procedures provide an invaluable data base. Confidence in this approach is also greatly enhanced by the overall agreement in compositions derived from a number of different workers in several areas (Table 2.1). The most comprehensive studies have been carried out by two groups working in the Canadian Shield. Although there is some overlap, these workers covered essentially different areas. Notwithstanding this, the upper crustal estimates by these two groups are in remarkable agreement (Fig. 2.1). The weight of the evidence strongly indicates a granodioritic average composition. The mean density (2.67 g cm^{-3}) of a large area of the Canadian Shield is also consistent with a granodioritic composition [9].

The third approach to this problem is to allow the geological processes of sedimentation to produce an average sample of the exposed crust. This procedure is also fraught with difficulty because of possible chemical biases.

Table 2.1. Average chemical composition of large areas of the continental crust.

	1	2	3	4
SiO_2	66.0	64.93	65.3	70.4
TiO_2	0.6	0.52	0.53	0.4
Al_2O_3	15.3	14.63	15.9	14.9
FeO	4.8	3.97	4.4	2.9
MnO	0.1	0.068	0.08	0.05
MgO	2.4	2.24	2.2	1.6
CaO	3.7	4.12	3.4	3.1
Na_2O	3.2	3.46	3.9	4.5
K_2O	3.5	3.10	2.87	2.0
P_2O_5	0.2	0.15	0.16	0.1
Tl		0.52	—	—
Ba		1070	730	795
Rb		110	—	85
Sr		316	380	530
Pb		17	18	—
La		32	71	55
Ce		65	—	65
Nd		26	—	—
Sm		4.5	—	—
Eu		0.94	—	—
Gd		2.8	—	—
Tb		0.48	—	—
Ho		0.62	—	—
Yb		1.5	—	—
Lu		0.23	—	—
Y		21	21	—
Th		10	10.8	—
U		2.5	1.5	—
Zr		240	190	135
Hf		5.8	—	—
Nb		26	—	4
Cr		35	76	<50
V		53	59	—
Sc		7	12	—
Ni		19	19	25
Co		12	—	35
Cu		14	26	35
Zn		52	60	45
Li		22	—	—
Ga		14	—	15

1. Baltic and Ukranian Shields and Basement of Russian Platform [6].
2. Canadian Shield [8].
3. Canadian Shield [4].
4. North-west Scotland Highlands [3].

However, we consider this approach to be more fruitful, since it can be carried out with much smaller resources than are needed for widescale sampling programme which, for logistical reasons, become somewhat rigid in their approach. The assumptions can be more readily tested, and problems which arise in interpretation also can be addressed readily. Moreover this approach tends to

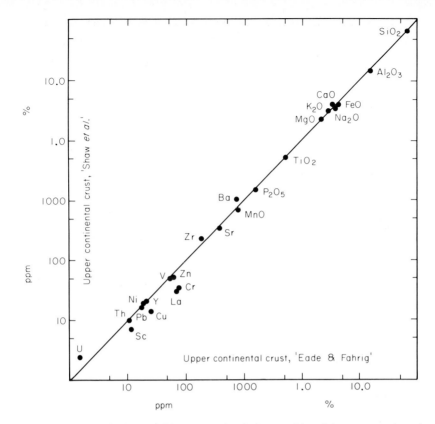

Fig. 2.1. Comparison of estimates of the average chemical composition of the upper continental crust by Eade & Fahrig [4] and Shaw et al. [8]. For the most part, these studies sampled different areas of the Canadian Shield. The agreement of these two estimates is excellent and indicates a granodioritic average composition.

lead to the uncovering of unanticipated scientific questions, such as the behaviour of elements in geological processes and secular variations in composition.

Two major sedimentary sampling processes have been employed. The first, suggested by Goldschmidt [10], was the use of glacial clays deposited in melt-water lakes adjacent to the Pleistocene ice front. The glacial ice-sheets act as a giant bulldozer, stripping the surface, grinding bedrock into rock flour, and depositing this ground, mixed and homogenized material in lakes. The processes of glacial erosion should, in theory, keep effects attributable to chemical weathering and alteration to a minimum. Glacial deposits are well documented in the geological record to about 2.3 Ae [11] and detailed investigation of such rocks should be a fruitful avenue of future research. A variation on this approach is to use the glacially derived wind-blown loess sediments [12]. The data from these are evaluated later in this chapter and also in Chapter 5.

The second major natural sampling process occurs during the weathering, erosion, transportation and deposition of most terrigenous clastic sedimentary rocks. Many complex processes are involved in comparison with the simple abrasion experienced during glacial erosion. Clearly, although the deposited sediments must ultimately bear a relationship to the composition of the exposed crust, the unravelling of this in detail depends on a wide variety of factors. The principal geochemical effects are due to solubility and removal of elements into the oceans where they may be stored for longer or shorter periods of time and ultimately redistributed into carbonates, evaporites and authigenic clay minerals. Elements which are relatively insoluble in natural waters and have very short residence times are transferred virtually quantitatively into terrigenous clastic sedimentary sequences. Those insoluble elements such as Zr which are strongly dispersed in clastic sediments, are of limited value in estimating crustal composition. Elements including the rare earth elements (REE), Th and Sc, which are less fractionated during sedimentary processes, are very useful for such estimates. Thus, the remarkable uniformity of REE patterns in post-Archean sedimentary rocks (see below) contrasts strongly with the considerable variability observed in igneous source rocks, and attests to the efficiency of mixing during transportation and sedimentation.

2.3 Some geological influences on sedimentary rock composition

Our ultimate purpose in examining the chemical composition of terrigenous clastic sedimentary rocks is to constrain the composition of the provenance or source rocks. By carefully selecting the sedimentary rocks which we analyse, to represent as wide and varied a provenance as possible, we might expect these compositions to be representative of the upper continental crust exposed to weathering. The sources of clastic sedimentary rocks can be divided into three main categories:

1 'Primary' igneous rocks (and their metamorphosed equivalents) derived from the mantle through one or more stages of partial melting.
2 'Secondary' igneous rocks derived from remelting of pre-existing crustal (mainly sedimentary) rocks.
3 Recycled sedimentary and metasedimentary rocks.

The various proportions of these sources in average sedimentary rocks is not well known but isotopic modelling suggests that the amount of 'primary' igneous material in a typical sediment is about 35% on average [13] although the amount is highly variable [e.g. see 12]. The provenance of many sediments is also dominated by recycled crustal material which necessarily implies large scale mixing of upper crustal rocks.

Sedimentation involves several complex chemical and physical processes, all of which may result in chemical fractionation among the various classes of sediments (e.g. carbonates, evaporites, sandstones, shales). The main processes involved in the sedimentary cycle include weathering, chemical and physical erosion, transportation, deposition, lithification (including diagenesis and possibly metamorphism) and in some cases uplift and recycling. Thus it is of critical importance to attempt to quantify the chemical effects of these processes on sedimentary rocks in order to isolate those elements which best preserve a record of the source rock composition in the sedimentary rocks of interest (in this case, shales).

2.3.1 Weathering

Well-documented studies of trace element behaviour during weathering are not abundant, although the general chemical environment is fairly well understood [14–17]. Weathering of typical upper crustal rocks can be divided into at least three major stages [15, 18].

1 *Early Stage*, which is dominated by the primary minerals and formation of amorphous phases, chlorite and complex clay minerals such as smectite, vermiculite, illite, with high cation-exchange capacity.
2 *Intermediate Stage*, in which the clay is dominated by complex clay minerals, particularly smectite, with lesser amounts of illite.
3 *Late Stage*, in which the clay fraction is dominated by kaolinite–gibbsite–quartz–ironoxide mineralogy. Ion exchange capacity is severely diminished in such systems. In extreme cases, bauxitic ores may form.

Deep chemical weathering occurs during sustained periods of tectonic quiescence and the formation of such highly weathered profiles requires times of the order of 10^7–10^8 years [e.g. 18]. High rates of mechanical erosion necessary to form clastic sediments generally follow periods of significant uplift [14] and accordingly it is inappropriate to view the bulk of clastic sediments as being derived from highly weathered sources. Thus, in considering the overall effects of weathering on shale composition, it is probably most appropriate to examine the early stage and, to a lesser extent, the intermediate stage of weathering.

Under such conditions, the alkali and alkaline earth elements are quite soluble although the larger cations, such as Rb, Cs, and Ba are commonly retained, adsorbed on clays [17]. Relatively immobile elements include the Al-group (Al, Ga), Ti-group (Ti, Zr, Hf), REE (including Y, Sc) and other high valency ions such as Th and Nb. First row transition metals tend to be relatively immobile although some Ni and V mobility is apparent [18]. Silicon is generally retained in quartz although dissolution of quartz during weathering is a common phenomenon. Effects on the REE will be discussed in greater detail below.

Virtually all studies of weathering have dealt exclusively with weathered profiles but the quantitative assessment of weathering effects on sedimentary rock composition is comparatively recent. Nesbitt & Young [19] have developed a chemical index of alteration (CIA) which is a useful index of the degree of weathering. Using molecular proportions:

$$\text{CIA} = [Al_2O_3/(Al_2O_3 + CaO^* + Na_2O + K_2O)] \times 100$$

where CaO^* is the amount of CaO in silicate minerals only (i.e. excluding carbonates and apatite). Since feldspar makes up in excess of 50% of the upper crust (with quartz comprising an additional 20%), this index effectively measures the degree of alteration of feldspars to clay minerals during weathering. A summary of typical CIA values in various sedimentary rocks and minerals is given in Table 2.2. The values vary from about 50 for unweathered upper crust to about 100 for severely weathered residual clays. Shales typically have intermediate values of about 70–75, indicating that weathering effects have not proceeded to the stage where alkali and alkaline earth elements are substantially removed from the clay minerals.

Table 2.2. Typical values of the chemical index of alteration (CIA) for the upper continental crust and various sedimentary rocks and minerals (adapted from [19]).

	CIA
Average upper continental crust	50
Pleistocene till (matrix)	50–55
Pleistocene glacial clays	60–65
Loess	55–70
Average shales	70–75
Amazon cone muds	80–90
Residual clays	85–100
Albite	50
Anorthite	50
K-feldspar	50
Muscovite	75
Illite	75–85
Montmorillonites	75–85
Beidellites	75–85
Chlorite	100
Kaolinite	100

Water rock partition coefficients. Whitfield and co-workers [20] have examined the partitioning of elements between upper crustal rocks and natural waters during the weathering and erosion of surficial rocks to form sediments. The partition coefficient for some element y between some natural water and the upper crust can be defined as:

$$K_y = X_y^W / X_y^{UC}$$

where X_y^W is the concentration of element y in some natural water (W) and X_y^{UC} is the concentration of element y in the upper continental crust (UC). For river water and sea water, these values have been found to be significantly affected by the electrostatic contribution to the element–oxygen bond energies (Q_{y-O}) such that:

$$Q_{y-O} = (x_y - x_O)^2 \text{ eV}$$

where x_y and x_O are the electronegativities of the element y and oxygen, respectively. Thus, the concentration of an element in natural waters is strongly controlled by solid state chemistry in addition to solution chemistry.

These findings have considerable bearing in understanding rock weathering, sea water compositions and the influence of compositional fluctuations of input (e.g. pollution) on ocean water chemistry. In addition, we can gain considerable insight into the relationships between sedimentary rocks and average upper crustal compositions. In this respect, it is particularly instructive to examine the partition coefficients between sea water, which is the ultimate sink of most dissolved material, and the upper crust (K_y^{SW}). These values, calculated from the data in Tables 2.3 and 2.15 (see below), are listed in Table 2.3, expressed as log K_y^{SW}. Whitfield [20] has also pointed out the strong correlation between residence time (τ) and K_y^{SW} for the various elements. In Fig. 2.2, it can be seen that the values of K_y^{SW} vary over almost nine orders of magnitude.

Table 2.3. Composition of natural waters and pelagic sediments, residence times (τ) and sea water–upper crust partition coefficients (K_y^{SW})*.

Element	River water (g g^{-1})	Ocean water (g g^{-1})	Pelagic clay (g g^{-1})	Pelagic carbonate (g g^{-1})	Pelagic siliceous (g g^{-1})	log τ	log K_y^{SW}
Li	3×10^{-9}	1.7×10^{-7}	5.7×10^{-5}			6.4	-2.1
Be	1×10^{-11}	2×10^{-13}	2.6×10^{-6}			1.8	-7.2
B	1×10^{-8}	4.5×10^{-6}	2.3×10^{-4}			7.2	-0.52
F	1×10^{-9}	1.3×10^{-6}	1.3×10^{-3}			5.9	—
Na	6.3×10^{-6}	1.08×10^{-2}	4.0×10^{-2}			8.3	-0.43
Mg	4.1×10^{-6}	1.29×10^{-3}	2.1×10^{-2}			7.7	-1.0
Al	5×10^{-8}	8×10^{-10}	8.4×10^{-2}			0.85	-8.0
Si	6.5×10^{-6}	2.8×10^{-6}	2.5×10^{-1}		4.7×10^{-1}	3.9	-5.0
P	2×10^{-8}	7.1×10^{-8}	1.5×10^{-3}			4.6	—
S	3.7×10^{-6}	9.0×10^{-4}	1.3×10^{-3}			8.7	—
Cl	7.8×10^{-6}	1.95×10^{-2}	2.1×10^{-2}			8.8	—
K	2.3×10^{-6}	3.99×10^{-4}	2.5×10^{-2}			7.1	-1.8
Ca	1.5×10^{-5}	4.13×10^{-4}	9.3×10^{-3}	4.0×10^{-1}		6.1	-1.9
Sc	4×10^{-12}	6.7×10^{-13}	1.9×10^{-5}			1.4	-7.3
Ti	3×10^{-9}	$<9.6\times10^{-10}$	4.6×10^{-3}			<2.2	<-6.5
V	9×10^{-10}	1.2×10^{-9}	1.2×10^{-4}			3.9	-4.7
Cr	1×10^{-9}	2×10^{-10}	9×10^{-5}			3.2	-5.2
Mn	7×10^{-9}	3×10^{-10}	6.7×10^{-3}			1.5	-6.3
Fe	4×10^{-8}	6×10^{-11}	6.5×10^{-2}			-0.16	-8.9
Co	1×10^{-10}	2×10^{-12}	7.4×10^{-5}			1.3	-6.7
Ni	3×10^{-10}	5×10^{-10}	2.3×10^{-4}			3.2	-4.6
Cu	7×10^{-9}	3×10^{-10}	2.5×10^{-4}			3.0	-4.9
Zn	2×10^{-8}	4×10^{-10}	2.0×10^{-4}	1.0×10^{-4}		3.1	-5.2
Ga	9×10^{-11}	2×10^{-11}	2×10^{-5}			2.9	-5.9
Ge	5×10^{-12}	5×10^{-12}	2×10^{-6}			3.3	-5.5
As	2×10^{-9}	1.7×10^{-9}	1.3×10^{-5}			5.0	-2.9
Se	6×10^{-11}	1.3×10^{-10}	1.7×10^{-7}			5.8	-2.6
Br	2×10^{-8}	6.7×10^{-5}	7×10^{-5}			8.9	—
Rb	1×10^{-9}	1.2×10^{-7}	1.1×10^{-4}			5.9	-3.0
Sr	7×10^{-8}	7.6×10^{-6}	1.8×10^{-5}	2.3×10^{-3}		6.7	-1.7
Y	4×10^{-11}	7×10^{-12}	4×10^{-5}			2.1	-6.5
Zr	—	3×10^{-11}	1.5×10^{-4}			2.2	-6.8
Nb	—	$<5\times10^{-12}$	1.4×10^{-5}			<2.4	<-6.7
Mo	6×10^{-10}	1.1×10^{-8}	2.7×10^{-5}			5.5	-2.1
Pd	—	—	4.0×10^{-9}*			—	—
Ag	3×10^{-10}	2.7×10^{-12}	1.1×10^{-7}			4.3	-4.3
Cd	1×10^{-11}	8×10^{-11}	3.0×10^{-7}	6.5×10^{-7}		4.9	-3.1
In	—	1×10^{-13}	8×10^{-8}			3.0	-5.7
Sn	4×10^{-11}	5×10^{-13}	3.0×10^{-6}			2.1	-7.0
Sb	7×10^{-11}	1.5×10^{-10}	1×10^{-6}			5.1	-3.1
I	7×10^{-9}	5.6×10^{-8}	$\geq 1\times10^{-5}$			≤ 6.6	—

Table 2.3 (*continued*)

Element	River water (g g^{-1})	Ocean water (g g^{-1})	Pelagic clay (g g^{-1})	Pelagic carbonate (g g^{-1})	Pelagic siliceous (g g^{-1})	log τ	log K_v^{SW}
Cs	2×10^{-11}	2.9×10^{-10}	6×10^{-6}			4.6	-4.1
Ba	2×10^{-8}	1.4×10^{-8}	2.3×10^{-3}			3.7	-4.8
La	4.8×10^{-11}	4.5×10^{-12}	4.2×10^{-5}			1.9	-6.8
Ce	7.9×10^{-11}	3.5×10^{-12}	8.0×10^{-5}			1.5	-7.3
Pr	7.3×10^{-12}	1.0×10^{-12}	1.0×10^{-5}			1.9	-6.9
Nd	3.8×10^{-11}	4.2×10^{-12}	4.1×10^{-5}			1.9	-6.8
Sm	7.8×10^{-12}	8.0×10^{-13}	8.0×10^{-6}			1.9	-6.8
Eu	1.5×10^{-12}	1.5×10^{-13}	1.8×10^{-6}			1.8	-6.8
Gd	8.5×10^{-12}	1.0×10^{-12}	8.3×10^{-6}			2.0	-6.6
Tb	1.2×10^{-12}	1.7×10^{-13}	1.3×10^{-6}			2.0	-6.6
Dy	7.2×10^{-12}	1.1×10^{-12}	7.4×10^{-6}			2.0	-6.5
Ho	1.4×10^{-12}	2.8×10^{-13}	1.5×10^{-6}			2.1	-6.5
Er	4.2×10^{-12}	9.2×10^{-13}	4.1×10^{-6}			2.2	-6.4
Tm	6.1×10^{-13}	1.3×10^{-13}	5.7×10^{-7}			2.2	-6.4
Yb	3.6×10^{-12}	9.0×10^{-13}	3.8×10^{-6}			2.3	-6.4
Lu	6.4×10^{-13}	1.4×10^{-13}	5.5×10^{-7}			2.3	-6.4
Hf	—	$<7 \times 10^{-12}$	4.1×10^{-6}			<3.1	<-5.9
Ta	—	$<2.5 \times 10^{-12}$	1×10^{-6}			<3.3	<-5.9
W	3×10^{-11}	1×10^{-10}	1×10^{-6}			4.9	-4.3
Re	—	4×10^{-12}	1×10^{-9}			6.5	-2.1
Ir	—	—	3.1×10^{-10}†			—	—
Au	2×10^{-12}	4.9×10^{-12}	$1.5 \times 10^{-}$†			6.2	-2.6
Hg	7×10^{-11}	1×10^{-12}	1×10^{-7}			3.9	—
Tl	—	1×10^{-11}	1.2×10^{-6}			3.8	-4.9
Pb	1×10^{-9}	2×10^{-12}	3.0×10^{-5}			1.7	-7.0
Bi	—	2×10^{-11}	5.5×10^{-7}			4.4	-3.8
Th	$<1 \times 10^{-10}$	6×10^{-14}	1.34×10^{-5}			0.53	-8.3
U	4×10^{-11}	3.1×10^{-9}	2.6×10^{-6}			6.0	-3.0

* Data sources
1 River water: Li [21]; Martin *et al.* [22] for (REE); Y=ch. norm. Ho
2 Ocean water: Broecker & Peng [23]; Li [21]; B. Ullman (pers. comm.); various sources for REE, Y
3 Pelagic sediments: Chester & Aston [24]; Li [21]; Heinrichs *et al.* [2]; Crocket & Kuo [25]; McLennan & Taylor [26]
   Note: Values for pelagic carbonate and siliceous sediments not reported if mass element in these sediments <10% of entire mass of element in average ocean sediments.
4 log τ: Sedimentation rates [27]: Terrigenous—1.73×10^{15} g.a^{-1}
   Carbonate—1.079×10^{15} g.a^{-1}
   Siliceous—0.172×10^{15} g.a^{-1}
   Ocean mass (excl. continental shelf water): 1.3×10^{24} g

2.3.2 Erosion

The agents of erosion include running water (streams, ground water), aeolian action, glacial action, and gravity [14]. In Table 2.4 the mass of material delivered to the ocean by the various flux agents are compared. Two observations stand out. First, the particulate load is greater than the dissolved load by more than a factor of 4 and secondly, the river flux is by far the most important for both particulate and dissolved material.

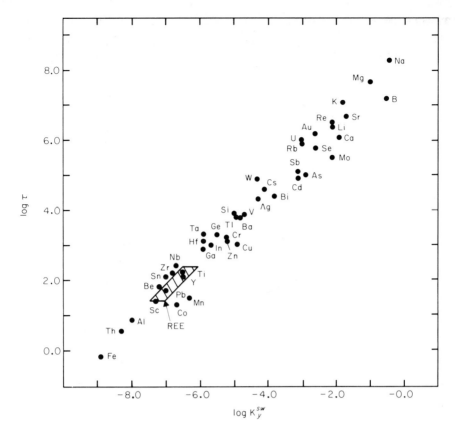

Fig. 2.2. Plot of residence time (expressed as log τ) against sea water–upper crust partition coefficient (expressed as log K_y^{SW}). Data from Table 2.3. Elements with the lowest residence times and sea water–upper crust partition coefficients (e.g. REE, Y, Sc, Ti, Zr, Hf, Al, Ga, Th, Nb, Sn, Be) have the greatest potential for being transferred into the clastic sedimentary record, thus preserving a record of upper crustal composition. For reasons discussed throughout Chapter 2, the distribution of REE, Y, Th and Sc in shales are considered to most faithfully reflect upper crustal abundances.

The rates of chemical erosion vary at present by a factor of only 3–4 among the various continents [14]. The magnitude of chemical denudation is closely related to rock types being eroded. Thus, about one-third of the chemical load delivered to the oceans results from the direct weathering of carbonate rocks. In contrast, mechanical denudation varies by a factor of more than 25 among the continents and is strongly correlated to elevation above sea-level and thus to tectonic activity.

2.3.3 Physical transport and deposition

The mechanical sorting of clastic particles during sedimentary transport is governed by certain laws of fluid flow. The critical properties of a particle are specific gravity, size and shape [29, 30]. In terms of understanding mineral fractionation, the concept of hydraulic equivalence is important. Particles which have hydraulic equivalence display identical hydraulic properties in the

Table 2.4. Present mass of material annually delivered to the oceans (excluding organic material) [14, 28].

	Mass ($\times 10^{15}$ g)	% Mechan.	% Total
A. Mechanical load			
Rivers	16.0	87	70
Glacial	2.07	11	9
Marine erosion	0.26	1	1
Aeolian	0.06	—	—
Total	18.4		

	Mass ($\times 10^{15}$ g)	% Dissolved	% Total
B. Dissolved load			
Rivers	3.9	91	17
Ground waters	0.4	9	2
Total	4.3		

Values quoted for density of 2.8 g cm^{-3}

transporting fluid (e.g. water or air) resulting from varying combinations of physical properties (i.e. size, shape, specific gravity).

A simple case is to examine the settling velocity of spherical particles in standing water. For a fluid regime with a Reynolds number (R) ≤0.5 particle settling velocity is governed by the Stokes Law equation which can be expressed:

$$V = [\Delta \varrho g / 18 \mu] d^2$$

where V is the settling velocity (cm s^{-1}), $\Delta \varrho$ is the particle–fluid density difference (g cm^{-3}), g is the acceleration due to gravity (cm s^{-2}), μ is the viscosity (poise) and d is the particle diameter (cm). In practice, Stokes Law only holds for particles with diameters ≤0.01 cm (fine sand). For larger particles, viscosity alone is not the controlling factor in the fluid resistance to the settling particles; inertia also plays an important role.

The relationship between settling velocity and grain size for spheres has been investigated experimentally [31] and is compared with the theoretical Stokes Law calculation in Fig. 2.3(a) for a sphere with constant density (2.50 g cm^{-3}) settling in pure water. Shown in Fig. 2.3(b) are the curves for spheres of differing density, one being typical of common framework grains such as quartz and feldspar ($\varrho = 2.5$–2.8 g cm^{-3}) and the other typical of heavy minerals ($\varrho > 2.8$ g cm^{-3}).

Theoretically, the effects of these processes on the chemical composition of sediments is not complicated. Clearly, the most important first order effect is separation on the basis of grain size. In Table 2.5, chemical analyses for various grain size fractions for modern sediments from the Barents Sea [32] and Gulf of Paria [33] and analyses for the average greywacke and argillite from an Ordovician turbidite sequence from New Zealand [34] are shown. In each case the chemical relationships, for the most part, can be explained by the concentration of clay minerals at the expense of quartz and feldspar in finer-grained fractions. This results in a decrease in silica and a corresponding increase in the abundances of most other elements. Several of the alkali and alkaline earth elements (e.g. Ca, Sr, Na, Ba) show an erratic distribution. These elements are greatly fractionated during weathering processes (see

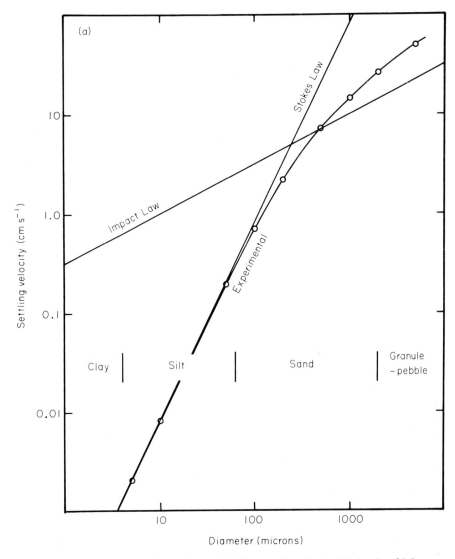

Fig. 2.3. (a) Plot of settling velocity against particle diameter for spheres with density of 2.5 g cm^{-3} settling in pure water at 20°C. Shown are the theoretical Stokes Law calculation and the experimentally determined relationship [31]. Note that the theoretical and experimentally determined relationships agree very well for particles smaller than fine sand. (b) Plot of experimentally determined settling velocities against particle diameter for spheres settling in pure water at 20°C [31]. Shown are the curves for densities of 5.0 g cm^{-3} (typical for heavy minerals) and 2.65 g cm^{-3} (approximate value for quartz and feldspar). It is clear from such a diagram that heavy minerals tend to be deposited in clastic sediments with an overall larger grain size.

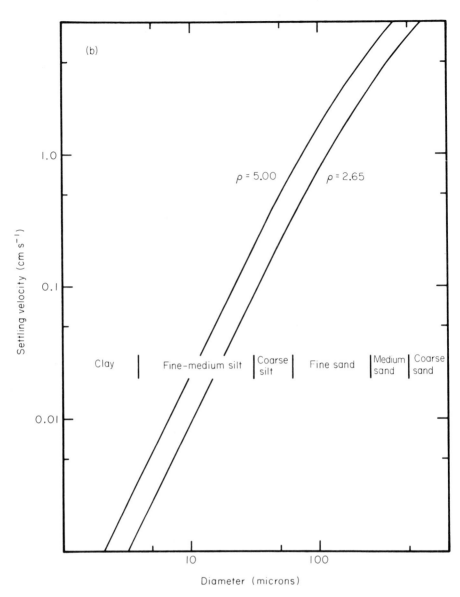

Fig. 2.3 (*continued*)

Section 2.3.1) and in the case of Ca and Sr, may be located as well in carbonate minerals. Of great importance is the observation that critical elemental ratios (e.g. La/Yb, Eu/Eu*, La/Th, La/Sc, Th/Sc) are not seriously affected (see Section 2.3.6).

Table 2.5 (*footnotes*)
 1. Surface sediments from the Barents Sea [32].
 2. Averages of 12 platform sands (3 only for SiO_2) and 7 basinal clays (3 only for SiO_2) from Gulf of Paria [33].
 3. Average Greenland Group (Ordovician) greywacke and argillite [34].
 *Includes H_2O^+ and H_2O^-.

Table 2.5. Chemical composition of grain-size fractions.

	Barents Sea[1]			Gulf of Paria[2]		Greenland Group[3]	
	Sand	Silt	Clay	Sand	Clay	Greywacke	Argillite
SiO_2	84.65	76.95	55.67	81.26	60.99	74.33	63.11
TiO_2	0.16	0.74	1.12	0.47	0.88	0.69	0.85
Al_2O_3	7.72	11.54	21.84	6.34	18.52	12.57	18.85
FeO	1.47	2.38	9.74	4.37	7.82	4.60	6.85
MnO	0.04	0.06	0.10	0.06	0.27	0.06	0.05
MgO	0.54	1.43	3.62	0.92	2.70	2.39	3.60
CaO	0.84	1.88	1.02	4.06	3.17	0.90	0.56
Na_2O	—	2.58	—	1.22	2.90	1.51	1.14
K_2O	2.01	2.32	3.97	1.18	2.60	2.76	4.80
P_2O_5	0.07	0.13	0.42	0.11	0.16	0.20	0.19
Σ	97.50	100.01	97.50	99.99	100.01	100.01	100.00
CO_2	—	2.21	0.80	2.17	1.92	—	—
LOI	—	—	—	3.61*	10.27*	2.84	4.08
Cs	—	—	—	1.4	9.0	6.3	9.0
Ba	—	769	671	301	371	409	702
Rb	—	65	173	47	150	136	204
Sr	—	280	177	147	203	64	56
Pb	—	17	53	22	23	—	—
La	—	34	66	—	—	33.7	36.3
Ce	—	61	121	—	—	60.5	65.1
Nd	—	—	—	—	—	38	36
Sm	—	—	—	—	—	7.61	8.07
Eu	—	—	—	—	—	1.23	1.32
Tb	—	—	—	—	—	0.76	0.81
Yb	—	—	—	—	—	2.99	3.14
Lu	—	—	—	—	—	0.46	0.52
ΣREE	—	—	—	—	—	168	179
La/Yb	—	—	—	—	—	11.3	11.6
Eu/Eu*	—	—	—	—	—	0.59	0.59
Y	—	34	28	—	—	—	—
Th	—	10	18	—	—	12.7	13.3
U	—	—	—	—	—	3.0	3.3
Zr	—	393	131	413	139	—	—
Hf	—	—	—	—	—	5.2	2.9
Th/U	—	—	—	—	—	4.2	4.0
La/Th	—	3.4	3.7	—	—	2.7	2.7
Cr	—	—	—	31	88	67	91
V	—	—	—	79	137	—	—
Sc	—	—	—	—	—	11.8	16.8
Ni	—	22	83	16	30	41	36
Co	—	—	—	7.7	11	10.4	16.0
Cu	—	20	96	7.1	18	—	—
Zn	—	49	245	—	—	—	—
Ga	—	—	—	7.1	20	—	—
La/Sc	—	—	—	—	—	2.9	2.2
Th/Sc	—	—	—	—	—	1.1	0.8
B	—	—	—	60	77	—	—

It is also apparent from Table 2.5 that Zr and Hf are enriched in the coarser fractions. This introduces us to a second major geochemical factor: heavy mineral fractionation. Due to a combination of resistance to weathering and high specific gravity, some minerals tend to be deposited in those sediments with an overall larger grain size (Fig. 2.3(b)). In sands and sandstones, heavy minerals typically constitute ≤0.1% although concentrations up to 1.0% are known [35]. Common heavy minerals are ilmenite, magnetite, garnet, zircon, hornblende, rutile and epidote. Although rare, cassiterite, tourmaline, apatite and monazite may be important locally [36]. From the point of view of the geochemical balance of trace elements, those most affected are Zr, Hf, Nb and Ta (zircon, rutile) and to a lesser extent Sn (cassiterite), B (tourmaline), Sr (apatite), REE and Th (zircon, garnet, apatite, monazite).

2.3.4 Ocean water chemistry

During weathering and erosion, considerable material is dissolved by natural waters and ultimately carried to the oceans, mainly by rivers. Thus the oceans represent an important link between the chemical compositions of the upper continental crust and sedimentary rocks. In this section, we will show how the distribution and controls of the various elements in sea water provide insight into the rationale of using the chemical composition of terrigenous clastic sedimentary rocks as an index of upper crustal composition.

Considerable advances in our understanding of marine chemistry have been made during the past two decades [23, 37]. In Table 2.3, we list an estimate of the average chemical composition of ocean water. Our data are mainly taken from a compilation of Broecker & Peng [23] with Th values from Li [21], and I values from W. Ullman (pers. comm.). An examination of the several recent compilations of sea water compositions indicate excellent agreement for the major and minor ionic species and better than 'order of magnitude' agreement for the trace and ultratrace elements. Thus, for our purposes the composition of sea water is reasonably well known.

Residence times. The residence time (τ) of an element in the ocean, a concept introduced by Barth [38], is given, for a steady-state system, as:

$$\tau_y = M_y/F_y$$

where M_y is the mass of element y in the oceans and F_y is the annual mean flux of element y through the oceanic reservoir. There is evidence that the oceans are not in steady-state [23, 39], although the open oceans, exclusive of the continental shelves approximate steady-state [21].

Elemental flux through the oceans can be measured either by river input [20, 23, 37, 40] or by rate of removal in deep-sea sediments [21, 41]. Li [21] has argued against the first approach on account of the effects of human influence on the distribution of some elements in river water and the complexities of estuarine mixing. Average river water compositions are very difficult to estimate [23] and accordingly we find Li's arguments persuasive.

The alternative approach is to use the composition of oceanic pelagic sediments as a measure of elemental flux. The abundance of most elements in pelagic carbonates and cherts is negligible except for Ca, Sr and Si. In Table 2.3, an estimate of the average composition of pelagic clay is given [21]. REE values are from McLennan & Taylor [26] with Y estimated from the chondrite-normalized REE diagrams. Also shown are Ca and Sr composition for pure biogenic carbonate and Si for pure biogenic siliceous material. Using the recently compiled sedimentation rates of Lisitsyn et al. [27], the residence times have been calculated and are also given in Table 2.3 (expressed as log τ). The residence times vary over some nine orders of magnitude from less than one year (Fe) to more than 10^8 years (e.g. Na).

Residence times calculated by the two methods (riverine input, pelagic deposition) are highly correlated but not perfectly concordant [21]. Residence times using river water are about one order of magnitude less for the highly soluble elements (e.g. Br, Cl, Na) and about one to two orders of magnitude more for the least soluble elements (e.g. REE, Al, Th, Sc).

2.3.5 Diagenesis and metamorphism

Erosion and transport of sediment generally occurs under fairly oxidizing conditions at more or less constant pH, pressure and temperature. Upon deposition, these parameters can change drastically and thus influence the composition of the sediment (see reviews in [42, 43]). As in the case of weathering, the more soluble alkali and alkaline earth elements are liable to movement and/or redistribution [44]. The anoxic conditions under which many sediments are deposited can have severe consequences; for example Fe and Mn, which are generally insoluble at surface conditions, may change oxidation state and become readily soluble and mobilized. Table 2.6 compares trace element data for oxic and anoxic sediments from the same basin (Oslo Fiord). As well as Mn (see above), the elements Cu, Mo, Pb, Zn, S and C are clearly enriched in the anoxic sediments. The reasons for enrichments are not entirely clear but incorporation in sulphide phases or adsorption on organic compounds are obvious possibilities [45]. Similarly, U also is enriched commonly in anoxic sediments [45]. The enrichment of U is almost certainly affected by the reduction of soluble U^{6+} to the relatively insoluble form U^{4+} and precipitation as $U(OH)_4$ or other appropriate compounds. Another common feature of diagenesis is enhanced solubility of silica [e.g. 35].

Element mobility (particularly for trace elements) during metamorphism of pelitic rocks is inadequately understood and no substantial advance has been made since the pioneering work on the Littleton Formation [46]. Shaw [46] found little change in the major elements beyond loss of water and carbon dioxide during regional metamorphism (up to about sillimanite grade). Trace elements also showed little evidence of mobility except that Li and Pb increased. Possible decreases in Ni and Cu were considerably less well defined. Ronov [47] also noted only minor geochemical effects during progressive regional metamorphism of sedimentary rocks. Under very high grades of

Table 2.6. Trace elements in oxic and anoxic sediments from the Oslo Fjord (in ppm except where noted).

	Anoxic	Oxic	Probable significant difference†
Ba	735	768	
Rb	149	162	
Sr	241	230	
Pb	148	94	*
Y	37	38	
Zr	197	235	
Cr	125	113	
V	181	186	
Ni	55	54	
Co	27	24	
Cu	133	54	*
Mn	5743	3990	*
Zn	571	342	*
P	1322	1305	
Mo	33	5	*
C_{org} (%)	3.58	2.27	*
S (%)	1.15	0.22	*

Anoxic sediment mean of 52 samples; oxic sediment mean of 62 samples. From compilation of Calvert [45].
† Differences are considered significant if >20%.

metamorphism, such as in the granulite facies, many elements, including K, Rb, Cs, Th, U are mobile, probably in fluid phases [48].

2.3.6 Sedimentary rocks as crustal samples.

It is instructive to examine Fig. 2.2 in assessing the overall effects of the various processes on the composition of clastic sedimentary rocks. The sea water–upper crust partition coefficient (K_y^{SW}) varies over eight orders of magnitude. Elements with very high K_y^{SW} and τ values (log $K_y^{SW} \geqslant -3$; log $\tau \geqslant 5$), including most alkali and alkaline earth elements, B and U, are strongly partitioned into natural waters and remain for long periods of time. Accordingly, such elements are not likely to reflect crustal abundances in fine-grained terrigenous clastic sedimentary rocks in any simple fashion. On the other hand, elements with very low K_y^{SW} and τ values (log $K_y^{SW} \leqslant -6$; log $\tau \leqslant 3$), including Ti-group (Ti, Zr, Hf), Al-group (Al, Ga), REE (including Y, Sc), Th, Nb, Sn, Be are strongly excluded from natural waters and remain in the oceans for times less than average oceanic mixing times ($\leqslant 10^3$ a). Consequently, it is likely that these elements are transferred almost quantitatively into

---

Table 2.7 (footnotes)
1. 54943, Fortescue Gp., Hamersley Basin (2.6–2.4 Ae)
2. 46436, Earaheedy Gp., Nabberu Basin (c. 1.7 Ae)
3. MI5, Mount Isa Gp. (c. 1.7 Ae)
4. A010, Pertatataka Fm., Amadeus Basin (c. 0.85 Ae)
5. SC8, State Circle Shale (Silurian)
6. PL6, Laurel Fm., Canning Basin (Carboniferous)
7. PL1, Poole Sst., Canning Basin (Permian)
8. PW5, Kockatea Shale, Perth Basin (Triassic)

Table 2.7. Chemical composition of some typical post-Archean shales from Australia [49, 50].

	1	2	3	4	5	6	7	8
SiO_2	57.98	69.25	69.66	67.39	68.01	66.03	67.20	59.52
TiO_2	0.81	0.64	0.79	0.86	0.75	0.74	0.77	1.03
Al_2O_3	23.28	18.76	15.55	16.95	17.12	16.94	19.85	23.75
FeO	6.51	5.99	6.20	6.82	5.90	4.80	5.53	8.54
MnO	0.08	0.02	0.04	0.09	0.23	0.05	0.04	0.16
MgO	4.33	0.56	2.60	2.05	2.73	2.01	1.62	1.81
CaO	0.44	0.03	0.51	0.27	0.22	5.11	0.39	0.57
Na_2O	0.66	0.75	0.42	1.01	0.66	0.46	0.18	0.95
K_2O	5.92	4.00	3.90	4.45	4.24	3.75	4.33	3.44
P_2O_5	—	—	0.32	0.12	0.15	0.11	0.08	0.23
Σ	100.01	100.00	99.96	100.01	100.01	100.00	99.99	100.00
LOI	5.88	3.75	4.29	5.57	3.12	7.31	4.34	5.50
Cs	13.4	4.65	16	17	13	18	17	10
Ba	879	1051	310	240	470	500	590	580
Rb	—	—	215	198	180	194	175	185
Sr	—	—	48	86	34	269	67	120
Pb	18.0	17.1	10	43	6.2	24	35	32
La	42.3	51.0	27	34	44	52	40	50
Ce	88.9	102.9	57	63	94	132	86	104
Pr	9.87	12.8	8.8	7.0	8.1	12	9.1	13
Nd	35.2	48.2	32	24	29	37	30	43
Sm	7.18	7.39	4.9	5.3	5.0	6.5	6.0	8.2
Eu	1.14	1.23	0.94	1.0	0.95	1.2	1.1	1.7
Gd	4.63	4.19	4.0	4.8	4.7	5.2	4.7	6.7
Tb	0.81	0.68	0.68	0.71	0.74	0.84	0.82	1.2
Dy	5.08	3.80	3.8	3.7	4.0	4.6	5.0	6.2
Ho	1.07	0.83	0.86	0.87	0.89	1.1	1.2	1.4
Er	3.01	2.15	2.3	2.6	2.5	3.2	3.6	3.8
Yb	3.03	2.10	2.4	2.5	2.4	2.8	3.6	3.7
ΣREE	203	238	145	149	196	258	192	242
La_N/Yb_N	9.4	16.4	7.6	9.2	12.4	12.6	7.5	9.2
Eu/Eu^*	0.60	0.67	0.65	0.61	0.60	0.63	0.63	0.70
Y	29.9	30.0	18	20	34	36	29	38
Th	15.8	19.9	12	15	16	15	18	19
U	3.58	2.62	2.7	2.6	2.7	2.5	2.9	3.3
Zr	148	189	219	202	155	210	245	185
Hf	4.28	4.59	4.1	4.3	3.0	3.1	4.9	4.3
Sn	7.48	5.08	3.4	4.3	2.6	3.6	3.9	3.9
Nb	15.3	12.2	13	16	18	23	24	30
Mo	1.34	0.35	—	—	—	—	—	—
W	1.32	0.90	0.75	1.1	0.76	0.71	1.1	0.94
Th/U	4.4	7.6	4.4	5.8	5.9	6.0	6.2	5.8
La/Th	2.7	2.6	2.3	2.3	2.8	3.5	2.2	2.6
Cr	395	151	67	79	99	61	78	110
V	190	69	55	105	110	70	125	100
Sc	21	17	13	19	20	15	19	21
Ni	211	19	29	40	49	31	36	54
Co	23	*	16	14	19	14	14	27
Cu	65	3.2	—	25	110	35	27	52
Ga	21	12	13	22	14	25	25	30
La/Sc	2.0	3.0	2.1	1.8	2.2	3.5	2.1	2.4
Th/Sc	0.75	1.2	0.92	0.79	0.80	1.0	0.95	0.90
Bi	0.84	0.25	0.19	0.40	0.09	0.30	0.39	0.23
B	28	98	95	210	96	75	56	41

clastic sedimentary rocks and hence give the best information regarding the source. This conclusion is further reinforced by the very low abundances of these elements in chemical sediments such as carbonates and evaporites. The remaining elements with intermediate values of K_y^{SW} and τ, including several first row transition elements, Si, Ba, Cs, Tl, may also give some information about source composition, although considerably more caution is warranted.

Many of the elements with low values of K_y^{SW} and τ in sedimentary rocks are unsuitable to use for inferring upper crustal abundances for other reasons. Some elements exhibit marked changes of solubility with changing oxidation state (Fe, Mn) or are strongly chalcophile (Pb). Other elements show little dispersion in common igneous rocks (Al, Ga). Another important factor, discussed briefly above, is the fact that some elements form major constituents in common heavy minerals (e.g. Zr, Hf, Sn) and accordingly may be fractionated in sedimentary processes.

We consider the distribution of REE, Y, Th, Sc and possibly Co to be the most useful for the purposes of determining upper crustal abundances from the composition of clastic sedimentary rocks. Both strongly incompatible elements (LREE, Th) and strongly compatible elements (Sc, Co) are represented. Their ratios provide an index of chemical differentiation (e.g. Th/Sc, La/Sc). In addition, certain key elemental ratios remain fairly constant or are predictable during igneous processes (Th/U, K/U, K/Rb).

In our analytical work, we have concentrated on the fine-grained terrigenous sedimentary rocks (mudstones, shales, etc.). This approach is used because any sedimentary mass balance calculation (see Table 1.5) is controlled to a large extent by the shale data for the elements under consideration (e.g. REE, Th, Sc). Complete high quality analytical data are still comparatively scarce for all sediment and sedimentary rock types. In Table 2.7 we present some typical shale analyses from post-Archean successions of Australia [49, 50]. In Table 2.8 are listed analyses for particulate matter from several modern rivers [51] (we have not reported elements which have been affected by pollution such as Cu, Mo, Sb, Pb, Zn). These analyses effectively represent the composition of the solid fraction being eroded from the continents. Finally, in Table 2.9, we list three estimates of average post-Archean shale. Included are averages from Clarke [52] and Krauskopf [53]. The first column is taken from our own work and differs from the others in that the carbonate and evaporite components are not included. It comprises essentially an estimate of average terrigenous shale.

To demonstrate the coherence of certain trace elements between the upper crust and clastic sediments, we plot several sets of data on ternary diagrams, Th–Hf–Co and La–Th–Sc (Fig. 2.4). Included are shale analyses from various post-Archean sequences in Australia [49, 50], Antarctica [34], New Zealand [34], geochemical sedimentary standards SCo-1, SGR-1, MAG-1 [54] as well as particulate matter from the world's rivers [51] and particulate matter from an eastern USA estuary [55]. The cogent features are the low dispersion of the sediment data with respect to these parameters and the good correspondence to the average upper crustal composition. On the Th–Hf–Co diagram, the

Table 2.8. Chemical composition of particulate matter from selected rivers [51].

	Amazon	Congo	Ganges	Garonne	Mekong
SiO_2	60.1	58.0	66.9	58.2	61.9
TiO_2	1.3	1.6	0.96	0.84	0.63
Al_2O_3	22.8	25.0	16.0	22.4	22.3
FeO	7.5	10.4	5.2	7.5	7.6
MnO	0.14	0.20	0.14	0.22	0.13
MgO	2.0	1.1	2.3	2.9	2.3
CaO	2.3	1.4	4.1	2.7	0.87
Na_2O	1.2	0.32	1.5	0.75	0.74
K_2O	2.3	1.6	2.7	4.1	3.0
Σ	99.6	99.6	99.8	99.6	99.5
Cs	13	6	8.2	—	21
Ba	700	790	490	815	600
Rb	138	60	116	—	190
Sr	309	61	—	164	92
La	48	50	42	44	48
Ce	112	90	98	93	93
Pr	—	—	—	8.2	8.5
Nd	—	—	48	36	47
Sm	9.7	—	9.7	6.2	5.4
Eu	1.8	1.6	1.2	1.1	1.5
Gd	—	—	—	6.1	5.3
Tb	—	1.6	0.7	0.9	0.9
Ho	—	—	—	0.9	0.9
Er	—	—	—	2.4	2.7
Tm	—	—	0.35	0.44	0.45
Yb	3.7	2.6	3.2	2.8	3.2
Lu	0.6	0.37	0.51	0.42	0.58
ΣREE	—	—	—	207	222
La_N/Yb_N	8.8	13.0	8.9	10.6	10.1
Eu/Eu*	0.57	0.47	0.51	0.54	0.85
Th	13	16.2	17.5	13	17
U	2.5	3	2.8	3.6	5.8
Hf	6.2	5.1	8.4	—	4.0
Ta	2.0	1.1	1.25	—	1.35
Th/U	5.2	5.4	6.3	3.6	2.9
La/Th	3.7	3.1	2.4	3.4	2.8
Cr	193	175	71	255	102
V	232	163	—	150	175
Sc	18	12	11.5	—	19.5
Ni	105	74	80	33	99
Co	41	25	14	39	20
Ga	19	25	—	16	28
La/Sc	2.7	4.2	3.7	—	2.5
Th/Sc	0.72	1.4	1.5	—	0.87
B	68	43	—	120	87

fine-grained sediment data appear to be offset from the upper crustal composition, away from the Hf apex. This is probably the result of heavy mineral fractionation with zircon being concentrated in coarser-grained sediments (see Section 2.3.3).

Table 2.9. Various estimates of average post-Archean shales.

	1	2	3
SiO_2	62.8	62.0	61.6
TiO_2	1.0	0.8	1.0
Al_2O_3	18.9	18.0	18.3
FeO	6.5	7.1	7.4
MnO	0.11	—	0.13
MgO	2.2	2.9	2.7
CaO	1.3	3.6	4.2
Na_2O	1.2	1.5	1.1
K_2O	3.7	3.8	3.4
P_2O_5	0.16	0.20	0.22
Σ	99.9	99.9	100.1
LOI (CO_2, H_2O, etc.)	6.0	9.2	—
Cs	15	—	5
Ba	650	—	580
Rb	160	—	140
Sr	200	—	450
Pb	20	—	20
La	38	—	40
Ce	80	—	50
Pr	8.9	—	5
Nd	32	—	23
Sm	5.6	—	6.5
Eu	1.1	—	1
Gd	4.7	—	6.5
Tb	0.77	—	0.9
Dy	4.4	—	4.5
Ho	1.0	—	1
Er	2.9	—	2.5
Tm	0.40	—	0.25
Yb	2.8	—	3
Lu	0.43	—	0.7
ΣREE	183.0	—	145
La_N/Yb_N	9.2	—	9.0
Eu/Eu^*	0.66	—	0.5
Y	27	—	30
Th	14.6	—	11
U	3.1	—	3.2
Zr	210	—	200
Hf	5.0	—	6
Sn	4.0	—	6
Nb	19	—	20
Mo	1.0	—	2
W	2.7	—	2
Th/U	4.7	—	3.4
La/Th	2.6	—	3.6
Cr	110	—	100
V	150	—	130
Sc	16	—	10
Ni	55	—	95

Table 2.9 (continued)

	1	2	3
Co	23	—	20
Cu	50	—	57
Zn	85	—	80
Li	75	—	60
Ga	20	—	19
La/Sc	2.4	—	4.0
Th/Sc	0.91	—	1.1
Bi	0.25	—	0.01
B	100	—	100

1. This work—represents terrigenous shale, excluding carbonate components.
2. Clarke [52].
3. Krauskopf [53].

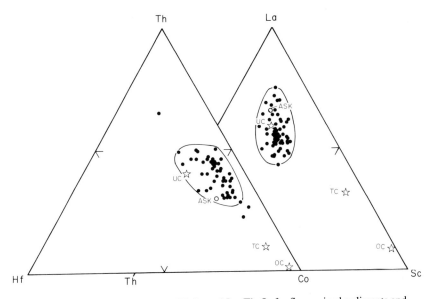

Fig. 2.4. Ternary diagram plots of Th–Hf–Co and La–Th–Sc for fine-grained sediments and sedimentary rocks of post-Archean age. Symbols as follows: UC—upper continental crust; TC—bulk continental crust; OC—average oceanic crust; ASK—Krauskopf's average shale [53]. Note the generally low dispersion and the similarity of the data to the present upper continental crust (UC) in comparison to more mafic compositions. The data are offset from the upper crustal composition, away from the Hf apex. This is probably a result of zircon concentration in sandstones (see Section 2.3.3). See text for data sources.

2.4 Rare earth elements in sedimentary rocks

2.4.1 The post-Archean record

Goldschmidt [56] was first to suggest that the homogenizing effects of sedimentary processes should result in nearly constant REE distributions in sedimentary rocks and that the pattern should reflect upper continental crust abundances. Although findings by other workers [57–60] have shown significant variability of REE distributions in some sedimentary environments, the original prediction of Goldschmidt has proven remarkably accurate [61]. Taylor [62] used the uniformity of REE in sedimentary rocks as a basis to construct an average crustal abundance table. Haskin [63, 64] suggested that a composite of mainly North American Palaeozoic shales (NASC) was representative of the upper continental crust. Re-analysis of the composite of European Palaeozoic shales (ES) [65] substantiated the original observations of Minami [66] and confirmed the similarity to NASC and average sediments of the Russian Platform [67]. An average of 23 post-Archean shales from Australia (PAAS) was reported by Nance & Taylor [49]. The values for the various composites and averages of shales from throughout the world are listed in Table 2.10 and displayed in Fig. 2.5 [68]. The remarkable uniformity of

Table 2.10. REE in various composites and averages of shales of post-Archean age (in ppm).

	PAAS (1)	NASC (2)	ES (3)
La	38	32	41.1
Ce	80	73	81.3
Pr	8.9	7.9	10.4
Nd	32	33	40.1
Sm	5.6	5.7	7.3
Eu	1.1	1.24	1.52
Gd	4.7	5.2	6.03
Tb	0.77	0.85	1.05
Dy	4.4	5.8	—
Ho	1.0	1.04	1.20
Er	2.9	3.4	3.55
Tm	0.40	0.50	0.56
Yb	2.8	3.1	3.29
Lu	0.43	0.48	0.58
ΣREE	183.0	173.2	204.0
La_N/Yb_N	9.2	7.0	8.4
Eu/Eu^*	0.66	0.70	0.70

1. Post-Archean average Australian shale [49]; average of 23 post-Archean shales from Australia.
2. North American shale composite [64]; composite of 40 North American Shales.
3. European shale composite [65]; composite of numerous European shales.

these patterns adds a great deal of support to the idea that typical shales reflect the composition of the exposed crust for the REE. The use of average analyses is favoured over a single analysis of a composite sample which may contain aberrant material and accordingly, we adopt PAAS as our average shale value.

REE abundances in a series of typical post-Archean fine-grained sedimentary rocks are tabulated in Table 2.7. Chondrite-normalized plots are given in Fig. 2.6. The similarity to the various shale composites and averages is apparent. The REE distribution in modern sedimentary environments is similar to that of the post-Archean shales (NASC, ES, PAAS). Typical analyses of suspended

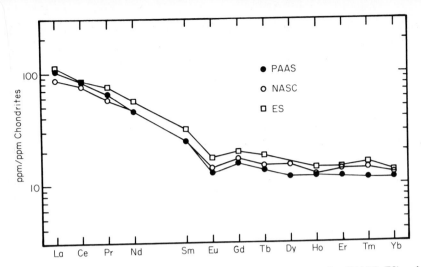

Fig. 2.5. Chondrite-normalized REE diagram of post-Archean shale composites (NASC, ES) and averages (PAAS). Data from Table 2.10. The uniformity of these patterns is particularly noteworthy and adds support to the view that sedimentary REE patterns reflect the upper continental crust exposed to weathering and erosion. The salient features of these patterns are the high abundances, relative to chondritic meteorites, light REE enrichment, fairly flat heavy REE patterns and, most importantly, the significant negative Eu-anomaly. Of the common igneous rocks, such patterns would best compare to granodioritic compositions. See Appendix 2 for chondrite normalizing factors and a definition of the europium anomaly.

particulate matter in some of the world's major rivers are listed in Table 2.8 [51]. The REE data are incomplete and somewhat scattered but it is clear that the patterns are very similar to PAAS (Fig. 2.7) although slightly enriched in total REE (about 1.1–1.4 times). The cause of the enrichment is probably related to the fine grain size of the river sediment. Young terrigenous sediments from the marine environment also show the same REE pattern as typical shales but again, enrichment in total REE is seen in some samples. These differences show that while most typical shales show the same relative REE characteristics (La_N/Yb_N, Eu/Eu^*; see Appendix 2), some caution is warranted in estimating average total REE abundances.

REE analyses of sandstones are less common than for shales. Chondrite-normalized plots for sand and sandstone of post-Archean age are given in Fig. 2.8. The REE patterns of these sandstones tend to have lower total REE abundances than shales, although like shales, the values are quite variable. A number of authors have noted the lower abundances in coarser-grained sedimentary rocks as compared to shales [49, 59, 63]. On the other hand, the shape of the REE patterns (Eu/Eu^*, La_N/Yb_N) is generally similar for sandstones and shales [49, 59, 69]. The REE abundances in quartz-rich sediments and sedimentary rocks (quartzites, orthoquartzites, etc.) are typically very low (Fig. 2.9). The shape of the pattern, however, is similar in a general way, to that of typical shales. Increased analytical uncertainty, inherent at lower concentrations, could explain some of the variability (particularly for samples analysed before 1970). The role of heavy minerals is also significant when clay fractions are minor ([59]; also see below).

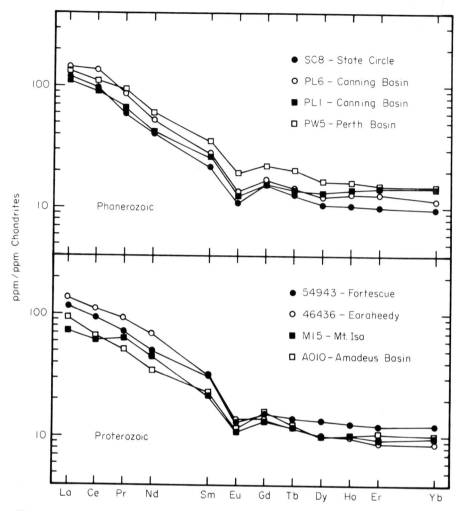

Fig. 2.6. Chondrite-normalized REE diagram of selected post-Archean shales from Australia. Data from Table 2.7. Note the uniformity of these patterns from sedimentary rocks of differing age and widely separated locations. Also note the similarity to the average and composite patterns of Fig. 2.5.

Most investigations have also reported rather low concentrations of REE in sedimentary carbonate rocks and minerals with typical total REE in marine calcite not exceeding 10 ppm [63, 70]. In general, carbonate REE patterns are similar to clastic sedimentary rocks. In marine carbonates, a distinct Ce depletion is common, reflecting the Ce depletion in sea water relative to the other REE. REE distributions in evaporites have not been studied in detail. Some incomplete average analyses of evaporites from Russia were characterized by very low abundances with total REE probably being less than about 10 ppm [71].

Although REE patterns in fine-grained sedimentary rocks are parallel to upper crustal abundances, they probably overestimate the absolute abundances by about 20% [72]. Three observations add support to this suggestion. Upper

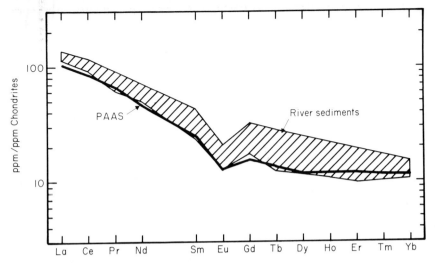

Fig. 2.7. Chondrite-normalized REE diagram showing the field occupied by modern river sediments. Data from Table 2.8. Also shown for comparison, as a solid heavy line, is PAAS. The river sediments have REE patterns which are generally parallel to PAAS although enriched in ΣREE. The enrichment is probably due to the very fine-grained nature of the sediment analysed.

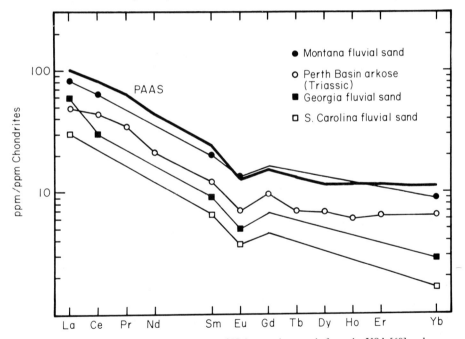

Fig. 2.8. Chondrite-normalized REE diagram of Holocene river sands from the USA [69] and a Triassic arkose from the Perth Basin, Australia [49]. PAAS is plotted as the solid heavy line. Note that the sands and sandstones have generally similar patterns to PAAS but are considerably depleted in ΣREE. In detail, sands and sandstones also tend to have more variable La/Yb ratios.

crustal estimates based on sampling programmes of shield areas give average La abundances of 32 ppm [8]. Fine-grained sedimentary rocks typically contain about 10–30% higher total REE than do associated coarser clastic sedimentary rocks (for example in greywacke–mudstone turbidite sequences). Clastics

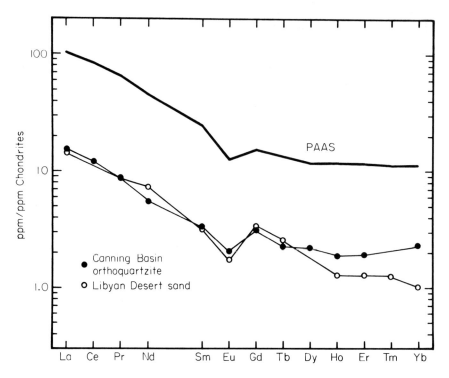

Fig. 2.9. Chondrite-normalized REE diagram of quartz-rich Libyan desert sand [63] and average of three Permian orthoquartzites from the Canning Basin, Australia [49]. PAAS is plotted as the solid heavy line. Quartz-rich sands and sandstones are greatly depleted in ΣREE compared with PAAS.

comprise about 70–80% of sedimentary rocks at the surface of the crust [14] but contain the bulk of the REE, carbonates and evaporites containing very low abundances (see below). We have performed mass balance calculations using available sedimentary REE data and known lithological proportions [e.g. 14] to estimate upper crustal abundances (Table 2.11). Within reasonable uncertainties, the results agree with those produced by simply reducing the average shale REE values by 20%.

Table 2.11. Mass balance of REE in sedimentary rocks.

	Shale	Sandstone	Carbonate	Evaporite	UC* (mass balance)	UC† (shale)
La	38	14	4.5	1.1	30	30
Sm	5.6	3.1	0.9	0.4	4.5	4.5
Eu	1.1	0.6	0.2	0.1	0.9	0.88
Gd	4.7	2.7	0.8	0.3	3.8	3.8
Yb	2.8	1.2	0.4	0.5	2.2	2.2
La/Yb	13.6	11.7	11.3	2.2	13.6	13.6
Eu/Eu*	0.66	0.63	0.72	0.88	0.67	0.65

*Upper crust using shale:sandstone:carbonate:evaporite=72:11:15:2 [14].
†Upper crust using PAAS reduced by 20%.

2.4.2 Mineralogical controls

The mineralogical location of the REE in sedimentary rocks is a question of some importance. Of the framework grains, quartz, feldspar and rock fragments are volumetrically the most important components. Quartz contains little or no REE (thus explaining the low abundances in sandstones), although mineral inclusions within quartz grains may have appreciable amounts. Feldspars are particularly important minerals in sediments. Some typical analyses of plagioclase and K-feldspars are listed in Table 2.12 and plotted on a chondrite-normalized diagram in Fig. 2.10. The notable features are the fairly low abundances and the positive Eu-anomalies. Eu-enrichment results from preferential substitution of Eu^{2+} for Sr^{2+}, because of its greater size in

Table 2.12. REE content of some common rock-forming minerals.

	A	B	C	D	E	F	G	H	I
La	—	—	—	—	—	—	—	86.0	—
Ce	0.569	5.94	0.442	22.5	1.36	0.264	2.20	127.3	20.0
Nd	0.365	7.23	0.645	27.5	0.252	0.0769	1.03	—	15.0
Sm	0.090	3.3	0.347	8.67	0.0200	0.0112	0.221	20.91	15.1
Eu	0.024	0.554	0.064	1.375	0.155	0.0821	0.0377	1.14	1.42
Gd	0.084	—	—	9.74	—	—	0.213	—	53.6
Tb	—	—	—	—	—	—	—	2.73	—
Dy	0.079	6.75	1.35	8.29	0.00552	0.0060	0.170	—	122
Er	0.046	4.04	1.40	4.18	0.00308	0.0029	0.0913	—	77.9
Yb	—	—	2.10	3.18	0.00301	0.0033	0.0792	14.48	70.3
Lu	0.0094	—	0.414	—	—	—	—	2.12	10.1
Ref.	[73]	[73]	[73]	[74]	[74]	[74]	[74]	[75]	[73]

A. Olivine, basalt
B. Clinopyroxene, andesite
C. Orthopyroxene, andesite
D. Hornblende, granodiorite
E. Plagioclase, granodiorite
F. Alkali feldspar, granodiorite
G. Biotite, granodiorite
H. Muscovite, granite
I. Garnet, dacite

comparison to the trivalent REE. The REE patterns of rock fragments will be highly variable and unpredictable on an *a priori* basis. In Table 2.12, we also list some typical REE analyses for other common igneous minerals and plot the data in Fig. 2.10. The main feature to be noted is the extreme variability in REE patterns. It is also important to point out that the REE content of an igneous mineral is controlled by the mineral-melt distribution coefficient and thus mineral REE patterns depend strongly on the bulk REE composition of the parent magma. Similarly, the REE content of metamorphic minerals is influenced by bulk rock REE abundances.

The importance of heavy minerals in sedimentary rocks, and the possible geochemical fractionation associated with heavy mineral segregation, has been

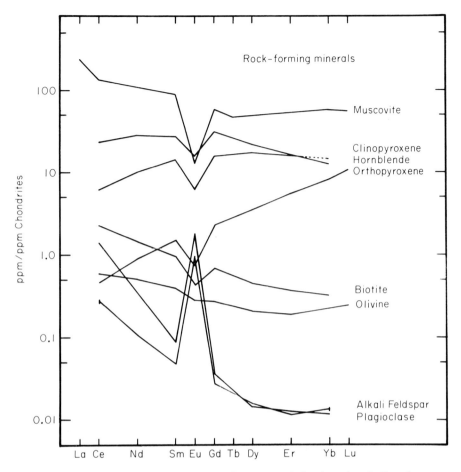

Fig. 2.10. Chondrite-normalized REE diagram of common rock-forming minerals. Data from Table 2.12.

discussed above. In Table 2.13, we present some typical REE analyses of minerals which may be present in heavy mineral suites. The patterns are plotted in Fig. 2.11. The important features to note are the high abundances and extreme HREE enrichment in the common heavy minerals zircon and garnet. Less common heavy minerals such as apatite may have high abundance but are less important in the overall REE mass balance in sediments. The REE mineral monazite is generally not abundant enough to play a dominant role although, in individual cases, concentrations of as little as 0.005% would have to be considered.

Zircon is probably the most important heavy mineral in terms of REE abundances. In fine-grained clastic sedimentary rocks, Zr typically is about 200 ± 100 ppm which is comparable to the upper crustal Zr abundance of 190 ppm (see Section 2.6). Typical Zr/Yb ratios are greater than 1000 in zircon (see Table 2.13). If a rock contains enough zircon to contribute 100 ppm Zr, then the zircon will also contribute $\leq 0.5 \times$ chondritic levels of the heavy REE. This is insignificant in relation to the REE content of shales which typically

Table 2.13. REE content of some common heavy minerals.

	A	B	C	D	E	F	G	H
La	—	—	—	119,000	—	—	—	10.3
Ce	42.3	20.0	509	195,000	66,560	152	3305	13.7
Pr	—	—	—	32,100	—	—	—	—
Nd	14.9	15.0	302	98,000	16,060	58.2	2680	—
Sm	5.40	15.1	52.9	24,500	1,260	9.45	655	1.86
Eu	1.27	1.42	15.2	635	133.3	3.38	165	0.11
Gd	17.4	53.6	—	14,700	460	8.15	564	—
Tb	—	—	—	1,960	—	—	—	0.16
Dy	56.9	122	31.7	7,710	118.4	5.67	470	—
Ho	—	—	—	1,400	—	—	—	—
Er	116	77.9	17.1	—	28.5	2.69	237	—
Yb	253	70.3	13.9	540	17.4	2.10	207	0.59
Lu	—	10.1	—	—	—	—	—	0.10
Ref.	[74]	[73]	[74]	[76]	[74]	[74]	[74]	[75]

A. Zircon, granodiorite
B. Garnet, dacite
C. Apatite, granodiorite
D. Monazite, granite
E. Allanite, granodiorite
F. Epidote, granodiorite
G. Sphene, granodiorite
H. Tourmaline, granite

contain 10–20 × chondritic Yb. In siltstones and sandstones, where REE abundances are lower and Zr concentrations may reach 400–500 ppm, REE patterns may be significantly affected.

The clay mineral fraction is an important carrier of the REE. Individual analyses of various clay minerals are unavailable but Cullers *et al.* [59] have examined the REE content of various grain size fractions (including <2 micron) of some unconsolidated late Palaeozoic sediments from central USA. In some cases, they also separated heavy minerals from silt and sand fractions to evaluate the overall effects of heavy mineral suites. The results of two particularly instructive samples are listed in Table 2.14. In Fig. 2.12, we plot the sand and silt data normalized to the <2 micron fraction. The finest, clay fraction (<2 micron) is useful for normalization in this case because it most closely approximates what we would expect for the REE content in shales derived from the same provenance. It is clear that there is a significant, although erratic, effect of heavy minerals on the HREE distribution of sand and, to a much lesser extent, silt fractions.

From these data and others [59, 77], the following conclusions are indicated:
1. the bulk of REE in clastic sedimentary rocks is present in the <2 micron and silt fractions. There is no correlation of REE content to clay mineralogy;
2. there is little or no difference in REE patterns in the grain sizes ranging from <1 micron to about 20 microns (medium-fine silt). The silt fraction tends to have lower total REE than the clay fraction (the shape of the pattern

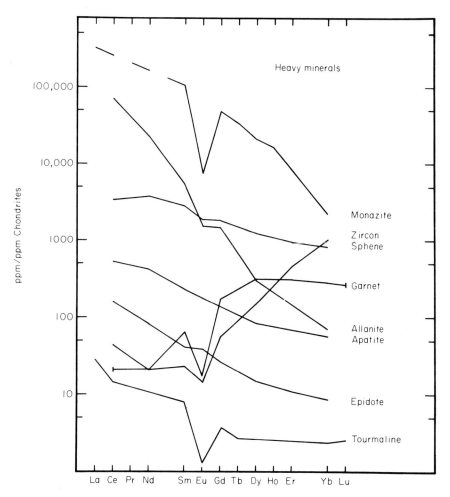

Fig. 2.11. Chondrite-normalized REE diagrams of common heavy minerals. Data from Table 2.13. Zircon is a heavy mineral of particular importance and is notable for its heavy rare earth element enrichment.

is similar). There is no obvious correlation between REE content and clay mineral to quartz ratio in the silt fraction;

3 the sand-size fraction tends to have lower total REE and less fractionated patterns (lower La/Yb) than silt and clay fractions;

4 when the heavy minerals are removed from the sand-size fractions, the effect is to lower the total REE and raise the La/Yb ratio. This suggests that the heavy mineral suite is HREE-enriched and high in total REE. The silt fraction appears to exhibit similar characteristics but the effects of heavy minerals are less pronounced;

5 the nature of the Eu-anomalies is similar in all of the size fractions considered.

2.4.3   Effects of weathering

Few studies are available on REE behaviour during weathering [57, 58, 78,

Table 2.14. Rare earth elements in various size fractions of sedimentary rocks [59].

	Whole rock	Sand	Sand less heavy minerals	Silt	Silt less heavy minerals	<2 micron	Weighted sum of fractions
Havensville shale—W2							
La	30.0	4.2	4.0	27.9	19.1	55.3	34.2
Ce	58.1	10.6	6.5	61.7	38.6	100	65.8
Sm	5.0	0.79	0.59	5.17	3.11	5.85	4.33
Eu	1.01	0.16	0.10	0.85	0.61	1.08	0.77
Tb	0.93	—	0.068	0.70	0.47	0.82	0.62
Yb	2.6	0.65	0.32	2.60	1.67	3.81	2.64
Lu	0.44	0.084	0.053	0.41	0.29	0.54	0.38
ΣREE	142	25	19	145	92	250	—
La_N/Yb_N	7.8	4.4	8.4	7.3	7.7	9.8	—
Eu/Eu*	0.59	—	0.59	0.53	0.61	0.59	—
Weight % of rock		24.7		28.1		45.8	
Okaloosa shale—ELK6AII							
La	7.3	4.4	3.0	31.5	22.6	27.8	8.05
Ce	15	9.2	6.3	73.1	48	61.6	17.7
Sm	1.53	0.98	0.45	7.48	4.31	4.62	1.78
Eu	0.29	0.17	0.077	1.42	0.78	0.80	0.32
Tb	0.25	0.20	0.05	1.19	0.59	0.61	0.32
Yb	1.50	1.25	0.31	4.5	2.22	2.2	1.62
Lu	0.23	0.20	0.052	0.84	0.36	0.40	0.27
ΣREE	40	25	14	178	115	135	—
La_N/Yb_N	3.3	2.4	6.5	4.7	6.9	8.5	—
Eu/Eu*	0.58	0.50	0.60	0.59	0.59	0.57	—
Weight % of rock		86.0		10.2		3.8	

79]. It is generally agreed that there is an overall enrichment of REE in weathering profiles but the nature and magnitude of such enrichments are difficult to judge since most studies to date have not considered volume changes or have had to assume an element, other than the REE, to be completely immobile during weathering. Whether fractionation occurs during weathering is of importance to sedimentary REE studies.

During incipient to moderate stages of weathering of a granodiorite a progressive enrichment of REE (up to 100–200% over parent rock) with particular enrichment of HREE (up to 300%) was noted [78]. Extremely weathered residual clay material (mainly kaolinite and illite) in fractures, however, displayed a complementary pattern (Fig. 2.13). Nesbitt [78] argued that REE were mobilized in response to changes in soil and general water chemistry and that the REE fractionation was mineralogically controlled. The complementary REE patterns of weathered granodiorite and residual clays suggests that there is not a significant net loss of REE from the weathering profile.

The intensity of weathering effects are very much related to the rock type [57]. Thus rocks with unstable mineralogy under surface conditions, such as

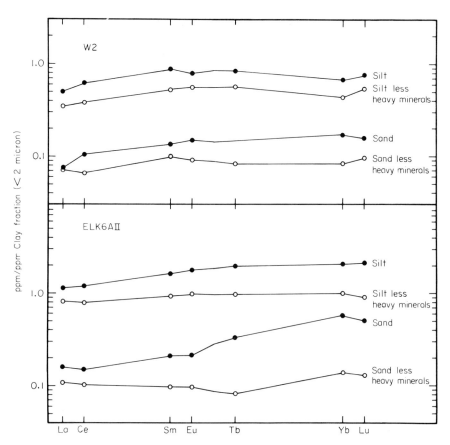

Fig. 2.12. Rare earth element diagram of various sand size fractions of two sediment samples (W2, ELK6AII), normalized to the <2 micron fraction. Data from Table 2.14.

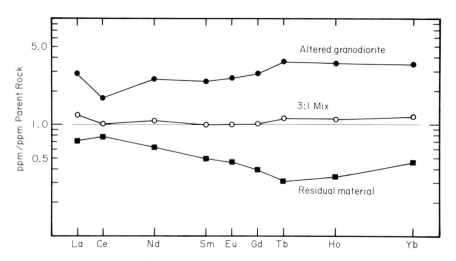

Fig. 2.13. Rare earth element diagram of weathered materials from a granodiorite, normalized to the unweathered parent rock. Shown is the most altered granodiorite and fracture filling residual material. The REE pattern of a 3:1 mix of residual material and altered granodiorite is essentially equivalent to the unweathered parent rock. (Adapted from [78].)

ultramafic and basaltic rocks may show considerably more effect than rocks with relatively stable mineralogy, such as granites and granodiorites.

It is generally agreed that the transport of REE into sedimentary basins is primarily a result of mechanical rather than chemical transport ([49, 50, 62–65, 69, 71–73, 78]; also note low REE abundances in natural waters). Thus, it is unlikely that redistribution of REE within weathering profiles would have a significant effect on REE patterns in well mixed sedimentary rocks. The remarkable uniformity of REE distribution patterns in shales is an obvious testimonial to this conclusion.

2.4.4 Effects of diagenesis and metamorphism

Pliocene–Miocene sediments from the Gulf Coast, Louisiana, taken from a deep well (1.81–4.77 km depth), have REE patterns similar to typical post-Archean sedimentary rocks [77] but were slightly to moderately enriched in total REE. Total REE and La_N/Yb_N decreased by about 20–30% for bulk samples and clay fractions with depth. The magnitude of the Eu-anomalies, which are similar to post-Archean sedimentary rocks, remained constant. The changes in REE characteristics were not easily correlated with mineralogical and other geochemical changes associated with diagenesis. Thus, the changes in REE characteristics were mainly controlled by provenance rather than diagenesis.

A large, and somewhat confusing literature exists on the question of REE mobility during metamorphic and hydrothermal alteration. Much of the debate turns on apparent evidence of mobility of REE during low grade metamorphism. The apparent evidence for some REE mobility during these processes contrasts with the general evidence for lack of mobility of the REE at higher metamorphic grades. Workers claiming such mobility, during spilitization, include Floyd [80] and Hellman & Henderson [81]. These workers demonstrated that changes in REE patterns may occur, but have not identified any regular or systematic cause. The observed mobility during zeolitization can be attributed to the breakdown of residual glass, enriched particularly in LREE [82, 83]. Selective loss of, or redistribution of LREE during zeolite facies metamorphism thus finds a reasonable explanation by this mechanism, without appealing to exotic processes or to alleged elemental mobility. Hydrothermal experiments on a glassy tholeiite at temperatures of 150–600°C and water/rock mass ratios of 1–125 showed that the REE were immobile even when basalt was altered totally to clay [84]. On the other hand, Staudigel & Hart [85] noted removal (but not fractionation) of REE from basalt glass during alteration to palagonite. They argued that significant movement of REE only occurred at the basalt–sea water interface where water/rock ratios are extremely high. The results of this study have been confirmed by Michard et al. [86]. A number of experimental studies [87] have also shown that REE have very large silicate melt–aqueous fluid and mineral–aqueous fluid distribution coefficients, suggesting extremely large water/rock ratios would be necessary to cause significant changes in REE patterns during metamorphism.

2.4.5 Origin and significance of Eu-anomalies in sedimentary rocks (see Appendix 2)

Understanding the origin of the depletion in Eu, relative to the other chondrite-normalized REE in clastic sedimentary rocks is fundamental to any interpretations of crustal composition and evolution. The most significant observation in this regard is that virtually all post-Archean sedimentary rocks (sandstones, mudstones, carbonates) are characterized by Eu depletion of approximately comparable magnitude. The only important sedimentary rock types which do not have Eu depletion are some of the first cycle volcanogenic sediments deposited in fore-arc basins of island-arcs and derived mainly from andesites which, not unnaturally, reflect the parent rock patterns [88]. No common sedimentary rock is *characterized* by Eu enrichment. Similarly, river and sea water also show Eu depletion [e.g. 22, 89]. This is strong evidence that the upper continental crust must be similarly depleted in Eu based on mass balance.

Common igneous rocks derived from the mantle (e.g. MORB, island-arc basalts and andesites) are not characterized by Eu-anomalies of any sort [90]. Those which have been noted can usually be readily attributed to late stage addition or removal of feldspar [90] or hydrothermal alteration [91]. Accordingly, if the continental crust is derived from the mantle, then it follows that the bulk continental crust is not anomalous with respect to Eu. Thus, any suggestion that Eu-anomalies seen in post-Archean clastic sedimentary rocks are due to oxidation–reduction processes during weathering or the breakdown of feldspar is unfounded since there is no significant upper crustal reservoir with the complementary Eu enrichment. The only tenable explanation is that the Eu depletion in sedimentary rocks, and hence the upper continental crust, is due to chemical fractionation within the continental crust, related to production of K-rich granitic rocks which typically possess negative Eu-anomalies. The residual material in the lower continental crust would thus contain the complementary enrichment in Eu, the magnitude of which would depend on the relative proportions of the upper and lower crust [72, 92]. Any Eu^{2+} released during weathering will be oxidized to Eu^{3+}, and hence behave like the other trivalent REE. The presence of an EU anomaly is thus the signature of earlier events in a more reducing igneous environment than now exists in the upper crust.

2.5 Other upper crustal samples

2.5.1 Loess and crustal composition

About 10% of the land surface of the Earth is covered with loess of Pleistocene age. The blanketing deposits may in places exceed 100 m in thickness, but generally are a few metres thick, comprised of silt-sized (30–50 microns) angular particles, principally quartz and feldspar. Their origin was originally the cause of much debate, puzzling geologists as eminent as Lyell [93]. The aeolian origin of loess was originally proposed by von Richthofen [94] and this hypothesis has survived. A remaining controversy concerns the origin of the high carbonate contents occasionally encountered in loess. Ancient loess deposits are rare, probably due to the difficulty of preservation [95]. The origin of loess by aeolian transport from glacial outwash deposits, particularly during cold dry climatic regimes, appears well established. The uniformity of grain size in loess is presumably due to wind transport, while the angularity of the grains attests to a glacial, rather than a desert origin. Glacial erosion grinds up rocks, a task carried out less efficiently by geochemists. This combination of widespread production of silt-sized rock flour, and its transport by wind over tens to hundreds of kilometres provides geochemists with a natural sampling of comparatively unweathered material from the exposed crust. Can loess be used to provide an average composition of the upper crust?

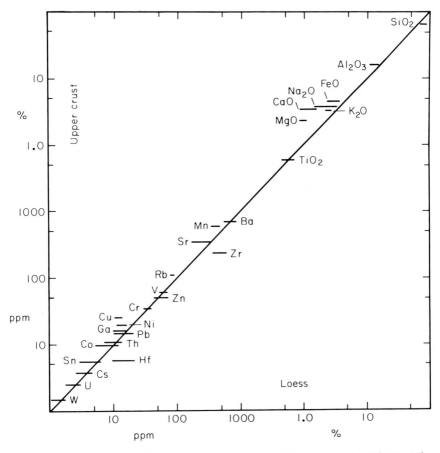

Fig. 2.14. Comparison diagram for the chemical composition of the upper continental crust and the range of unweathered and non-carbonate loess samples (adapted from [12]). On average, the loesses are enriched in silica and depleted in most other major elements in comparison with the upper crust. Zr and Hf are significantly enriched in the loesses.

Studies on the chemical and isotopic composition of loess have attempted to answer this question and to see whether loess can provide useful insights into the evolution of the continental crust [12]. Loess has generally a very similar major element composition, expressed on a carbonate-free basis, although some local provenance effects are apparent [97]. Figure 2.14 shows a comparison of loess with that of the upper continental crust. The principal difference is that loess has a higher silica content, causing the other constituents to fall below the crustal average. This effect is presumably due to the behaviour of quartz during glacial erosion, resulting in its concentration in the 30–50 micron size range. The trace element abundances show a general similarity to upper crustal averages (Figs 2.14, 2.15). They are slightly lower due to the diluting effect of quartz. Zr and Hf are enriched, like silica, relative to upper crustal estimates indicating that zircon, like quartz, is being concentrated in loess. Soluble elements such as Cs, Ba, Pb and Rb are not depleted in loess. Chalcophile elements, such as Cu, Tl and Ni are low, possibly reflecting a low abundance in the uppermost part of the continental crust.

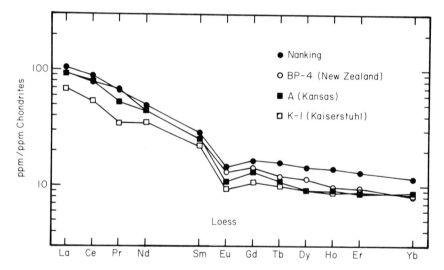

Fig. 2.15. Chondrite-normalized REE diagram of selected loess samples from around the world [12]. The patterns are all very similar and bear a strong resemblance to PAAS. The Kaiserstuhl sample is carbonate-rich and this is reflected in lower ΣREE abundances resulting from dilution of the carbonate minerals. The Banks Peninsula sample has slightly lower Eu/Eu* than the other samples reflecting derivation from the New Zealand greywacke terrains.

The REE data for the New Zealand loesses show slightly lower Eu depletion, reflecting their derivation from greywackes (Fig. 2.15). The other loess from North America, Europe and China show very uniform REE patterns with Eu depletions equivalent to those of shales. The uniformity of the REE patterns for the widely scattered loess deposits and their similarity to PAAS, ES and NASC indicates that loess is providing the same information on REE abundances as other clastic sediments. These two independent sampling techniques thus provide identical REE patterns, reinforcing the concept that both processes (glacial erosion and normal erosion and deposition) are providing an average sample of the upper continental crust.

Information on model ages for loess is presented in Chapter 5 during the discussion on the uniformity of crustal composition though time. Loess Nd model ages are generally younger than about 1500 m.y., reflecting their derivation from younger orogenic areas. These ages overlap those of modern sediments. The uniformity of the REE patterns in the loesses indicates that additions to the upper continental crust over that period of time have not changed in composition.

2.5.2 Tektites

The chemical composition of meteorite impact glasses bears a close resemblance, except for the most volatile elements, to that of the country rocks in which the impact crater was excavated, and sometimes bears the signature of a small contribution from the impacting body (e.g. excess Ni, Ir). The origin of tektites, which are the results of much larger scale events than the typical kilometre-sized impact craters, was long the subject of controversy. Extraterrestrial origins were ruled out both by the analysis of returned lunar samples and by the lack of exposure to cosmic radiation, which precludes origins from outside the Earth–Moon system. The composition of tektites bears an unmistakable terrestrial signature, recognized in early studies [98]. All subsequent geochemical and isotopic investigations have served to reinforce this conclusion [99, 100]. Ivory Coast tektites are linked to the Bosumtwi crater, Ghana, whereas the Bohemian and Moravian modavites are most likely derived as splashed melt from Tertiary sands forming the surface layer at the Ries meteorite impact structure, Germany. The sources of the Texan bediasites (K–Ar age 34 m.y.) and the South-east Asian and Australian strewn fields remain to be discovered. This latter occurrence, about 700 000 years old, extends from Hainan to Tasmania with the source area most probably in the region of Vietnam, Thailand or South China [101]. The composition of this group is quite close to that of loess or subgreywacke, and is more silica-rich than the average composition of the upper crust [102]. The REE patterns are identical to those of PAAS [103]. Nd and Sr isotopic studies [100] confirm an origin by melting of sedimentary rocks. The Australasian group were derived from sediments laid down about 250 m.y. ago, while the age of the derivation from mantle sources was 1.15 Ae [100].

The chemical and isotopic data are thus both consistent with derivation of tektites by large-scale meteorite or cometary impact into continental crustal sediments or continental shelf sediments.

2.6. The chemical composition of the upper crust

2.6.1 The rare earth elements

The uniform distribution of REE in most fine-grained clastic sedimentary rocks is very well documented. It is also generally agreed that the average shale

REE pattern must represent that of the upper crust. We have suggested that the average upper crustal pattern is parallel to average shale but lower in absolute abundances due to the presence of sediments with much lower REE abundances (sandstones, carbonates, evaporites). The REE concentrations in the hydrosphere are so low that they are not a significant reservoir (Table 2.3). One can calculate a simple mass balance among various sedimentary reservoirs to show that the upper crust is about 20% lower in ΣREE when compared with average shale (Table 2.11).

The general uniformity of REE patterns in post-Archean sedimentary rocks can be seen in Figs 2.5–2.7 and 2.15 where representative patterns from many sequences of widely varying age are plotted. Absolute abundances vary somewhat but the shapes of the patterns are remarkably similar. The values selected for the REE abundances in the upper crust are identical to those published by Taylor & McLennan [72] and Taylor [104] and are listed in Table 2.15. The average upper crust has an estimated Sm/Nd ratio of 0.173. The upper crustal REE pattern provides a guide for evaluating the abundance levels for the other elements. The pattern is close to that of typical granodiorites, which most other studies suggest is close to the average composition of the upper crust.

2.6.2 The heat producing elements (K, Th, U)

There are two independent approaches to obtain these values. The first

Table 2.15. Chemical composition of the upper continental crust. (See text for references.)

	%		NORM	
SiO_2	66.0	Q	15.7	
TiO_2	0.5	Or	20.1	
Al_2O_3	15.2	Ab	13.6	
FeO	4.5	Di	6.1	
MgO	2.2	Hy	9.9	
CaO	4.2	Il	0.95	
Na_2O	3.9			
K_2O	3.4			
	99.9			

Li	20 ppm	Ni	20 ppm	In	50 ppb	Tm	0.33 ppm
Be	3 ppm	Cu	25 ppm	Sn	5.5 ppm	Yb	2.2 ppm
B	15 ppm	Zn	71 ppm	Sb	0.2 ppm	Lu	0.32 ppm
Na	2.89%	Ga	17 ppm	Cs	3.7 ppm	Hf	5.8 ppm
Mg	1.33%	Ge	1.6 ppm	Ba	550 ppm	Ta	2.2 ppm
Al	8.04%	As	1.5 ppm	La	30 ppm	W	2.0 ppm
Si	30.8%	Se	0.05 ppm	Ce	64 ppm	Re	0.5 ppb
K	2.80%	Rb	112 ppm	Pr	7.1 ppm	Ir	0.02 ppb
Ca	3.00%	Sr	350 ppm	Nd	26 ppm	Au	1.8 ppb
Sc	11 ppm	Y	22 ppm	Sm	4.5 ppm	Tl	750 ppb
Ti	3000 ppm	Zr	190 ppm	Eu	0.88 ppm	Pb	20 ppm
V	60 ppm	Nb	25 ppm	Gd	3.8 ppm	Bi	127 ppb
Cr	35 ppm	Mo	1.5 ppm	Tb	0.64 ppm	Th	10.7 ppm
Mn	600 ppm	Pd	0.5 ppb	Dy	3.5 ppm	U	2.8 ppm
Fe	3.50%	Ag	50 ppb	Ho	0.80 ppm		
Co	10 ppm	Cd	98 ppb	Er	2.3 ppm		

involves direct measurement. The data of Shaw [8] and Fahrig & Eade [4] are in agreement, except for U where the value of Fahrig & Eade is about 40% lower. The higher value is considered the most reliable because it gives values of K/U = 10^4 and Th/U = 3.4.

The sedimentary data provide an entirely independent estimate. McLennan et al. [105] noted a correlation between La and Th in fine-grained sedimentary rocks (Fig. 2.16). They suggested that this was due to similar

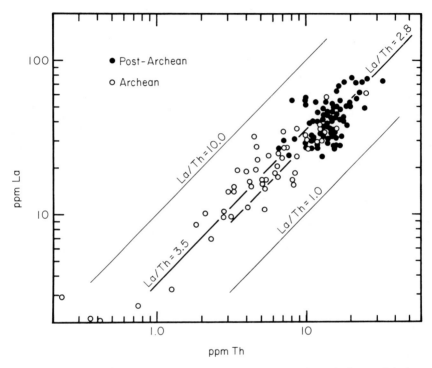

Fig. 2.16. Plot of La against Th for fine-grained sedimentary rocks of post-Archean and Archean age. There is a good correlation between these elements, resulting from similar behaviour during sedimentary processes. Note that Archean sedimentary rocks have lower abundances of La and Th and higher La/Th ratios, on average. The importance of this is discussd in Chapters 5 and 7.

behaviour for REE and Th in sedimentary processes and pointed out that an upper crustal Th value could be estimated. The most recent value for average La/Th ratios in sedimentary rocks is 2.8 ± 0.2. For an upper crustal La value of 30 ppm, a Th value of 10.7 ppm is derived. If we assume an upper crustal Th/U ratio of 3.8, U = 2.8 ppm and for K/U = 10^4, K = 2.8%. This totally independent method agrees with the measured values to within about 2% for K and Th and to within about 10% for U, and is adopted here. The contribution of these abundances to the measured surface heat flow is discussed in Chapter 3. Estimates of the thickness of the upper crust, assessed here at 10 km, are given in Section 2.6.5.

2.6.3 Rubidium and strontium

The upper crustal K/Rb ratio is well established at about 250. Shaw [8] has estimated Rb abundances, from direct measurement at 110 ppm. Using the K value derived from the sediment data (2.8%), and K/Rb = 250, Rb = 112 ppm. Thus agreement among the various estimates is very good, to within 10%. We adopt a value of 112 ppm for Rb.

The K/Rb ratio for the upper crust is very close to the value in Type I carbonaceous chondrites (K/Rb = 248). This agreement is probably fortuitous. K/Rb ratios for the bulk earth are probably between 320 and 390 (see Chapter 11), implying that Rb is depleted relative to less volatile K. During partial melting in the mantle, Rb will be partitioned more readily than K into the melt phase, and so will eventually be concentrated relative to K in the upper crust as a consequence of successive partial melting and fractional crystallization. Accordingly the ratio K/Rb will decrease from bulk mantle values during formation of the upper crust, and the agreement between chondritic and upper crustal K/Rb is accidental. Cesium exhibits this behaviour to a more extreme degree. Rb/Cs for the upper crust is about 30, but is only 12 in chondrites.

The value of Sr is not quite as well known. Shaw [8] and Fahrig & Eade [4] give values 316 ppm and 380 ppm, respectively, based on direct sampling. These agree only to within about 20%. An intermediate value of 350 ppm is adopted. Thus, the Rb/Sr ratio for the upper crust is estimated at 0.32, but has at least a 10% uncertainty.

2.6.4 Other elements

The compilations published by Taylor & McLennan [72] and Taylor [104] were modified by more recent data from the values published by Taylor [106]. These compilations have been amended here by the addition of more recent data, and include estimates for elements not listed since the 1964 compilation [62], particularly for the Pt group elements, and new abundance estimates for other elements. The changes to the previous tables are not extensive, and it appears that a consensus is being reached on the average abundances in the upper crust (Table 2.15).

The data for Zn, Cd, Tl, Pb and Bi are from the new estimates by Heinrichs et al. [2]. The value for gallium is derived from the Ga/Al ratio of 0.21×10^{-3}. The Ta value comes from the crustal Nb/Ta ratio of 11.6. The abundance of Sc is derived from the La/Sc ratio in 67 sediments of 2.7 and from the Th/Sc ratio of 1.0 in the same suite. Data for Au, Pd and Ir are from Shaw et al. [8]. The tin value, which represents one of the few significant changes, is from Taylor & McLennan [107]. Updated values for Li, Be, B, Ba, Ge, As, Se, Zr, Mo, Ag, In, Sb, W and Re are from the compilations in the *Handbook of Geochemistry* [108].

The values for the major elements are mostly unchanged from the previous compilations. They are based on the large scale sampling programmes in the Canadian Shield, and are representative of the upper crust, as distinct from total crustal estimates.

2.6.5 Heat flow and the thickness of the upper crust

The thickness of the upper crustal layer may be estimated from the abundances of the heat producing elements, K, U and Th, and the heat flow data. A crust about 30 km thick with the abundances listed in Table 2.15 would produce the observed continental heat flow of 60 mW/m^{-2} [109]. This does not allow for any contribution from the mantle, which accounts for about 60% of the observed heat flow [110].

The linear relationship between heat flow and surface heat production [111] is usually given as

$$q_s = q^* + A_s b$$

where q_s = measured heat flow
A_s = heat production of surface rocks
b = thickness of heat producing layer
q^* = 'reduced heat flow' from beneath the crust

The depth parameter b does not of course uniquely define the depth distribution of the heat producing elements. This is usually considered to be one of two extremes. The exponential model assumes that heat production decreases exponentially from the surface. In this model, the linear relationship between heat production and heat flow is preserved. The second or step model assumes that the heat production remains constant down to depth b. The distribution of K, U and Th with depth will generally lie between these two models, and is likely to be complex in detail. For example, in the upper crustal section exposed in the Vredefort structure, South Africa, U and Th showed a general exponential decrease with depth, but the distribution of K remained constant to depths of 12 km (Fig. 2.17) [112]. In the near-surface environment, uranium is highly mobile, due to meteoric water circulation. Both K and U are likely to be mobile in magmatic and metamorphic fluid phases while Th should be more stable [113]. There appear to be regional differences among heat flow provinces [113]. Caution must be exercised in relating heat flow measurements and radioactive element distributions on too fine a scale. Several studies have shown that groundwater flow on scales of tens of kilometres may perturb heat flow on a regional scale. Accordingly, much complexity is expected, and b varies from 3.8 km for the Bohemian Massif to 16 km for England and Wales [114].

In this book, we are concerned with an overall average. An average value of 10 km ±3.5 is derived from the review by Morgan [114] and is adopted here. This is taken as the average maximum thickness of the upper crust, and is the zone within the crust where the bulk of the K, U and Th are concentrated.

2.6.6 Mineralogy of the upper crust

Various estimates have been made of the mineralogical composition of the upper crust. These should be compatible with the estimates of chemical compositions. In Table 2.16 we present four such estimates. The first (column

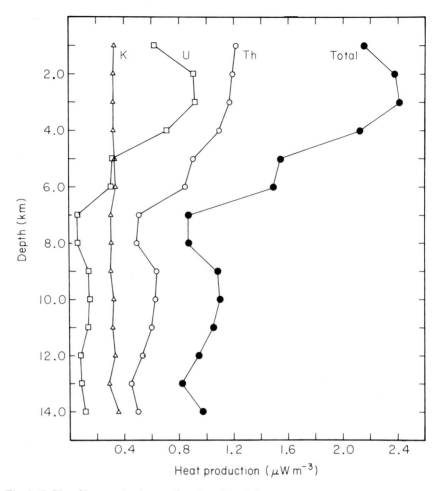

Fig. 2.17. Plot of heat production as a function of depth for a crustal profile through the Vredefort structure [112]. Data shown for K, Th, U and total heat production. Note that U, Th and total heat production show a general exponential decrease with depth but K heat production remains constant.

A) gives the average mineralogy for plutonic rocks in the crust from Wedepohl [115]. The second (column B) is the calculation of the mesonorm [17] for the average Canadian Shield composition. A revised average is given in column C [17] along with an estimate of the mineralogy of the exposed continental crust. This, effectively a surface estimate, includes the large areas of plateau basalts and other extrusive rocks. The principal difference, important in weathering studies, is the significant amount of glass present in such rocks. This estimate is given in column D.

2.7 Summary

1 Estimates of the composition of the upper continental crust have been made by large-scale sampling programmes and from analyses of sedimentary

Table 2.16. Mineralogical composition of the upper crust (in volume per cent) (adapted from [17]).

	A	B	C	D
Plagioclase	41	39.3	30.9	34.9
K-feldspar	21	8.6	12.9	11.3
Quartz	21	24.4	23.2	20.3
Glass	0	0	0	12.5
Amphibole	6	0	2.1	1.8
Biotite	4	11.2	8.7	7.6
Muscovite	0	7.6	5.0	4.4
Chlorite	0	3.3	2.2	1.9
Pyroxene	4	0	1.4	1.2
Olivine	0.6	0	0.2	0.2
Oxides	2	1.4	1.6	1.4
Others	0.5	4.7	3.0	2.6

A. Mineral composition of plutonic rocks (Wedepohl [115]).
B. Mesonorm of Canadian Shield average (Shaw et al. [8]).
C. Mineral composition of upper crust assuming metamorphic/intrusive ratio of 17/9 (Nesbitt & Young [17]).
D. Mineral composition of exposed continental crust surface [17].

rocks, utilizing the natural sampling processes of erosion and deposition to provide an average.

2 The factors to be considered in using sedimentary rocks to provide an upper crustal average include weathering, solubility, erosion, physical transport, element residence time in the ocean, deposition, diagenesis and metamorphism. An evaluation of these factors leads us to conclude that only a few chemical elements, in fine-grained terrigenous clastic sediments, preserve a direct relationship between their abundance in the upper crust and their presence in the sedimentary record. These include the REE, Th and Sc.

3 In shales, the REE are concentrated mainly in the <2 micron fractions. There is no significant difference in the REE patterns between the finest fractions (<1 micron) and the silt-sized (20 micron) fraction. Heavy minerals contribute only a minor proportion of the REE but tend to be enriched in heavy REE.

4 Although some fractionation of REE may take place within weathering profiles, there appears to be no selective loss of REE during weathering.

5 Diagenesis and metamorphism have little relative effect on REE patterns. The REE appear to be truly immobile elements.

6 REE patterns in post-Archean fine-grained clastic sediments (shales) have $La_N/Yb_N = 9.2$ and $Eu/Eu^* = 0.66$. These values are taken to be representative of the upper crust. From mass balance calculations, the absolute abundances of REE in the upper crust are about 20% lower than in average shales.

7 The depletion in Eu is not due to surficial processes, but is inherited from igneous precursors, where some of the Eu was present as Eu^{2+}.

8 The wind-blown sediment, loess, has REE patterns parallel to those of shales, reinforcing the concept that this pattern is typical of the upper crust.

9 Table 2.15 provides our estimate for the composition of the upper continental crust, for 62 elements. Key abundances and ratios include:

$SiO_2 = 66\%$; K/Rb = 250; Rb/Cs = 30; K = 2.8%; U = 2.8 ppm; Th = 10.7 ppm; Rb/Sr = 0.32; Sm/Nd = 0.17; $La_N/Yb_N = 9.2$; $Eu/Eu^* = 0.65$.

10   From the abundances of the heat-producing elements K, U and Th, and the relationship between surface heat production and heat flow, the thickness of the upper crust is assessed as 10 km.

Notes and references

1   Poldevaart, A. (1955) *Chemistry of the Earth's Crust.* GSA Spec. Pap. 62, 119.
2   Heinrichs, H. *et al.* (1980) Terrestrial geochemistry of Cd, Bi, Tl, Pb, Zn and Rb. *GCA*, **44**, 1519.
3   Bowes, D.R. (1972) Geochemistry of Precambrian crystalline basement rocks, north-west highlands of Scotland. *24th IGC, Sect. 1*, 97.
4   Fahrig, W.F. & Eade, K.E. (1968) The chemical evolution of the Canadian Shield. *CJES*, **5**, 1247; Eade, K.E. & Fahrig, W.F. (1971) Chemical evolutionary trends of continental plates—a preliminary study of the Canadian Shield. *GSC Bull.*, **179**; (1973) Regional, lithological and temporal variation in the abundances of some trace elements in the Canadian Shield. *GSC Pap.* **72–46**.
5   Holland, J.G. & Lambert, R.St.J. (1972) Major element chemical composition of shields and the continental crust. *GCA*, **36**, 673.
6   Ronov, A.B. & Yaroshevsky, A.A. (1969) Chemical composition of the Earth crust. *In:* The Earth's Crust and Upper Mantle. *AGU. Mono.* **13**, 37.
7   Ronov, A.B. & Migdisov, A.A. (1971) Geochemical history of the crystalline basement and the sedimentary cover of the Russian and North American platforms. *Sedimentology*, **16**, 137.
8   Shaw, D.M. *et al.* (1967) An estimate of the chemical composition of the Canadian Precambrian Shield. *CJES*, **4**, 829; (1976) Additional estimates of continental surface Precambrian Shield composition in Canada. *GCA*, **40**, 73.
9   Gibb, R.A. (1968) The densities of Precambrian rocks from northern Manitoba. *CJES*, **5**, 433.
10 Goldschmidt, V.M. (1933) Grundlagen der quantitativen Geochemie. *Fortschr. Mineral. Krist. Petrog.*, **17**, 112.
11 Hambrey, M.J. & Harland, W.B. (eds) (1981) *Earth's Pre-Pleistocene Glacial Record.* Cambridge University Press.
12 Taylor, S.R. *et al.* (1983) Geochemistry of loess, continental crustal composition and crustal model ages. *GCA*, **47**, 1897.
13 Veizer, J. & Jansen, S.L. (1979) Basement and sedimentary recycling and continental evolution. *J. Geol.*, **87**, 341.
14 Garrels, R.M. & Mackenzie, F.T. (1971) *Evolution of Sedimentary Rocks.* Norton.
15 Chesworth, W. (1973) The residua system of chemical weathering: a model for the chemical breakdown of silicate rocks at the surface of the earth. *J. Soil Sci.*, **24**, 69; (1977) Weathering stages of the common igneous rocks, index minerals and mineral assemblages at the surface of the earth. *J. Soil Sci.*, **28**, 490; (1980) The haplosoil system. *AJS*, **280**, 969.
16 Nesbitt, H.W. *et al.* (1980) Chemical processes affecting alkalis and alkaline earths during continental weathering. *GCA*, **44**, 1659.
17 Nesbitt, H.W. & Young, G.M. (1984) Prediction of some weathering trends of plutonic and volcanic rocks based on thermodynamic and kinetic considerations. *GCA*, **48**, 1523.
18 Kronberg, B.I. *et al.* (1979) The chemistry of some Brazilian soils: element mobility during intense weathering. *Chem. Geol.*, **24**, 211; Kronberg, B.I. & Nesbitt, H.W. (1981) Quantification of weathering, soil geochemistry and soil fertility. *J. Soil Sci.*, **32**, 453.
19 Nesbitt, H.W. & Young, G.M. (1982) Early Proterozoic climates and plate motions inferred from major element chemistry of lutites. *Nature*, **299**, 715.
20 Whitfield, M. (1979) The mean oceanic residence time (MORT) concept—a rationalisation. *Mar. Chem.*, **8**, 101; Whitfield, M. & Turner, D.R. (1979) Water–rock partition coefficients and the composition of seawater and river water. *Nature*, **278**, 132; Turner, D.R. &

Whitfield, M. (1979) Control of seawater composition. *Nature*, **281**, 468; Turner, D.R. *et al.* (1980) Water–rock partition coefficients and the composition of natural waters—a reassessment. *Mar. Chem.*, **9**, 211.
21. Li, Y-H. (1982) A brief discussion on the mean oceanic residence time of elements. *GSA*, **46**, 2671.
22. Martin, J.M. *et al.* (1976) Rare earth element supply to the ocean. *JGR*, **81**, 3119.
23. Broecker, W.S. & Peng, T.-H. (1982) Tracers in the sea. *Lamont-Doherty Geol. Obs.*
24. Chester, R. & Aston, S.R. (1976) The geochemistry of deep-sea sediments. *In:* Riley, J.P. & Chester, R. (eds), *Chemical Oceanography*, Vol. 6 (2nd Ed.). Academic Press. 281.
25. Crocket, J.H. & Kuo, H.Y. (1979) Sources for gold, palladium and iridium in deep-sea sediments. *GCA*, **43**, 831.
26. McLennan, S.M. & Taylor, S.R. (1981) Role of subducted sediments in island-arc magmatism: constraints from REE patterns. *EPSL*, **54**, 423.
27. Lisitsyn, A.P. *et al.* (1982) The relation between element influx from rivers and accumulation in ocean sediments. *Geochem. Int.*, **19**, 102.
28. Milliman, J.D. & Meade, R.H. (1983) World-wide delivery of river sediment to the oceans. *J. Geol.*, **91**, 1.
29. Friedman, G.M. & Sanders, J.E. (1978) *Principles of Sedimentology*. Wiley; Selley, R.C. (1976) *An Introduction to Sedimentology*. Academic Press.
30. Bagnold, R.A. (1966) An approach to the sediment transport problem from general physics. *USGS Prof. Pap.* 422-I.
31. Gibbs, R.J. (1971) The relationship between sphere size and settling velocity. *JSP*, **41**, 7.
32. Wright, P.L. (1974) The chemistry and mineralogy of the clay fraction of sediments from the southern Barents Sea. *Chem. Geol.*, **13**, 197.
33. Hirst, D.M. (1962) The geochemistry of modern sediments from the Gulf of Paria—I. The relationship between the mineralogy and the distribution of major elements. *GCA*, **26**, 309; (1962) The geochemistry of modern sediments from the Gulf of Paria—II. The location and distribution of trace elements. *GCA*, **26**, 1147.
34. Nathan, S. (1976) Geochemistry of the Greenland Group (early Ordovician), New Zealand. *NZJGG*, **19**, 683.
35. Pettijohn, F.J. (1975) *Sedimentary Rocks* (3rd Ed.). Harper & Row.
36. Brenninkmeyer, B.M. (1978) Heavy minerals. *In:* Fairbridge, R.W. & Bourgeois, J. (eds), *The Encyclopedia of Sedimentology*. Dowden, Hutchinson & Ross. 400.
37. Holland, H.D. (1978) *The Chemistry of the Atmosphere and Oceans*. Wiley; Schopf, T.J.M. (1980) *Paleooceanography*. Harvard University Press.
38. Barth, T.W. (1952) *Theoretical Petrology*. Wiley. More recently, Yuan-Hui Li has pointed out that most standard texts provide a definition of residence time which does not correspond to the original, and preferred, meaning (Li, Y-H. (1977) Confusion of the mathematical notation for defining the residence time. *GCA*, **41**, 555).
39. Li, Y-H. (1981) Ultimate removal mechanisms of elements from the ocean. *GCA*, **45**, 1659.
40. Goldberg, E.D. *et al.* (1971) Marine chemistry. *In:* Radioactivity in the Marine Environment. *NSF*, 137.
41. Goldberg, E.D. & Arrhenius, G.O.S. (1958) Chemistry of Pacific pelagic sediments. *GCA*, **13**, 153.
42. Price, N.P. (1976) Chemical diagenesis in sediments. *In:* Riley, J.P. & Chester, R. (eds), *Chemical Oceanography*. Vol. 6 (2nd Ed.). Academic Press. 1.
43. Curtis, C.D. (1977) Sedimentary geochemistry: environments and processes dominated by involvement of an aqueous phase. *Phil. Trans. Roy. Soc.*, **A286**, 353.
44. The control of illite/smectite stability on the redistribution of potassium during diagenesis and burial metamorphism is a classic example in this regard. (Hower, J. *et al.* (1976). Mechanism of burial metamorphism of argillaceous sediment: I. Mineralogical and chemical evidence. *GSA Bull.*, **87**, 725.)
45. Calvert, S.E. (1976) The mineralogy and geochemistry of near-shore sediments. *In:* Riley, J.R. & Chester, R. (eds), *Chemical Oceanography*. Vol. 6 (2nd Ed.). Academic Press. 187.
46. Shaw, D.M. (1954) Trace elements in pelitic rocks. Part I: variations during metamorphism. *GSA Bull.*, **65**, 1151; (1954) Trace elements in pelitic rocks. Part II: geochemical relations. *GSA Bull.*, **65**, 1167; (1956) Geochemistry of pelitic rocks. Part III: major elements and general geochemistry. *GSA Bull.*, **67**, 919.

47 Ronov, A.B. *et al.* (1977) Regional metamorphism and sediment composition evolution. *Geochem. Int.*, **12**, 90.
48 Heier, K.S. (1973) A model for the composition of the deep continental crust. *Fortschr. Min.*, **50**, 174; (1978) The distribution and redistribution of heat producing elements in the continents. *Phil. Trans. Roy. Soc.*, **A288**, 393; (1979) The movement of uranium during higher grade metamorphic processes. *Ibid.*, **A293**, 413; Tarney, J. & Windley, B.F. (1977) Chemistry, thermal gradients and evolution of the lower continental crust. *J. Geol. Soc. Lond.*, **134**, 153.
49 Nance, W.B. & Taylor, S.R. (1976) Rare earth element patterns and crustal evolution—I. Australian post-Archean sedimentary rocks. *GCA*, **40**, 1539.
50 McLennan, S.M. (1981) *Trace Element Geochemistry of Sedimentary Rocks: Implications for the Composition and Evolution of the Continental Crust.* Ph.D. Thesis, Australian National University.
51 Martin, J.-M. & Meybeck, M. (1979) Elemental mass-balance of material carried by major world rivers. *Mar. Chem.*, **7**, 173.
52 Clarke, F. (1924) The data of geochemistry. *USGS Bull.*, **770**.
53 Krauskopf, K.B. (1967) *Introduction to Geochemistry.* McGraw-Hill.
54 McLennan, S.M. & Taylor, S.R. (1980) Geochemical standards for sedimentary rocks: trace element data for U.S.G.S. standards SCo-1, MAG-1 and SGR-1. *Chem. Geol.*, **29**, 33.
55 Sigleo, A.C. & Helz, G.R. (1981) Composition of esturine colloidal material: major and trace elements. *GCA*, **45**, 2501.
56 Goldschmidt, V.M. (1938) Geochemische verteilungsgesetze der elemente IX. Die mengen-verhaltnisse der elemente und atom-arten: Norske Videnskaps—Akad. Skr. M.N.Kl., **4**, 1.
57 Ronov, A.B. *et al.* (1967) Geochemistry of the rare earths in the sedimentary cycle. *Geochem. Int.*, **4**, 1; (1972) Trends in rare earth distribution in the sedimentary shell and in the earth's crust. *Geochem. Int.*, **9**, 987.
58 Cullers, R.L. *et al.* (1975) Rare earth distributions in clay minerals and in the clay-sized fraction of Lower Permian Havensville and Esbridge shales of Kansas and Oklahoma. *GCA*, **39**, 1691.
59 Cullers, R.L. *et al.* (1979) Rare-earths in size fractions and sedimentary rocks of Pennsylvanian–Permian age from the mid-continent of the USA. *GCA*, **43**, 1285.
60 Dypvik, H. & Brunfelt, A.O. (1976) Rare-earth elements in Lower Paleozoic epicontinental and eugeosynclinal sediments from the Oslo and Trondheim regions. *Sedimentology*, **23**, 363.
61 The incomparable role now held by the rare-earth elements in geochemistry and cosmochemistry no doubt would have comforted Bertram Boltwood. In a letter to Ernest Rutherford dated December, 1905, he stated, 'In point of respectability your radium family will be a Sunday school compared with the (rare-earth elements), whose (chemical) behaviour is simply outrageous. It is absolutely demoralizing to have anything to do with them' (Badash, L. (1969) Rutherford & Boltwood: *Letters on Radioactivity.* Yale University).
62 Taylor, S.R. (1964) Abundance of chemical elements in the continental crust: a new table. *GCA*, **28**, 1273.
63 Haskin, L.A. *et al.* (1966) Rare earths in sediments. *JGR*, **71**, 6091.
64 Haskin, L.A. *et al.* (1968) An accurate procedure for the determination of rare earths by neutron activation. *J. Radioanal. Chem.*, **1**, 337.
65 Haskin, M.A. & Haskin, L.A. (1966) Rare earths in European shales: a redetermination. *Science*, **154**, 507.
66 Minami, E. (1935) Gehalte an seltenen Erden in europäishen und japaneschen Tonschiefern. *Nach. Gess. Wiss. Goettingen, Z, Math-Physik KL.IV*, **1**, 155.
67 Balashov, Y.A. *et al.* (1964) The effect of climate and facies environment on the fractionation of the rare earths during sedimentation. *Geochem. Int.*, **10**, 951.
68 Chondrite factors used for normalizing rare earth element diagrams are listed in Appendix II.
69 Basu, A. *et al.* (1982) Rare earth elements in the sedimentary cycle: a pilot study of the first leg. *Sedimentology*, **29**, 739.
70 Jarvis, J.C. *et al.* (1975) Rare earths in the Leadville Limestone and its marble derivates. *Chem. Geol.*, **16**, 27; Parekh, P.P. *et al.* (1977) Distribution of trace elements between carbonate and non-carbonate phases of limestone. *EPSL*, **34**, 39.
71 Ronov, A.B. *et al.* (1974) Regularities of rare-earth element distribution in the sedimentary shell and in the crust of the earth. *Sedimentology*, **21**, 171.

72 Taylor, S.R. & McLennan, S.M. (1981) The composition and evolution of the continental crust: rare earth element evidence from sedimentary rocks. *Phil. Trans. Roy. Soc.*, **A301**, 381.
73 Schnetzler, C.C. & Philpotts, J.A. (1970) Partition coefficients of rare-earth elements between igneous matrix material and rock-forming mineral phenocrysts—II. *GCA*, **34**, 331.
74 Gromet, L.P. & Silver, L.T. (1983) Rare earth element distribution among minerals in a granodiorite and their petrogenetic implications. *GCA*, **47**, 925.
75 Arniaud, D. *et al.* (1984) Geochemistry of Auriat granite (Massif Central, France). *Chem. Geol.* **45**, 263.
76 Lee, D.E. & Bastron, H. (1967) Fractionation of rare-earth elements in allanite and monazite as related to geology of the Mt. Wheeler mine area, Nevada. *GCA*, **31**, 339.
77 Chaudhuri, S. & Cullers, R.L. (1979) The distribution of rare-earth elements in deeply buried Gulf coast sediments. *Chem. Geol.*, **24**, 327.
78 Nesbitt, H.W. (1979) Mobility and fractionation of rare earth elements during weathering of a granodiorite. *Nature*, **279**, 206.
79 Duddy, I.R. (1980) Redistribution and fractionation of rare-earth and other elements in a weathering profile. *Chem. Geol.*, **30**, 363.
80 Floyd, P.A. (1977) Rare earth element mobility and geochemical characterisation of spilitic rocks. *Nature*, **269**, 134.
81 Hellman, P.L. & Henderson, P. (1977) Are rare earth elements mobile during spilitisation? *Nature*, **267**, 38.
82 Frey, F.A. *et al.* (1974) Atlantic ocean floor: geochemistry and petrology of basalts from Legs 2 and 3 of the Deep Sea Drilling Project. *JGR*, **79**, 5507.
83 Wood, D.A. *et al.* (1976) Elemental mobility during zeolite facies metamorphism of the Tertiary basalts of eastern Iceland. *CMP*, **55**, 241.
84 Menzies, M. *et al.* (1979) Experimental evidence of rare earth element immobility in greenstones. *Nature*, **282**, 398; Hajash, A. (1984) Rare earth element abundances and distribution paterns in hydrothermally altered basalts: experimental results. *CMP*, **85**, 409.
85 Staudigel, H. & Hart, S.R. (1983) Alteration of basaltic glass: mechanisms and significance for the oceanic crust-seawater budget. *GCA*, **47**, 337.
86 Michard, A. *et al.* (1983) Rare-earth elements and uranium in high-temperature solutions from East Pacific Rise hydrothermal vent field (13°N). *Nature*, **303**, 795.
87 Cullers, R.L. *et al.* (1973) Experimental studies of the distribution of rare earths as trace elements among silicate minerals and liquids and water. *GCA*, **37**, 1499; Zielinski, R.A. & Frey, F.A. (1974) An experimental study of the partitioning of a rare earth element (Gd) in the system diopside-aqueous vapour. *GCA*, **38**, 545; Flynn, R.T. & Burnham, C.W. (1978) An experimental determination of rare earth partition coefficients between a chloride containing vapor phase and silicate melts. *GCA*, **42**, 685.
88 Nance, W.B. & Taylor, S.R. (1977) Rare earth element patterns and crustal evolution—II. Archean sedimentary rocks from Kalgoorlie, Australia. *GCA*, **41**, 225; Bhatia, M.R. & Taylor, S.R. (1981) Trace element geochemistry and sedimentary provinces: a study from the Tasman Geosyncline, Australia. *Chem. Geol.*, **33**, 115.
89 Elderfield, H. & Greaves, M.J. (1982) The rare earth elements in seawater. *Nature*, **296**, 214.
90 Bence, A.E. *et al.* (1980) Basalts as probes of planetary interiors: constraints on the chemistry and mineralogy of their source regions. *PCR*, **10**, 249; Basaltic Volcanism Study Project (1981) *Basaltic Volcanism on the Terrestrial Planets*. Pergamon.
91 Sun, S.-S. & Nesbitt, R.W. (1978) Petrogenesis of Archean ultrabasic and basic volcanics: evidence from rare earth elements. *CMP*, **65**, 301.
92 Taylor, S.R. *et al.* (1983) Residual lower continental crustal compositions. *LPS XIV*, **781**.
93 Lyell complained that 'the more that I have studied the subject, the more difficult I have found it to form a satisfactory theory'. (Lyell, C. (1834) Observations on the loamy deposit called 'loess' of the basin of the Rhine. *Edinburgh New Phil. J.*, **17**, 118.)
94 Richthofen, Baron F. von (1882) On the mode of origin of the loess. *Geol. Mag.*, **9**, 293. This classic paper is well worth reading, if only to appreciate the keen observational power of Baron Ferdinand. At times, this talent took the form of a finely honed wit, such as in his view of Howorth's theory of volcanic origin for loess: 'If the author of the volcanic theory of the loess had devoted the same admirable industry, with which he has ... written the history of the Mongols, to the personal observation of the soil on which their wanderings and warfare (took) place, ... he would hardly have ventured to adopt views ... which could be pronounced

at an early ... stage of geological science, but are long since abandoned' (p. 395). We note that his observational ability must have been a family trait as it reappeared with somewhat different results in Manfred, the Red Baron.

95 Edwards, M.B. (1979) Late Precambrian glacial loessites from north Norway and Svalbard. *JSP*, **49**, 85.
96 It may be noted in passing that the base of the extensive Chinese loess deposits is dated at about 2.4 million years, consistent with a glacial origin. (Heller, F. & Liu, T.-S. (1982) Magnetostratigraphical dating of loess deposits in China. *Nature*, **300**, 431.)
97 For example, loess from the South Island of New Zealand has high Na_2O/K_2O ratios consistent with derivation from greywackes.
98 Spencer, L.J. (1933) Origin of tektites. *Nature*, **131**, 117; Taylor, S.R. (1962) The chemical composition of australites. *GCA*, **26**, 685.
99 Taylor, S.R. (1973) Tektites: A post-Apollo view. *Earth Sci. Rev.*, **9**, 101.
100 Shaw, H.F. & Wasserburg, G.J. (1982) Age and provenance of the target materials for tektites as inferred from Sm–Nd and Rb–Sr systematics. *EPSL*, **60**, 155.
101 Taylor, S.R. (1969) Criteria for the source of australites. *Chem. Geol.*, **4**, 451.
102 Taylor, S.R. & Kaye, M. (1969) Genetic significance of the chemical composition of australites. *GCA*, **33**, 1083.
103 Taylor, S.R. & McLennan, S.M. (1979) Chemical relationships among irghizites, zhamanshinites Australasian tektites and Henbury impact glasses. *GCA*, **43**, 1551.
104 Taylor, S.R. (1982) *Planetary Science: A Lunar Perspective*. LPI.
105 McLennan, S.M. *et al.* (1980) Rare earth element—thorium correlations in sedimentary rocks, and the composition of the continental crust. *GCA*, **44**, 1833.
106 Taylor, S.R. (1977) Island arc models and the composition of the continental crust. *AGU Ewing Ser.* **I**, 325.
107 Taylor, S.R. & McLennan, S.M. (1983) Geochemical application of spark-source mass spectrography—IV. The crustal abundance of tin. *Chem. Geol.*, **39**, 273.
108 Wedepohl, K.H. *et al.* (eds.) (1969 *et seq.*) *Handbook of Geochemistry*. Springer-Verlag.
109 Units of terrestrial heat flow are expressed as $mW\ m^{-2}$. $41.8\ mW\ m^{-2} = 1\ \mu cal\ cm^{-2}\ s^{-1}$ or 1 HFU (heat flow unit).
110 Vitorello, I & Pollack, H.N. (1980) On the variation of continental heat flow with age and the thermal evolution of continents. *JGR*, **85**, 983.
111 Birch, F. *et al.* (1968) Heat flow and thermal history in New York and New England. *In:* Zen, E. *et al.* (eds), *Studies of Appalachian Geology*. Interscience. 437.
112 Nicolaysen, L.O. *et al.* (1981) The Vredefort radioelement profile extended to supracrustal strata at Carletonville, with implications for continental heat flow. *JGR*, **86**, 10653; also see Hart, R.J. *et al.* (1981) Radioelement concentrations in the deep profile through Precambrian basement of the Vredefort structure. *JGR*, **86**, 10639.
113 Jaupart, C. *et al.* (1981) Heat flow studies: constraints on the distribution of uranium, thorium and potassium in the continental crust. *EPSL*, **52**, 328.
114 Morgan, P. (1984) The thermal structure and thermal evolution of the continental lithosphere. *PCE*, **15**.
115 Wedepohl, K.H. (1969) *The Handbook of Geochemistry*. Vol. 1. Springer-Verlag. 247.

3

Models of Total Crustal Composition

3.1 Boundary conditions from the upper crustal composition

The composition of the presently exposed upper crust was given in Section 2.6. Various limitations were noted in the factors involved in the calculation, but the estimates from the separate methods of establishing crustal abundances converge in a reasonable manner consistent with the geological evidence. To a first approximation, the composition is close to that of the plutonic igneous rock, granodiorite. The upper crust is not siliceous enough to resemble true granite, nor so silica-poor that it resembles diorite or tonalite. The REE patterns, Th and Sc abundances in sedimentary rocks are consistent with abundances observed in granodiorite and form one boundary on bulk crustal compositions.

The abundances of K, U and Th are, however, high enough so that a 30 km thickness of this composition would generate the observed crustal heat flow, without any mantle contribution. Accordingly, the total thickness of the upper crust must be less than this. Numerous complicating factors enter these calculations. Th, U and K are probably concentrated in the last melts to crystallize from granitic liquids and hence will be more abundant in the upper levels of granitic batholiths. Most models envisage an exponential or steplike decrease in the contents of K, U, and Th with depth although, as noted in Section 2.6.5, these models are only an approximation to reality (see Sections 2.6.5 and 3.2). Upper crustal thicknesses between 8 km (20%) and 13 km (30%) are usually specified. In Section 2.6.5 we adopt a value of 10 km as the thickness of the upper crust, based on the heat-flow parameters. This is about 25% of the total crust.

The bulk composition of the crust thus depends heavily on models for the composition of the generally inaccessible lower crust and on mechanisms for producing the crust from the mantle. Various approaches are possible. One is to select terrains thought to be representative of the deep crust. This approach has become increasingly fraught with difficulties as our understanding of the heterogeneous nature of the lower crust has grown (e.g. Section 1.2). A second line of attack on the problem is to try to establish the process responsible for crustal growth, and identify its products. These will accordingly provide the bulk composition for the crust. Many of the complexities involved are discussed throughout this book. Does the crust grow by lateral accretion of island-arcs? Is underplating a significant mechanism for growth? Can we distinguish chemically between andesitic volcanism as a growth mechanism, and growth by intrusion of diorites or tonalites?

Among the constraints adopted here is the requirement that the bulk crustal composition be able to produce, by intra-crustal melting, the presently observed granodioritic upper crust. This restriction seems inherently reasonable. It is generally agreed that the granitic rocks typical of the upper crust are produced by such a mechanism (Section 9.5 [1, 2]) and we adopt this concept throughout this book.

3.2 Heat flow and radioactive element abundances

The heat flow data form one of the better constraints on the composition of the bulk crust. The number of heat flow measurements in continental areas has increased dramatically in the past three decades. Birch [3] listed 43 values. This number grew to 2808 in 1980 out of a global total of 7217 [4]. Two important empirical relationships have been established. The first was the discovery [5] of a linear relationship between heat flow and heat production in rocks at the surface of the earth (see Section 2.6.5). This relationship has enabled the identification of heat flow provinces (see Chapter 5), and also allows the effective thickness of the upper crust (10 km) to be calculated (see Section 2.6.5).

The second important finding was the recognition [6] that the component of heat flow which arises in the crust is about 40% of the total, and that 60% of the heat flow is derived from the mantle. There are three principal features of the observed heat flow. A relatively high heat flow, averaging about 72 mW m^{-2} is typical of Mesozoic and younger rocks. The period from the mid-Palaeozoic to the base of the Proterozoic is marked by rather uniform heat flow averaging 51–54 mW m^{-2}. Archean sites have lower heat flows of about

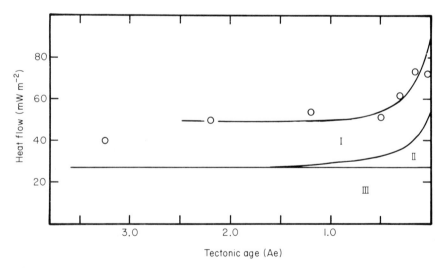

Fig. 3.1. The variation in continental heat flow with age, showing the three principal components. I represents the radiogenic heat component generated in the crust. II represents the transient thermal perturbation associated with tectonic activity. It is mostly gone within 300–500 m.y. III represents the mantle contribution, which averages 27 mW m^{-2}. Note that the heat flow in the continental crust is relatively constant from about 500 m.y. to over 2 Ae, but decreases in the Archean. (Adapted from [4].)

41 mW m^{-2}. Three components are recognized as contributing toward the observed heat flow (Fig. 3.1) [6, 7].

(a) A tectonothermal source contributes about 30% (27 mW m^{-2}) of the heat flow in young orogenic areas. This decays with a time constant of about 350 m.y., and is effectively zero in Lower Palaeozoic rocks.

(b) A background heat flux of about 27 mW m^{-2} derived from mantle sources.

(c) A component, which averages about 23 mW m^{-2}, is produced by radioactive elements in the crust. There is no simple relationship between the crustal component and age [7] and the mean heat flow (24±8 mW m^{-2}) for Phanerozoic provinces from this source is about the same as that for the Proterozoic provinces of 23±9 mW m^{-2}.

The identification of the crustal component of the observed heat flow enable us to calculate the bulk crustal abundances of K, U and Th. We assume a value of 23 mW m^{-2}, a crustal thickness of 40 km, a density of 2.8 g cm^3, and K/U and Th/U ratios of 10^4 and 3.8 respectively. The K, U, and Th values from this calculation are K=0.9%, U=0.9 ppm and Th=3.4 ppm. These values constitute a valuable constraint on bulk crustal compositions for many other elements.

3.3 The andesite model

It is a reasonable assumption that the continental crust must be derived ultimately from the mantle by some common geological process. A durable candidate has been calc-alkaline volcanism typical of orogenic regions [8]. The model has been a favourite, since it provides large volumes (at least one-half km^3 per year; see Chapter 10) of material averaging about 60% silica, and has the Lyellian advantage of being an observable process, agreeable to the more strict proponents of uniformitarianism. Suggestions that the continents grew by geological processes associated with orogenic activity were made by many workers [9].

The derivation of the continental crust from various sources was discussed by Taylor [8]. An assessment of the geophysical, geological and geochemical evidence at that time indicated both that continents grew throughout geological time and were derived from the mantle. Igneous activity in orogenic zones, both at island arcs and in cordilleran regions provided the only presently observable viable source, both from volume and compositional considerations.

Andesitic volcanism is the most spectacular and visible manifestation of orogenic igneous activity, so that the process became labelled as the 'Andesite' model. However, it embraced all aspects of plate margin igneous activity, noting that 'continental areas grow mainly by the addition of andesites and associated calc-alkaline rocks in orogenic areas' [8]. Thus recent models [10] which distinguish underplating by intrusion of tonalites in Andean-type margins as a principal mechanism for continental growth, are only minor variations on the original theme.

Since andesitic rocks represented the mean composition of igneous rocks in orogenic areas, the average composition of present-day orogenic andesites was accordingly used, on uniformitarian grounds, to establish the bulk composition of the continental crust [8, 11]. Although considerable progress has been made since the model was originally proposed only slight modifications to the crustal averages have been needed. However, it has become clear that the bulk of the crust has probably grown by a mechanism other than that responsible for present-day igneous activity in orogenic areas. A principal new fact is that the bulk (75%) of the continental crust was in place by the end of the Archean, 2500 m.y. ago (see Chapter 10). Igneous activity in the Archean appears not to have produced much typical andesitic material. These interesting and spectacular examples of igneous activity are thus a less important factor in continental growth than previously thought.

However, one of our theses is that such igneous activity in orogenic regions constitutes the principal mechanism for present-day continental growth (and so accounts for perhaps 25% of the crust). Recent reviews of island-arc rocks, their occurrence and genesis have been provided by Gill [12] and Thorpe [13]. Compositions of typical arc and other igneous rocks of interest are given in Tables 3.1 and 3.2, and typical REE patterns of island-arc rocks in Fig. 3.2. The mechanism of the addition of new material to the present crust by orogenic processes may be volcanic (resulting also in the production of volcanic sediments) or by intrusion of diorites, tonalites and related lithologies. The compositional distinction between these two processes is minor, reinforcing the suggestion that underplating models involving tonalite intrusions are a variant of the andesite model.

Because of the possibility of crustal contamination, recycling of sediments and intracrustal melting to produce ignimbrites, for example, modellers have tended to place more attention on island-arcs, where mantle derivation is more

Table 3.1. Average composition of common igneous rocks.

	1	2	3	4	5	6	7	8
SiO_2	54.2	59.0	62.6	66.3	67.2	71.9	72.0	73.4
TiO_2	0.94	0.78	0.74	0.67	0.55	0.29	0.31	0.30
Al_2O_3	17.6	17.4	16.8	15.9	16.0	15.5	14.5	13.5
FeO	8.34	6.63	5.57	4.40	4.04	2.1	2.76	2.21
MgO	5.05	3.54	2.85	1.76	1.77	0.81	0.72	0.50
CaO	8.88	7.08	5.53	4.32	3.89	2.94	1.86	2.00
Na_2O	3.12	3.50	3.70	3.97	3.81	4.96	3.72	3.74
K_2O	1.10	1.44	2.10	2.57	2.78	1.52	4.11	4.24
Σ	99.2	99.4	99.9	99.9	100.0	100.0	100.0	99.9
n	636	945	97	1031	885	16	2485	593

1. Basaltic andesite (Orogenic) [14].
2. Andesite (Orogenic) [14].
3. Tonalite [15].
4. Dacite (Orogenic) [16].
5. Granodiorite [15].
6. Trondhjemite. Amîtsoq trondhjemitic gneisses [17].
7. Granite [15].
8. Rhyolite (Orogenic) [16].

Table 3.2. Typical REE data for island arc volcanic rocks. Average data from the Sunda arc, Indonesia [18]. Data in ppm.

	1	2	3	4
La	9.5	15.3	19.4	35.1
Ce	23.0	34.8	40.9	68.0
Pr	2.9	4.0	4.9	7.9
Nd	13.0	17.4	19.4	33.8
Sm	2.8	4.0	3.5	7.3
Eu	1.0	1.3	1.0	2.0
Gd	2.9	3.8	3.3	6.4
Tb	0.54	0.66	0.53	0.99
Dy	3.5	3.9	3.3	5.9
Ho	0.80	0.87	0.73	1.4
Er	2.3	2.5	2.2	4.0
Yb	2.4	2.5	2.1	3.8
ΣREE	65	92	102	178
Eu/Eu*	1.07	1.02	0.95	0.95

1. Andesite, Tholeiitic series.
2. Basaltic andesite, Calc-alkaline series.
3. Andesite, Calc-alkaline series.
4. High-K andesite.

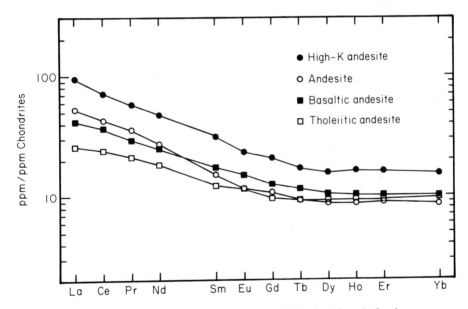

Fig. 3.2. Typical REE patterns for island-arc volcanic rocks. This suite is from the Sunda arc, Indonesia [18]. Data from Table 3.2.

secure, than on continental margins. The average composition of erupted material at island-arcs has been much debated, but a consensus appears to be emerging, based on statistical treatment, that the average silica content is about 57–58% silica [19]. The average continental crustal composition based on the andesite model is given in Table 3.3. The Th/U ratio (2.5) used in previous estimates [11] was based heavily on that of intraoceanic arcs. This

Table 3.3. The andesite model composition for the bulk crust (adapted from [11]).

	%
SiO_2	58.0
TiO_2	0.8
Al_2O_3	18.0
FeO	7.5
MgO	3.5
CaO	7.5
Na_2O	3.5
K_2O	1.5
Σ	100.3

Li	10 ppm	Ni	30 ppm	Eu	1.1 ppm
Be	1.5 ppm	Cu	60 ppm	Gd	3.6 ppm
Na	2.60%	Ga	18 ppm	Tb	0.64 ppm
Mg	2.11%	Rb	42 ppm	Dy	3.7 ppm
Al	9.50%	Sr	400 ppm	Ho	0.82 ppm
Si	27.1%	Y	22 ppm	Er	2.3 ppm
K	1.25%	Zr	100 ppm	Tm	0.32 ppm
Ca	5.36%	Nb	11 ppm	Yb	2.2 ppm
Sc	30 ppm	Cs	1.7 ppm	Lu	0.30 ppm
Ti	4800 ppm	Ba	350 ppm	Hf	3.0 ppm
V	175 ppm	La	19 ppm	Pb	10 ppm
Cr	55 ppm	Ce	38 ppm	Th	4.8 ppm
Mn	1100 ppm	Pr	4.3 ppm	U	1.25 ppm
Fe	5.83%	Nd	16 ppm		
Co	25 ppm	Sm	3.7 ppm		

ratio is much lower than the generally accepted crustal ratio of 3.8. These modern arcs may be sampling somewhat depleted mantle, since their Th/U ratios are similar to those of mid-ocean ridge basalt (MORB). Many calc-alkaline volcanic rocks in continental or continental margin zones do have Th/U close to 3.8 [12]. Whether this is due to crustal contamination or to recycling of material is uncertain. These problems with Th/U ratios illustrate a difficulty with the 'andesite' model. The mantle is being steadily depleted in the highly incompatible elements due to continental crust formation, so that present-day igneous rocks of mantle origin may not be the best candidates to represent the bulk composition of a crust of which about 75–80% was probably already extracted from the mantle by 2500 m.y. ago. Certainly, it is difficult to account for the crustal Th/U ratios, using present-day orogenic andesites as a source.

3.3.1 The Ni and Cr problem

Nickel and chromium abundances present other problems for the andesite model. The average nickel content of island-arc rocks is about 25 ppm. Chromium averages about 60 ppm [12, 13] (Fig. 3.3). Although the production of island-arc magmas is still enigmatic, a considerable school of thought [12] believes that extensive removal of Ni and Cr due to crystal fractionation has taken place before eruption. Alternatively, island-arc magmas

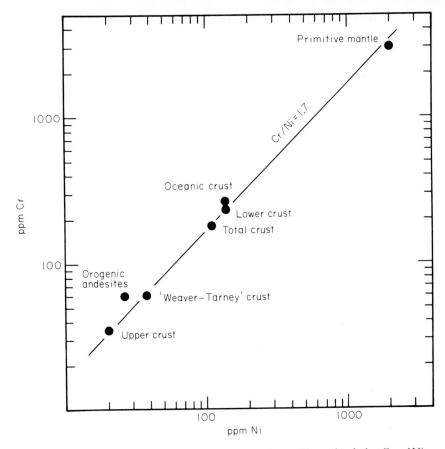

Fig. 3.3. Chromium and nickel abundances in the mantle and crust, illustrating the low Cr and Ni values predicted by the andesite model and the Weaver–Tarney model (Section 3.3.1). Our estimate for average lower crustal values (Section 4.7), primitive mantle (Table 11.3) and the basaltic oceanic crust (Table 11.6) are also shown.

are being produced from source materials depleted in these elements in previous melting events. Values for upper crustal rocks (20 and 35 ppm respectively, Table 2.15) are not very much less than the abundances in present-day andesites. These values seem to be too low to account for the bulk crustal abundances. It is reasonable to predict that both these elements will remain in residual phases in the lower crust during partial melting episodes to produce the upper crustal granitic rocks. Accordingly, one would expect Ni and Cr to be concentrated in the lower crust and bulk crustal estimates for these elements should be 2 or 3 times higher than the measured upper crustal values for these elements. This is supported by the observed abundances of Ni and Cr in xenoliths of lower crustal origin in which levels of 200–400 ppm are commonly observed [20].

In summary, the andesite model fails to provide enough Ni and Cr to account for the observed upper crustal abundances, and another source for these elements is required. There are also problems with Th/U ratios, as noted earlier.

3.4 Other primary mantle sources

In this section, we briefly consider other possible mantle sources for continental crustal material. Present-day igneous activity which produces voluminous eruptive material in tectonic environments other than plate margins is generally basaltic in character. We exclude here the large volumes of ignimbritic eruptives, which are almost certainly derived from partial melting within the crust. The most voluminous eruptive rocks, represented by MORB, the lesser amounts of intra-plate alkali basalts, and the voluminous plateau basalts, do not have the appropriate chemical composition from which to build the continental crust. The most fundamental difficulties are that not enough K, U and Th is provided to account for the heat flow, and that it is not possible to generate sufficient volumes of granodiorite by intracrustal melting to provide for the presently observed upper crust. Likewise, the seismic velocity data are too low for an overall basic composition. If the continental crust had a predominantly basaltic composition, then large amounts of such continental crust would of course have existed from the earliest times in apparent conflict with observations both on the Earth and other planets (see Chapter 12).

All estimates of continental crustal bulk composition, whether based on geophysical, geological or geochemical parameters, agree that the composition is more silica-rich than that of basalt. The consensus is that the bulk crust has a silica content of about 60% (Table 3.4). Production of continental crust at the present day by intrusion of diorites or tonalites of mantle derivation, is sometimes referred to as 'underplating' [10]. The average tonalite silica content is 62.6% (Table 3.1), so that tonalitic crusts are only a little more siliceous than andesite in bulk composition. Thus underplating models [e.g. 10] are only a modified version of the andesite model and suffers from the

Table 3.4. Various estimates of the composition of the bulk continental crust.

	1	2	3	4	5	6	7	8	9	10
% SiO_2	61.9	63.9	60.2	57.8	61.9	62.5	63.8	64.8	58.0	57.3
% TiO_2	1.1	0.8	1.0	1.2	0.8	0.7	0.7	0.51	0.8	0.9
% Al_2O_3	16.7	15.4	15.6	15.2	15.6	15.6	16.0	16.1	18.0	15.9
% FeO	6.9	6.1	7.1	7.6	6.2	5.5	5.3	4.8	7.5	9.1
% MgO	3.5	3.1	3.9	5.6	3.1	3.2	2.8	2.7	3.5	5.3
% CaO	3.4	4.2	5.8	7.5	5.7	6.0	4.7	4.6	7.5	7.4
% Na_2O	2.2	3.4	3.2	3.0	3.1	3.4	4.0	4.4	3.5	3.1
% K_2O	4.2	3.0	2.5	2.0	2.9	2.3	2.7	2.0	1.5	1.1
ppm Th	—	—	—	—	—	—	—	5.1	4.8	3.5
ppm U	—	—	—	—	—	—	—	1.3	1.25	0.91
ppm Cr	—	—	—	—	—	—	—	61	55	185
ppm Ni	—	—	—	—	—	—	—	39	30	105

Average continental crust estimates from:
1. Goldschmidt [21]
2. Vinogradov [22]
3. Taylor [23]
4. Pakiser & Robinson [24]
5. Ronov & Yaroshevsky [25]
6. Holland & Lambert [26]
7. Smithson [27]
8. Weaver & Tarney [10]
9. Taylor & McLennan [28]
10. This book

same limitations in explaining the Th/U crustal ratios and the Cr and Ni deficiency inherent in that model. Nevertheless, as noted later, addition of igneous rocks in orogenic zones is probably the principal mechanism for present day crustal growth. In Chapter 9, we conclude that such mechanisms are responsible for perhaps 20–25% of the bulk crustal composition, and that the bulk of the continental crust was derived by other igneous processes.

Are silica-rich igneous rocks a possible primary source for continental growth? Dacites which average 66% silica are much less voluminous than andesites. They typically comprise only 15% of the volcanic products of island arc regions and are balanced by low silica basalts and basaltic andesites [14]. The rhyolite data include many, probably a majority, derived from intra-crustal melting. This composition is indistinguishable from that of granites and so represents 'upper' crustal compositions, too rich in K to be representative of the bulk crust. This is a general problem for bulk crustal compositions with more than about 60% silica. The amounts of K, U and Th produce too much radiogenic heat. The very small amounts of silica-rich differentiates of basic magma are produced on too minute a scale to be viable candidates. The production of continental material is likely to be much less efficient than, for example, the generation of basaltic magma, which occurs in voluminous amounts on all terrestrial-type planets [29]. We conclude that present-day igneous activity in orogenic regions, averaging about 60% silica is the only major candidate for continental growth at present. We now proceed to seek candidates for continental growth in the Archean.

3.5 The Archean 'bimodal' basic-felsic igneous model

The Archean crust is dominated by the so-called bimodal basic-felsic suite of igneous rocks (Chapter 7). One component consists of basalts, generally similar to MORB, characterized by flat REE patterns with abundances enriched about ten times over those of chondritic meteorites. The other dominant rocks are sodium-rich plutonic or extrusive felsic rocks (tonalites-trondhjemites-felsic volcanics). All these are characterized by steep linear LREE enriched-HREE depleted patterns, typically ranging from $100\times$ chondritic for La to about $5\times$ chondritic for ytterbium (Fig. 3.4). The origin of these felsic rocks is generally attributed to partial melting of a garnet-bearing or eclogitic parent material, since the REE patterns are consistent with the presence of garnet in the source. This evidence is indicative of pressures exceeding 15 kbar and hence mantle depths. Archean tectonic models are discussed in Chapters 7 and 9 but do not resemble present day arc environments. They involve partial mantle melting to produce basalt, or komatiite, followed by sinking, conversion to garnet-bearing rocks or eclogite, and remelting at mantle depths to produce the felsic igneous suite.

The REE composition of the Archean crust can be modelled fairly successfully by appropriate mixtures of the bimodal suite (Fig. 3.4). A reasonable model, which complies with heat flow constraints, is a 2:1 mix of basic and felsic end members. The derivation of this model is discussed in

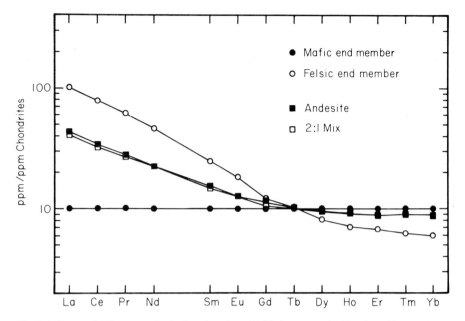

Fig. 3.4. Average REE patterns for Archean basalts and felsic igneous rocks (data from Chapter 7). The pattern resulting from a 2:1 mix of these components is very close to that of average andesite, illustrating that the REE patterns do not distinguish between island-arc and mixing models.

detail in Chapter 7 (see especially Section 7.10.2). The upper Archean crust composition can also be modelled by an equal mixture of basic and felsic rocks (Chapter 7). In some regions (e.g. Yellowknife area of the Slave Province) felsic rocks are more abundant, comprising perhaps two thirds of the exposed crust. As noted in Chapter 10 on the growth rate of the crust, the massive addition of crustal material from mantle sources which occurred between about 3.2 and 2.5 Ae may well have been principally derived from the mechanism which produced the bimodal suite. In this scenario, the composition of perhaps 75–80% of the crust is thus made up of the bimodal basic-felsic suite, in about 2:1 proportions. This has implications for deep crustal structure. Although extensive reworking, metamorphism and extraction of partial melt may have so modified the original material that remnants are difficult to recognize, nevertheless, the frequency of basic compositions among xenoliths derived from the lower crust [20] indicates that a substantial basic component is present (see Chapter 4).

3.6 A proposed bulk crustal composition

Except for the heat producing elements, estimates of total crustal composition will remain strongly model dependent until some independent methods of arriving at the composition of the whole crust can be established. In this section we present our estimate, which is limited by various constraints, and heavily influenced by our model for crustal growth. We also utilize information about the nature and composition of the Archean crust for which the details are to be found in Chapter 7 and of crustal evolution and growth outlined in Chapters 9 and 10 [30].

The constraints which we consider important for models of bulk crustal composition are:

1 it must meet the heat flow constraints, with the bulk of the K, U and Th concentrated in the upper crust;
2 the bulk crust must be capable of generating the upper crustal granodiorites by partial melting;
3 contributions to crustal composition from island-arc volcanism (the andesite model) are restricted to crustal growth since the Archean;
4 about 75% of crustal growth had occurred by 2500 m.y. ago, and the bulk of the crust was emplaced by mechanisms not necessarily related to present-day plate tectonic regimes.

We propose a model for the composition of the bulk crust which is comprised of 75% of the Archean crustal composition (derived in Chapter 7) and 25% of the 'andesite' model composition, representing the relative contributions of Archean and post-Archean crustal growth processes. Values for 62 elements are listed in Table 3.5. In Fig. 3.5, we compare this model with that of the 'Andesite Model' (Table 3.3). The most important differences are:

1 the present model is considerably enriched in Cr and Ni and more modestly enriched in the other ferromagnesian elements;

Table 3.5. Composition of the bulk continental crust (see text for sources).

	%	NORM	
SiO_2	57.3	Q	6.6
TiO_2	0.9	Or	6.5
Al_2O_3	15.9	Ab	26.2
FeO	9.1	An	26.2
MgO	5.3	Di	8.7
CaO	7.4	Hy	24.1
Na_2O	3.1	Il	1.7
K_2O	1.1		
Σ	100.1		

Li	13 ppm	Ni	105 ppm	In	50 ppb	Er	2.2 ppm
Be	1.5 ppm	Cu	75 ppm	Sn	2.5 ppm	Tm	0.32 ppm
B	10 ppm	Zn	80 ppm	Sb	0.2 ppm	Yb	2.2 ppm
Na	2.30%	Ga	18 ppm	Cs	1.0 ppm	Lu	0.30 ppm
Mg	3.20%	Ge	1.6 ppm	Ba	250 ppm	Hf	3.0 ppm
Al	8.41%	As	1.0 ppm	La	16 ppm	Ta	1.0 ppm
Si	26.77%	Se	0.05 ppm	Ce	33 ppm	W	1.0 ppm
K	0.91%	Rb	32 ppm	Pr	3.9 ppm	Re	0.5 ppb
Ca	5.29%	Sr	260 ppm	Nd	16 ppm	Ir	0.1 ppb
Sc	30 ppm	Y	20 ppm	Sm	3.5 ppm	Au	3.0 ppb
Ti	5400 ppm	Zr	100 ppm	Eu	1.1 ppm	Tl	360 ppb
V	230 ppm	Nb	11 ppm	Gd	3.3 ppm	Pb	8.0 ppm
Cr	185 ppm	Mo	1.0 ppm	Tb	0.60 ppm	Bi	60 ppb
Mn	1400 ppm	Pd	1.0 ppb	Dy	3.7 ppm	Th	3.5 ppm
Fe	7.07%	Ag	80 ppb	Ho	0.78 ppm	U	0.91 ppm
Co	29 ppm	Cd	98 ppb				

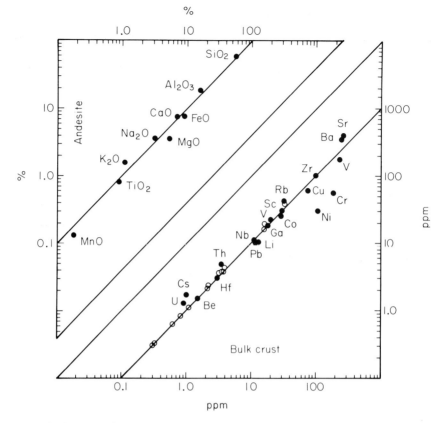

Fig. 3.5. A comparison of the bulk crustal abundances (Table 3.5) with those predicted previously by the andesite model (Table 3.3). The principal differences are the lower abundances of highly incompatible elements (e.g. K, U, Th, Cs) and higher abundances of Cr and Ni in the present estimate. Open symbols represent REE.

2 the present model is less enriched in the highly incompatible elements such as Cs, K, Th and U.

For a number of trace elements, information for Archean data is lacking and we have used bulk crustal estimates from the following sources. The data for Zn, Cd, Tl and Bi are from the crustal estimates by Heinrichs *et al.* [31]. The Sn estimate is from Taylor & McLennan [32]. The gallium abundance is calculated from the Ga/Al ratio of 0.21×10^{-3}. The value for K is 0.91%. Uranium is estimated from $K/U=10^4$ and thorium from $Th/U=3.8$, yielding values of 0.91 ppm U and 3.5 ppm Th. Estimates for a number of the less abundant elements are compiled from the *Handbook of Geochemistry* [33]. These include Li, Ge, As, Se, Mo, Ag, In, W, Re and Au.

In Section 3.2, we discussed the constraints from the heat-flow data. These indicate K, U and Th abundances of 0.90%, 0.90 ppm and 3.4 ppm respectively (for a crustal heat-flow contribution of 23 mW m^{-2}). These values are essentially identical to those calculated here (K=0.91%; U=0.91 ppm; Th=3.5 ppm). Our proposed composition generates a crustal heat flow component of 23 mW m^{-2}

3.6.1 Some comparisons

Various other estimates for the bulk crust are listed in Table 3.4. Many of these are of historical interest only, or do not distinguish clearly between the upper and the bulk crust. However, Weaver & Tarney [10] have proposed a recent model for the crust comprising a 12 km thick upper crust with the composition given by Taylor & McLennan [28], underlain by 6 km of 'middle' crust and 18 km of 'lower' crust. They adopt, for the lower crust, a composition based on the average Lewisian granulite. For the 'middle' crust of their three layer model, they adopt the average Lewisian amphibolite-facies composition [34]. The resulting bulk crustal composition (Table 3.3) has a silica content of 65%. The K_2O content is 2.0%, Th=5.1 ppm and U=1.3 ppm. This provides a total heat flux of 35 mW m^{-2}. This value is about 50% higher than the crustal heat flow component of 23 mW m^{-2} [7, 35]. Such models face three problems. The crustal heat flow produced by this composition is high, the silica abundance takes no account of the extensive evidence from the xenolith data for low silica compositions in the deep crust (see Chapter 4), and the Ni and Cr abundances are too low. The heat flow constraint seems

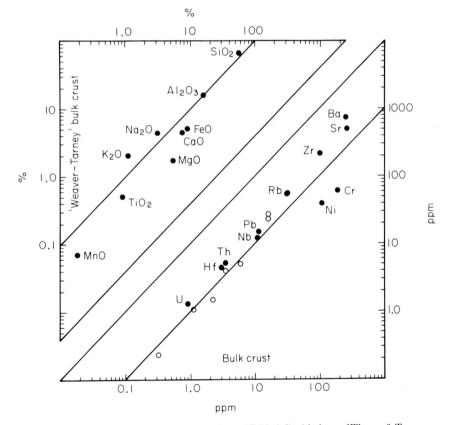

Fig. 3.6. A comparison of the bulk crustal abundances (Table 3.5) with those of Weaver & Tarney [10]. The latter model shows strong enrichment of incompatible elements (e.g. K, U, Th, LREE, Zr, Ba, Na) and depletion of Mg, Fe, Ti, Mn, Ni and Cr compared with the composition proposed in this book. Open symbols represent REE.

particularly serious and accordingly we conclude that the bulk crust must be less enriched in silica and the heat-producing elements than envisaged by Weaver & Tarney [10]. In Fig. 3.6, we compare our estimate with that of Weaver & Tarney [10]. Substantial differences exist for many elements.

Smithson et al. [36] also propose a more silica-rich bulk crust than that adopted here. From their interpretation of the mean seismic velocity crustal data, they estimate a bulk crustal composition of 64% silica and 2.7% K_2O, close to that of the upper crust. Values for U and Th are not given, but reasonable values for $K/U = 10^4$ and $Th/U = 3.8$ provide estimates of 2.2 ppm U and 8.4 ppm Th for the bulk crust. These values provide a crustal heat flow component of the same magnitude as the total observed heat flux. Such models face considerable petrological and geochemical difficulties since they imply that granodioritic and granitic compositions are effectively derived from the mantle. Attempts to resolve the heat flow dilemma by upward concentration of K, U and Th imply mantle derivation of silica rich but K, U and Th poor magmas on a scale sufficient to build the continental crust. In this context, Pakiser & Robinson [24] made an earlier estimate of crustal composition, using the mean seismic velocity data. They found values of 58% silica for the bulk crustal average, similar to those adopted here.

3.7 Summary

1 The composition of the bulk crust is model dependent, but it must meet the heat-flow constraints and must be capable of generating the upper crust by intra-crustal partial melting. Its composition must also be derived from the mantle by reasonably common geological processes.

2 The bulk crust at present generates a heat flow of 23 mW m^{-2} which, for a 40 km thick crust, is consistent with K=0.9%, U=0.9 ppm and Th=3.4 ppm.

3 The 'Andesite' model, which derives the continental crust by accretion of orogenic zone volcanic rocks (extrusive and intrusive) meets many of these requirements, but fails to provide appropriate Th/U ratios and enough Cr and Ni. However, it can account for present-day crustal growth, and is considered to have been the process by which the continental crust has grown since early Proterozoic time.

4 Underplating models, involving intrusion of tonalites and diorites at Cordilleran margins are difficult to distinguish chemically from the andesite model and form a variant of that model.

5 As discussed later, perhaps 75% of the crust was in place by about 2500 m.y. ago. Accordingly, the bulk of the continental crust was derived from the mantle by Archean-style igneous processes. Of these, the most important seems to have been the production of the bimodal basic-felsic suite.

6 We propose a model composition for the bulk continental crust in which 75% was produced by Archean-style igneous processes (Archean crustal composition) and 25% by additions from island-arc igneous activity (andesite model) since the early Proterozoic. Abundances for 62 elements are given in Table 3.5. Values for key elements and ratios include: SiO_2=57.3%; Cr=185

ppm; Ni=105 ppm; K=0.91%; U=0.91 ppm; Th=3.5 ppm; K/U=10^4; Th/U=3.8; Sm/Nd=0.22; Rb/Sr=0.12; La$_N$/Yb$_N$=4.9; Eu/Eu*=1.0. This composition provides a crustal heat flow component of 23 mW m^{-2}.

Notes and references

1 Tuttle, O.F. & Bowen, N.L. (1958) Origin of granite in the light of experimental studies in the system NaAlSi$_3$O$_8$– KAlSi$_3$O$_8$–SiO$_2$–H$_2$O. *GSA Mem.*, **74**.
2 Wyllie, P.J. (1977) Crustal anatexis: An experimental view. *Tectonophysics*, **43**, 41.
3 Birch, F. (1954) The present state of geothermal investigations. *Geophysics*, **19**, 645.
4 Pollack, H.N. (1980) The heat flow from the Earth: A review. *In:* Davis, P.A. & Runcorn, S.K. (eds), *Mechanisms of Continental Drift and Plate Tectonics.* Academic Press. 183.
5 Birch, F. *et al.* (1968) Heat flow and thermal history in New York and New England. *In:* Zen, E. *et al.* (eds), *Studies of Appalachian Geology.* Interscience. 437.
6 Pollack, H.N. & Chapman, D.S. (1977) On the regional variation of heat flow, geotherms and thickness of the lithosphere. *Tectonophysics*, **38**, 279.
7 Morgan, P. (1984) The thermal structure and thermal evolution of the continental lithosphere. *PCE*, **15**, Chap. 2.
8 Taylor, S.R. (1967) The origin and growth of continents. *Tectonophysics*, **4**, 17.
9 e.g. Rubey, W.W. (1951) Geologic history of sea water. *GSA Bull.*, **62**, 1111; Wilson, J.T. (1952) Orogenesis as the fundamental geological process. *Trans. AGU*, **33**, 444; Gilluly, J. (1955) Geologic contrasts between continents and ocean basins. *GSA Spec. Pap.*, **62**, 7.
10 Weaver, B.L. & Tarney J. (1984) Major and trace element composition of the continental lithosphere. *PCE*, **15**, Chap. 7.
11 Taylor, S.R. (1977) Island arc models and the composition of the continental crust. *AGU Ewing Series*, **I**, 325.
12 Gill, J.B. (1981) *Orogenic Andesites and Plate Tectonics.* Springer-Verlag.
13 Thorpe, R.S. (ed.) (1982) *Orogenic Andesites and Related Rocks.* Wiley, New York.
14 Ewart, A. (1982) *In:* Thorpe, R.S. (ed.), *Andesites.* Wiley. 25.
15 Le Maitre, R.W. (1976) The chemical variability of some common igneous rocks. *J. Petrol.*, **17**, 589.
16 Ewart, A. (1979) A review of the mineralogy and chemistry of Tertiary–Recent dacitic, latitic, rhyolitic, and related salic volcanic rocks. *In:* Barker, F. (ed.), *Trondhjemites, Dacites and Related Rocks.* Elsevier. Chap. 2.
17 McGregor, V.R. (1979) Archean gray gneisses and the origin of the continental crust: evidence from the Godthab region, West Greenland. *Ibid.* Chap. 6.
18 Whitford, D.J. (1979) Spatial variations in the geochemistry of Quaternary lavas across the Sunda Arc in Java and Bali. *CMP*, **70**, 341.
19 Ewart, A. (1976) Mineralogy and chemistry of modern orogenic lavas—some statistics and implications. *EPSL*, **31**, 417.
20 Arculus, R.J. *et al.* (1984) Eclogites and granulites in the lower continental crust. *Proc. Eclogite Symposium, France* (in press).
21 Goldschmidt, V.M. (1933) Grundlagen der quantitativen geochemie. *Fortsch. Min. Krist. Petrog.*, **17**, 112.
22 Vinogradov, A.P. (1962) Average contents of chemical elements in the principal types of igneous rocks of the Earth's crust. *Geochemistry*, **7**, 641.
23 Taylor, S.R. (1964) Abundance of elements in the continental crust: a new table. *GCA*, **28**, 1273.
24 Pakiser, L.C. & Robinson, R. (1967) The composition of the continental crust as estimated from seismic observations. *AGU Mono.*, **10**, 620.
25 Ronov, A.B. & Yaroshevsky, A.A. (1969) Chemical composition of the Earth's crust. *AGU Mono.*, **13**, 37.
26 Holland, J.G. & Lambert, R.St.J. (1972) Major element chemical composition of shields and the continental crust. *GCA*, **36**, 673.
27 Smithson, S.B. (1978) Modelling continental crust: structural and chemical constraints. *GRL*, **5**, 749.

28 Taylor, S.R. & McLennan, S.M. (1981) The composition and evolution of the continental crust: rare earth element evidence from sedimentary rocks. *Phil. Trans. Roy. Soc.*, **A301,** 381.
29 *Basaltic Volcanism on the Terrestrial Planets.* Pergamon Press. (1981).
30 In writing scientific books, unlike novels, it seems impossible to avoid drawing on conclusions reached in later chapters.
31 Heinrichs, H. *et al.* (1981) Terrestrial geochemistry of Cd, Bi, Tl, Pb, Zn and Rb. *GCA*, **44,** 1519.
32 Taylor, S.R. & McLennan, S.M. (1983) Geochemical application of spark source mass spectrometry. IV. The crustal abundance of tin. *Chem. Geol.*, **39,** 273.
33 Wedepohl, K.H. *et al.* (1969 *et seq.*) *Handbook of Geochemistry.* Springer-Verlag.
34 Weaver, B.L. & Tarney, J. (1981) Lewisian gneiss geochemistry and Archaean crustal development models. *EPSL*, **55,** 171.
35 Vitorello, I. & Pollack, H.N. (1980) On the variation of continental heat flow with age and the thermal evolution of the continents. *JGR*, **85,** 983.
36 Smithson, S.B. *et al.* (1981) Mean crustal velocity: a critical parameter for interpreting crustal structure and crustal growth. *EPSL*, **53,** 323.

4

The Lower Crust

4.1 Sampling problems

One of the principal regions of the Earth about which we know least is the lower portion of the continental crust. In Chapter 2 we assessed the thickness of the upper crust at about 10 km. The lower crust, as discussed here, comprises the lower 25–30 km or about 75% of the total crust. The composition of the upper continental crust, exposed to weathering and erosion, is well established as being of granodioritic composition (Chapter 2). For the lower crust we lack those large scale natural sampling processes (such as production of clastic sediments) which have simplified the task of arriving at upper crustal compositions. Available crustal samples comprise xenoliths in volcanic pipes and restricted outcrops of granulite terrains, which are inferred to be of lower crustal origin from recorded temperature and pressure conditions (i.e. 600–900°C, 5–10 kbars).

Geological observations are clearly constrained by inadequate sampling. Exposed granulite sections, which reveal very complex histories, may be incomplete or atypical or non-random exposures of a highly heterogeneous lower crust. In this context, xenoliths, if truly random, may be less biased samples of the lower crust (see Section 4.3). Seismic velocity studies in the lower crust, if these can be correlated with rock type, provide a way of establishing the overall lower crustal composition, analogous to the use of REE in clastic sedimentary rocks in the upper crust. However, the highly heterogeneous nature of the deep crust as revealed, for example, by the COCORP data (Section 1.2) may limit the use of this technique in arriving at a mean composition. One characteristic feature of the lower crustal composition, in the models adopted in this book, is that it should contain regions of residual composition resulting from the extraction of the upper crust. This follows from the conclusion that production of granodiorite and granite is due to intra-crustal melting, in agreement with most petrological and isotopic evidence [1]. A corollary is that the bulk crustal composition must be capable of generating the upper crustal composition by variable degrees of partial melting. One major compositional constraint is that the present heat-flow data limits the amount of K, U and Th in the lower crust to very low values in comparison with upper crustal values. In this chapter, we are concerned principally with the question of the overall composition of the lower crust and its relationship to bulk crustal compositional models rather than details of individual exposures. The probable complexity and heterogeneity of the lower crust, however, make all generalizations hazardous.

Detailed information on the structure of the lower crust is only now becoming available (Section 1.2). Seismic studies reveal many lateral and vertical variations. The Conrad discontinuity, at about 10–20 km, appears to be present in some areas (Fig. 4.1), but not in others. Many of the individual reflecting horizons lack lateral continuity. Apparently, the structure is complex in detail, perhaps matching the complexity of the upper crust. Occasionally there is variation in thickness as illustrated in Fig. 4.2. The existence of many low angle thrust sheets has recently been demonstrated (Section 1.2). This field of research is very active and current views are in a state of flux. Until the results of the new seismic experiments are available, it seems pointless to speculate in a book which deals principally with the geochemical data.

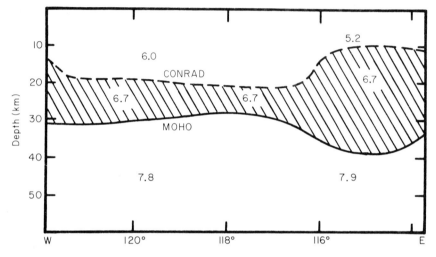

Fig. 4.1. Cross-section of the continental crust along the California–Oregon–Nevada–Idaho border line, showing the inferred presence of the Conrad discontinuity, separating a lower crustal region with $Vp=6.7$ km s^{-1} (adapted from [2]). Values on the figure are Vp in km s^{-1}.

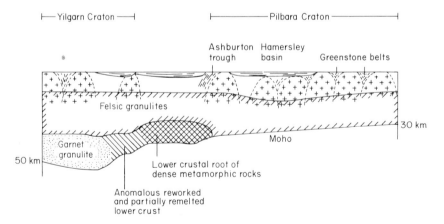

Fig. 4.2. A schematic north–south section of crust in Western Australia, showing thicker crust (40–50 km) underlying the Yilgarn craton in contrast to the thinner crust (28–33 km) of the Pilbara region (adapted from [3]). (See also discussion in Section 10.3.1.)

Until very recently, the Mohorovičić discontinuity (abbreviated hereafter to Moho) was considered to be a discrete boundary, uniform laterally, and so indicative of a chemical rather than a phase change. The view of the Moho as a sharp boundary, however, is almost certainly incorrect. Seismic evidence now suggests that the change in seismic velocities from crustal to mantle values extends over several kilometres in some places. The change in seismic velocities between crust and mantle often occurs in a series of steps extending over 5 km or so so that the structure at the base of the crust is laminated [4]. What are these structures; are they cumulate zones, imbricate thrust sheets, lenses of metasediments or slices of oceanic crust? Further complexity may result in regions of high geothermal gradients due to the difficulties of equating rock types with measured seismic velocities. There is much current uncertainty about the nature of the crust–mantle transition and so of the deepest portions of the lower crust.

4.2 The granulite facies terrains

Granulite terrains form our most accessible samples of materials which have been subjected to temperatures and pressures typical of those in the lower crust. Accounts of granulite facies terrains abound in the literature, and only some of the more crucial regions are discussed here. Extensive reviews and discussions of these terrains exist [5–7], and will not be repeated here. A particularly useful review of deep crustal sections has been given by Fountain & Salisbury [7]. They describe five regions: Ivrea zone, Italy; Pikwitonei subprovince, Canada; Frazer and Musgrave Ranges, Western Australia; and the Kasila zone of Sierra Leone, where granulite facies terrains have been upthrust and exposed as a result of continental collision processes. The uplift of denser lower crust along the thrust zone, and the depression of the less dense upper crust produces characteristic Bouguer gravity anomalies across these zones. The granulite facies rocks are dominated by mafic to intermediate compositions. To these examples, we may add the Kapuskasing structural zone of northern Ontario [8]. These zones have the advantage over xenoliths and other granulite facies terrains since they expose a section through the crust. Much work is currently in progress in such terrains [e.g. 8] and it is premature to discuss them in detail. At Kapuskasing, the crustal section extends to a depth of only about 25 km, and so the deeper crust may not be represented in this and other upthrust sections.

A large amount of apparently contradictory geochemical information appears in accounts of granulite terrains. Some are depleted in alkali elements, Th, and U, some appear to have undergone granitic melt extraction while others show no change in bulk composition compared to lower grade rocks. Anhydrous mineralogy typifies the granulite facies. Major changes in composition can occur in the upper amphibolite facies, before the onset of granulite facies metamorphism. These changes include formation and removal of granitic melts, and dehydration reactions, involving breakdown of hornblende and micas. Fyfe [9] has pointed out the importance of major removal of H_2O

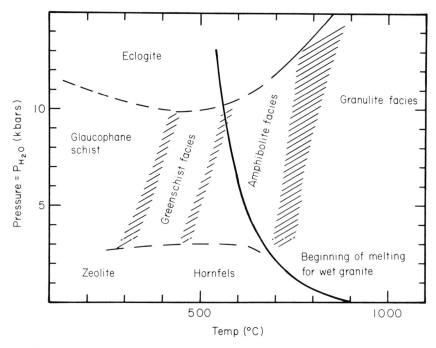

Fig. 4.3. Pressure and temperature relationships for metamorphic facies. The beginning of melting curve for wet granite illustrates that melting under hydrous conditions, with the formation of granitic magmas, is initiated in the amphibolite facies. Granulite facies metamorphism may accordingly develop in compositions from which a melt phase has already been extracted. (Adapted from [10].)

and granitic melts at the upper amphibolite grade of metamorphism, allowing the development of the anhydrous mineralogy typical of the granulite facies. True granites are common in amphibolite-facies rocks, but uncommon in granulite-facies zones (Fig. 4.3).

Heier [11] has argued that substantial loss of Rb, Cs, U and Th occurs during the loss of water accompanying granulite facies metamorphism. The mechanism of removal of elements during this process is envisaged as follows: 'In amphibolite facies gneisses the elements U, Th, Rb and Cs are probably mainly present in the lattices of micas and K-feldspar (Rb, Cs) and epidote or zircons (U, Th)... The micas tend to break down in granulite facies and to form K-feldspar... which do not concentrate Rb or Cs... Similarly, epidote minerals are not stable in granulite facies. A number of breakdown reactions involving epidote have been proposed but neither of the reaction products have suitable lattices for the incorporation of Th and U' [11, p. 437]. The escape of these elements is attributed to loss of the H_2O vapour phase. This will result in a high residual CO_2 partial pressure as is shown by the ubiquitous presence of CO_2 in fluid inclusions in granulite facies rocks [12, 13].

Thus, in dealing with the composition of the lower crust, two processes of element depletion must be considered. The first involves dehydration reactions associated with the breakdown of hydrous minerals which may selectively remove elements such as Rb and U. How effective this mechanism is on a broad scale may be questioned. How far the elements will travel depends on many factors. Uranium, which is very mobile should far outstrip the much less soluble Th so that high Th/U ratios (and high ^{208}Pb) should be characteristic of the lower crust. Much research remains to be done on the solubility of elements under high partial pressures of CO_2.

The second mechanism by which elements are removed from the lower crust is by partial melting and physical removal of the melt phase with or without portions of the restite. This process, responsible for the production of granodiorite and granite, and hence for typical upper crustal compositions, operates at lower temperatures than granulite facies metamorphism. As noted by Winkler [14, p. 312] 'melting begins, independently of the amount of free H_2O available, at the surprisingly low temperatures of 650° to 700°C' (Fig. 4.3). The enrichment of REE in the upper crust, and the development of the europium anomaly must be due to melt-liquid rather than metamorphic reactions. In this context, the well-known scarcity of granitic rocks in medium and high-pressure granulite terrains may be noted. This accords with loss of the acidic compositions by removal of low melting point fractions during melting prior to granulite facies metamorphism. Such considerations help in understanding the complex geochemical relationships observed in granulite facies terrains. Processes operating in granulite facies rocks may include one or more of the following:

(a) development of granulite facies mineralogy in dry source rocks from which a granitic melt has been extracted previously during amphibolite facies metamorphism [9]. Biotite is a common residual phase during anatexis [14]. Therefore, much K, Rb and Cs may be retained in that mineral following extraction of a granodioritic or granitic melt:

(b) dehydration of source rocks with loss of an hydrous fluid phase, resulting in granulites depleted in alkalies and U [11]. The removal of H_2O during upper amphibolite facies metamorphism and partial melting leaves CO_2 as the dominant gas phase in granulite facies terrains [15, 16]. However, the heavy REE in crustal samples show no effects attributable to complexing and removal or enrichment of these elements by CO_2;

(c) dehydration without partial melting or loss of trace elements (e.g. Jequie complex, Brazil [17]);

(d) dehydration accompanied by loss of CO_2 (e.g. Southern India [12]);

(e) subsequent retrograde metamorphism to produce amphibolite facies mineralogy in which one or more of the above processes have operated [9].

Further complications may arise if upper crust is overthrust in a continent-continent collision, as for example in the Himalayas. In such cases, typical upper crustal sediments, for example, may be subjected to granulite facies metamorphism.

4.2.1 The Lewisian gneiss complex

This is one of the classic localities where a possible sample of the lower crust is exposed. It comprises a central block of granulite facies (Scourian) bounded by

Table 4.1. Composition of Lewisian granulite and amphibolite facies rocks [18].

	1 %	2 %		1 ppm	2 ppm
SiO_2	63.0	68.1			
TiO_2	0.55	0.35			
Al_2O_3	16.1	16.3			
FeO	5.47	3.31			
MgO	3.50	1.43			
CaO	5.76	3.27			
Na_2O	4.53	5.00			
K_2O	1.03	2.15			
Σ	100.0	99.9			
Ba	760	710	Th	0.42	8.4
Rb	11	74	Zr	200	190
Sr	570	580	Hf	3.6	3.8
Pb	13	22	Nb	5	6
La	22	36	Ta	0.56	0.45
Ce	44	69	Cr	88	32
Nd	18.5	30	Ni	58	20
Sm	3.3	4.4			
Eu	1.18	1.09			
Tb	0.43	0.41			
Tm	0.19	0.14			
Yb	1.2	0.76			
Y	9	7			

1. Mean of 244 Scourian granulite gneisses, Drumbeg.
2. Mean of 39 amphibolite facies gneisses, Rhiconich.

amphibolite-facies gneisses. The composition of this terrain (Table 4.1) has been selected by Weaver & Tarney [19] as representative of the lower crust. They chose the Lewisian granulite average composition as typical of the lowermost crust. The Lewisian amphibolite facies gneisses 'appear to represent an intermediate crustal level' [19] and their composition is selected as representative of their 'middle' crustal layer. Accordingly, the Lewisian gneiss complex occupies an important position in discussions of crustal genesis. The averages [18] are listed in Table 4.1 for the granulite and amphibolite facies. The area has long been the subject of controversy. Disagreement exists at present on the interpretation of the chemical data and there is no consensus whether a partial melt has been extracted from the terrain, whether a more selective loss of elements has occurred during granulite facies metamorphism or possibly whether the compositions are little modified. Pride & Muecke [20] note that the following observations support the partial melting model:

(a) anhydrous nature of Scourian complex;
(b) incompatible element depletion, relative to upper crustal abundances;
(c) very narrow ranges of mineral composition, consistent with equilibration with a melt phase;
(d) estimated temperature of 950°C is in range for maximum melt production [21];

(e) major element chemistry shows trends *unlike* those displayed by upper crustal igneous rock sequences. Pride & Muecke [20] claim that these trends are due to equilibration of gneisses with granitic melts;

(f) REE abundances are low compared to upper crust and have positive Eu anomalies. The Eu/Eu* versus SiO_2 relationship is the opposite of that in upper crustal sequences. The Scourian complex has an overall small Eu enrichment relative to chondritic abundances.

Hamilton *et al.* [22] state that partial melting models for the Scourian would 'leave a residue with less fractionated REE patterns than observed for most Lewisian granulite facies gneisses' and would 'tend to eradicate the isotopic evidence for earlier events'. Pride & Muecke [20], however, claim that 'crustal anatexis (unlike mantle melting) does not significantly alter the Sm/Nd ratio of the starting material'. In summary, they comment that 'a range of features in the Scourian complex, including the REE data, suggests an origin fundamentally distinct from normal upper crustal igneous and metamorphic processes. Although each line of evidence may be explained in terms of some other process(es), taken together, a formidable case can be made for the residual nature of the Scourian complex.' The estimated temperature of melting is 950°C, and extraction of a 20% partial melt is envisaged. Tarney and co-workers [18, 23, 24] disagree with this interpretation. One difficulty appears to lie in comparing the compositions of the amphibolite facies terrains with those of the granulite facies. The averages reported by Weaver & Tarney [18] are claimed to show that 'overall there is a rather close correspondence between the major element and REE chemistry of Lewisian amphibolite-facies and granulite-facies gneisses'. Although we are not familiar with the field relationships, the chemical data as presented appear to us to show very significant differences. Fe, Mg and Ca are about a factor of two higher in the granulite facies rocks, while Cr and Ni show even greater differences. These were not commented upon. Much of the chemistry could be consistent with a tonalite parent for the granulite facies rocks and a trondhjemite parent for the amphibolite-facies gneisses, with some loss of Rb and Th during granulite facies metamorphism. Attempts to infer element mobility or immobility by comparisons of possibly differing precursors seems a difficult task. The positive Eu anomaly in the granulite facies gneisses resembles those seen sometimes as a primary feature in trondhjemites, and is so interpreted here, rather than being residual in nature, following extraction of a granitic melt.

Although the Lewisian terrain is of intense interest as a sample of the lower crust, it cannot be representative in the sense that it is used by Weaver & Tarney [19]. The principal reason is that the crustal model employed by these authors has too high a concentration of the heat producing elements to be representative of the continental crust (see Section 3.6.1). The heat production for the Weaver–Tarney composition of 35 mW m^{-2} is about 50% higher than the value assigned to the continental crust of 23 mW m^{-2} [25]. The information from the COCORP programme (Section 1.2) of heterogeneities on a scale greater than expected does not encourage the view that any one granulite terrain will be typical of the lower crust.

4.2.2 Geochemical aspects of other granulite terrains

The complexity shown both in the field and in the interpretation of the Lewisian is matched by observations in other granulite terrains. A wide variety of observations are available and we cite only a few to illustrate the range. Thus Nesbitt [26], after studying the New Quebec and Adirondack granulites, concluded that mass balance calculations, involving major elements show that those granulite facies rocks could represent residues after extraction of reasonable volumes of granite with compositions near those of minimum melts. Schmid & Wood [27], who studied the Ivrea–Verbano zone in Northern Italy interpreted the change in composition with increasing metamorphic grade as consistent with removal of a granodioritic magma from an 'average pelitic gneiss'. Temperature and pressure estimates were 700–820°C and 9–11 kbars, compatible with P-T conditions for the formation of granodiorites.

Extremely high Th/U ratios in Archean granulites are known from Bahia, Brazil [28]. As Heier [11] noted, Th/U may be high in amphibolite facies rocks (≈ 8), but lower in granulite facies (≈ 2–4). Possibly uranium is lost in low grade metamorphism, but both Th and U are lost by partial melting in the upper amphibolite or granulite facies. The whole question of element depletion is clearly very complex, with both dehydration reactions and partial melting operating either together or separately. Allen [29], for example, in a study of granulites in the Arunta block, South Australia, noted that while K was depleted, Rb was not, in contrast to the requirement of the conventional depletion hypothesis.

In a wide survey of granulite terrains from Africa, Canada, India and the United States, depletion in Rb relative to Sr appeared to be the exception rather than the rule [30]; the frequent presence of biotite may account for this. Recently this study was confirmed and extended by Ben Othman *et al.* [31].

Lower crustal rocks, described as eclogites, have been described from Sauviat-sur-Vige, Massif Central, France [32]. These rocks with 48% SiO_2, 18% Al_2O_3, 7% FeO, 8% MgO and 11% CaO are enriched in Eu. The REE patterns (Fig. 4.4) do not show any evidence of equilibrium with garnet, the development of which is considered to be a secondary metamorphic feature. Although these rocks are interpreted as of primary igneous origin, with some cumulate phases to account for the bulk enrichment of Eu, they could well represent residual compositions following extraction of a partial melt.

Models for the derivation of the lower crust frequently appeal to underplating mechanisms. Such models often claim igneous-like variation trends of major elements as evidence for the existence of these subterranean processes. However, recognizable metasediments are not uncommon in such terrains and their compositions may fall along 'igneous' trends. Many lower crustal rock compositions fall on 'pseudo-differentiation' trends. This misleads workers into treating them as 'primary' igneous rocks of mantle origin, although alternative explanations (e.g that they represent residual compositions following extraction of a granitic melt) may fit the data equally well.

In summary, granulite facies rocks exposed at the surface provide evidence

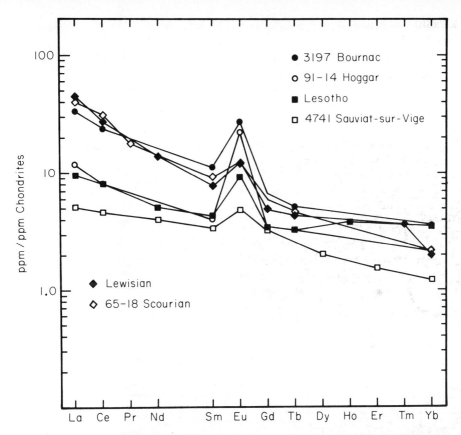

Fig. 4.4. REE patterns displaying enrichment in Eu from a wide variety of lower crustal sources including granulite facies terrains (Lewisian [33]; Scourian [20]; Sauviat-sur-Vige [32]) and xenolith localities (Lesotho [34]; Bournac [35, 36]; Hoggar [37]).

both of complex geochemical behaviour and for the extremely complex nature of the lower crust in detail [e.g. 6].

4.3 Xenoliths

A second major source of lower crustal samples are the xenoliths from volcanic pipes, and kimberlites. These are the only samples available in areas where there is an extensive cover of upper crustal rocks. They possess a wider range of chemical and isotopic characteristics than appears in surface outcrops of lower crustal rocks. Although they are random samples, snatched from their environment and transported quickly to the surface in times of the order of a day, this is not necessarily a disadvantage. They are not subjected to retrogressive metamorphism except perhaps within the pipe itself [38] or to long exposures to lower temperatures and pressures which have afflicted most of the lower crustal terrains now exposed to view. They are probably the least equivocal samples of the lower crust to which we have access.

As is the case with the granulite terrains, adequate assessment of the xenolith data requires book length treatment, and we can only give a few highlights here.

Among the most extensive studies of lower crustal xenoliths are the investigations by Dupuy and co-workers [35, 36, 37]. Over 2000 xenoliths from the Bournac pipe, France, have been investigated and a lower crustal composition established for that area. The Moho is at 27–28 km beneath the Massif Central at Bournac, so that the lower crustal depths may be more

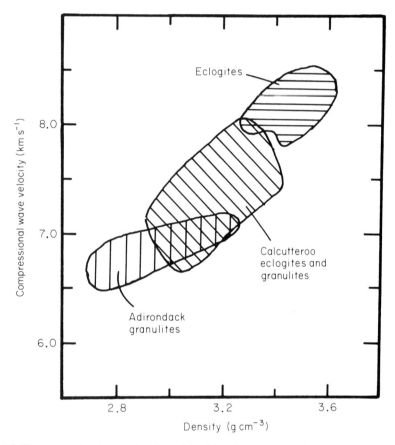

Fig. 4.5. The compressional wave velocities and densities of lower crustal eclogites and granulites from Calcutteroo, South Australia [40], overlap the fields for Adirondack granulites and eclogites from several localities [42]. (Adapted from [41].)

typical of intermediate crustal levels in other regions. Nevertheless many interesting conclusions emerge from the study of the granulitic xenoliths. Firstly, there does not seem to be a general depletion in LIL elements (K, Rb, Cs, U, Th) as advocated by many workers [e.g. 11]. The vertical zonation of elements at Bournac is related to the mineralogy and major element chemistry and does not appear to be due to a preferential upward migration of elements such as Rb, U or Th. The lower crust at Bournac appears to consist of heterogeneously banded granulite facies rocks, with much variation locally. The amount of 'restite' in the Bournac sample appears to be small (1%) with the remainder comprising essentially unmelted granulites. About 8–10 km of 'granitic' rocks forms the upper crust at Bournac, including granites derived by anatectic processes. Presumably there are unsampled restites at depth.

A second major occurrence are the xenoliths from the Lesotho kimberlite pipes [34, 39]. These samples are considered to have crystallized at temperatures in the range 660–830°C at pressures of 9–13 kbar (i.e. they come from depths of between 22 and 36 km). Typically, they are comprised of 20–40% garnet, 20–40% clinopyroxene and plagioclase. Their major and trace element compositions show much variation. These rocks have high Al_2O_3 and CaO. The bulk rocks commonly have Eu enrichment, which is reflected in the individual minerals, including garnet. The absence of any effects on the REE patterns due to garnet is interpreted as indicating that the bulk rock enrichment in Eu was established prior to crystallization of garnet in granulite facies metamorphism. Although regarded by some as cumulates, metamorphosed in the lower crust [39], their compositions are also consistent with that of residual material following extraction of a partial melt.

Extreme heterogeneity characterizes suites of xenoliths from SE Australia described by Arculus *et al.* [40]. Many examples of basic compositions are present but the chemistry is unlike that of modern basalts. For example, samples with the same major element chemistry may have very diverse REE patterns. Conversely samples with diverse mineralogy display similar REE patterns. None of the samples show heavy REE depletion characteristic of melting within the garnet stability field. It is clear that in most cases, the present phase assemblages post-date the establishment of the chemical composition. The most important finding is the plethora of basic compositions, in a region where the Moho displays a laminated character. Compressional wave velocity and density measurements on these xenoliths (Fig. 4.5) show probable *in situ* values in the range 7.0–7.7 km s^{-1} for V_p [41]. These values fall between the earlier measured values for silicic granulites and mafic eclogites [42], and correlate well with measured V_p for the lower crust in this part of Eastern Australia [4].

A suite of granulite xenoliths from central Hoggar, Algeria [37], shows a number of chemical features typical of xenoliths from other areas. Many samples show Eu enrichment, and these occur in samples referred to as 'meta-igneous' and 'metasedimentary'. About two thirds of the samples have enrichment in europium relative to the other REE. The composition of these samples is similar to those from the Bournac pipe and other areas.

A compilation of representative xenolith samples which show Eu enrichment is given in Table 4.2. Figure 4.4 shows some of the typical REE patterns. Typically the compositions are high in alumina. An interesting question is why such compositions are frequently encountered as xenoliths, but are uncommon in typical upper crustal rocks. If Eu enrichment was equally common in the upper and lower crust, then random sampling should give a similar frequency of Eu enrichment for both. This is not the case.

Table 4.2. Basic compositions in the lower continental crust.

	1	2	3	4	5	6	7
SiO_2	50.6	52.2	52.6	57.5	49.5	48.5	48.7
TiO_2	0.77	1.3	1.2	0.96	0.70	0.55	1.02
Al_2O_3	17.9	19.5	24.5	19.5	18.5	19.7	16.5
FeO	8.8	9.2	5.9	5.7	6.6	5.5	11.8
MgO	8.2	5.9	2.0	2.8	8.3	8.7	5.51
CaO	9.4	8.7	7.9	6.6	14.5	14.8	13.8
Na_2O	3.3	2.7	4.0	4.9	1.6	2.0	2.34
K_2O	1.0	0.55	1.7	0.9	0.4	0.26	0.34
Σ	100.0	100.1	99.8	98.9	100.1	100.0	100.0
Eu/Eu*	2.4	1.46	3.14	1.69	1.31	1.50	—
Nd_N/Sm_N	1.20	—	—	—	1.29	1.26	—

1. Lesotho garnet granulite xenoliths PHN 1670, 2533, 2582 and L.13 [34, 39].
2. Granulite xenolith 3199, Bournac [36].
3. Granulite xenolith 3197, Bournac [36].
4. Scourian granulite 65-16 [20].
5. Sauviat-sur-Vige metagabbro 4736 [32].
6. *Ibid.*, 4737 [32].
7. Kapuskasing mafic gneiss (average of 3) [43].

Examples of Eu enrichment in upper crustal rocks, or in igneous rocks in general, are rare. The common occurrence of lower crustal xenoliths enriched in Eu implies a substantial Eu enrichment in their source region. Although the presence of 'cumulate' plagioclase is often advanced to account for these ubiquitous Eu enrichments, this raises many problems. What mechanism is responsible for producing plagioclase-rich cumulates in the lower crust? Why are they missing from the upper crust? Where are the extensive suites of igneous rocks complementary to these cumulates? Although their major element composition resembles that of high-Al basalts, these eruptive rocks do not typically show enrichment in europium [44].

4.4   Basic or acidic lower crusts?

Many suites of crustal derived nodules in kimberlitic and other types of explosive volcanism are typically basic in character (e.g. Southern Africa, south-eastern Australia, Colorado Plateau and Front Ranges, South Central France, North-west Africa and West Germany). Many exposed sections of deeper crust are dominated by basic granulite [7]. Some seismic profiles in regions where basic materials appear as xenoliths show that a velocity gradient, rather than a sharp break, occurs between the lower crust and upper mantle. These profiles are compatible with the transformation at progressively deeper levels in the crust of relatively feldspar-rich to garnet-rich basic compositions. Furthermore, measured seismic velocities of the xenoliths coincide with inferred velocities in the depth range 25–45 km of about 7.0–8.0 km s^{-1}, which would imply basic material.

In general, medium and high pressure granulite facies terrains (e.g. Scourian) have a less silica-rich composition than that of the upper crust. They have lower abundances of Th, U and K, low Rb and high K/Rb ratios, and low initial and present-day $^{87}Sr/^{86}Sr$. Anorthosites are common, but granites are rare [20]. The Weaver & Tarney [19] three layer crustal model (12 km of upper crust, underlain by 6 km of 'middle' crust and 18 km of 'lower' crust) takes the composition of the middle crust to be that of the Lewisian amphibolite grade rocks and that of the lower crust to be that of average Lewisian granulites. This model has the advantage of dealing with observable terrains, but produces a silica rich lower crust. This value, combined with a granodioritic upper crust, produces a bulk crustal composition with too much K, U, and Th to be consistent with the heat flow data.

Rare earth element (REE) patterns in xenoliths span the range from relative light REE enrichment through unfractionated to light REE depleted. Garnet does not appear to have been involved as a residual phase during the formation of these rock types. Characteristically, lower crustal granulite terrains have high Sr, Ba and Pb abundances relative to light REE, depletions in U and Rb, but erratic and occasionally high Th contents.

The differences between the somewhat more basic composition suggested by the xenoliths and the more acidic compositions occurring in granulite facies terrains present a particular problem in attempting to decide on lower crustal compositions. This dichotomy may reflect differences in buoyancy with the less dense portions of the crust being preferentially upthrust by tectonic

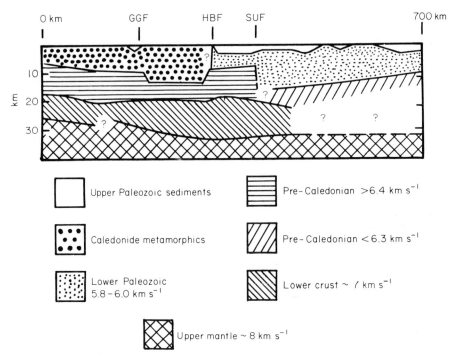

Fig. 4.6. A proposed cross-section through the crust and upper mantle in Northern Britain. SUF=Southern Uplands Fault; HBF=Highland Boundary Fault; GGF=Great Glen Fault. (Adapted from [45].)

forces. Such a structure for the lower crust, with basic granulites overlain by quartzo-feldspathic gneisses, is suggested [45] in the British Isles (Fig. 4.6). If the lower crust is chemically zoned in this manner, this certainly rules out attempts to use the more siliceous exposed granulite terrains as being representative of the entire lower crust.

The interpretation of the mean seismic velocity data by Smithson et al. [46] provides a bulk crust with 64% silica and 2.7% K_2O. These values are close to upper crustal estimates and imply similar lower crustal values. It seems difficult to reconcile such estimates with the heat flow constraints and the evidence both from xenoliths and granulite terrains (e.g. Ivrea) for more basic lower crustal compositions. Smithson et al. [46] note that present measurements of mean crustal seismic velocities were 'not acquired in a manner designed to attack this problem' (of crustal composition). Earlier estimates by Pakiser & Robinson [47] concluded that the seismic data were consistent with a bulk crust of andesitic composition (58% silica), more consistent with our estimates.

4.5   Nd and Sr isotopic studies

The close coherence between Sm and Nd even under the conditions of granulite facies metamorphism allows neodymium isotopic systematics to provide information about the lower crust. The initial work on high grade terrains has been restricted almost exclusively to Archean felsic granulites [22, 48]. These studies have been used to date the extraction from the mantle of the igneous precursors of the granulite-grade rocks. Large, negative values for

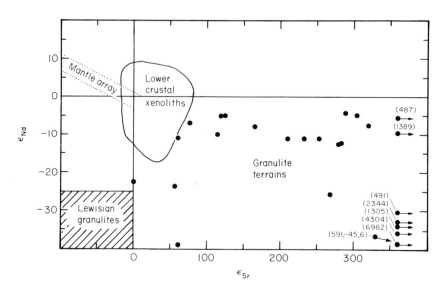

Fig. 4.7. Data from granulite terrains plotted on an ε_{Nd}–ε_{Sr} diagram. There is a very wide scatter particularly in the direction of high $^{87}Sr/^{86}Sr$ ratios, indicating prior residence in high Rb environment or metasomatic activity. The Lewisian gneiss field is unique. The field for lower crustal xenoliths from Fig. 4.8 is also shown. (Adapted from [31].)

$\varepsilon_{Nd}(0)$ (<-20) taken to be representative of the lower crust have also been recognized.

A survey of Sm–Nd and Rb–Sr systematics in felsic granulites has been undertaken [31] in an attempt to resolve the time interval between generation of new crust from the mantle, and internal crustal differentiation to produce granulite facies metamorphism and granites (Fig. 4.7). The time interval can be large and appears to be greater for younger granulites. Only a few of the granulites studied are depleted in Rb, probably because of the presence of biotite. Although the authors [31] claim that the $^{147}Sm/^{144}Nd$ ratio is constant at about 0.1121, there is a factor of three between the lowest and highest values. In contrast, we suggest that there is some fractionation of Sm and Nd during the melting event which produces the granitic upper crust. This study considered mainly acid or intermediate granulites, which are not necessarily representative of the lower crust as discussed earlier (Sections 4.3, 4.4).

McCulloch et al. [49] have questioned the use of Archean felsic granulites as representative of present-day lower crust both on the grounds of age and

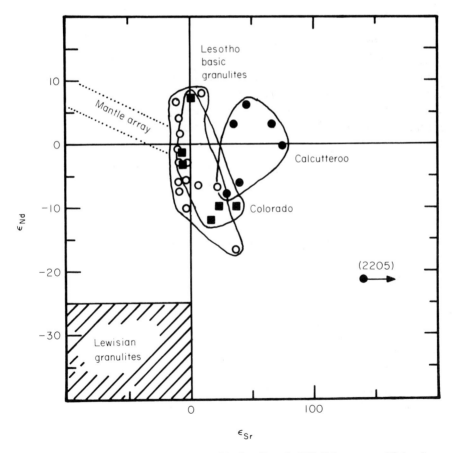

Fig. 4.8. ε_{Nd}–ε_{Sr} diagram for lower crustal xenoliths from Lesotho [39], Calcutteroo and Colorado [40] showing the much more restricted range of ε_{Nd} and ε_{Sr} compared with granulite terrains shown in Fig. 4.7, and the unique position of the Lewisian data.

because most studies have ignored the presence of intermediate, basic and ultramafic lithologies.

There are few isotopic analyses of lower crustal xenoliths and this represents one of the largest gaps of knowledge in our understanding of the continental crust. Most of the available data are plotted on an $\varepsilon_{Nd}(0)-\varepsilon_{Sr}(0)$ diagram in Fig. 4.8. From these data and the granulite data (see above), we can draw three major conclusions:

1 the lower crust is heterogeneous with respect to age and/or $^{143}Nd/^{144}Nd$, $^{87}Sr/^{86}Sr$ initial ratios;
2 there is considerable fractionation of Sm–Nd as well as Rb–Sr between the upper and lower crust;
3 the lower crust has higher ε_{Sr} and probably higher ε_{Nd} than indicated by the Lewisian felsic granulites.

Many of the xenoliths also display a correlation between $^{143}Nd/^{144}Nd$ and Sm/Nd. A regression of the data yield ages of 2470 m.y. for South Australia and 1400 m.y. for South Africa. In some cases, a similar age is derived from the Rb–Sr systematics (e.g. South Australia). These ages probably reflect the last equilibration event, possibly an intracrustal partial melting event.

4.6 Electrical conductivity

Most conventional geological wisdom suggests that the lower crust is dry. Granulite terrains, originating at pressures and temperatures appropriate to the lower crust, possess a dry mineralogy. Fluid inclusions in granulites commonly contain CO_2, rather than H_2O. Surface rocks possess electrical resistivity values typically of the order of 10^5 Ω.m. It is therefore surprising that some interpretations of electrical conductivity measurements indicate that a layer deep within the crust may be highly conducting, with resistivities in the range of 1–100 Ω.m. If some portion of the lower crust possesses such high conductivities, then only the presence of a conductor such as water or perhaps a melt layer could explain this observation. Accordingly this apparent discrepancy is of great geological and geochemical significance.

Do the geophysical measurements uniquely specify the presence of a conducting layer in the lower continental crust? Measurements in many parts of the world have apparently detected such conducting layers. Figure 4.9 shows a map with 23 such locations. Generally the conducting layer is reported to lie between 15 and 25 km depth, so that it apparently occurs in the upper portions of the lower crust, which as defined here begins at a depth of 10 km.

Measurements at Timmins, Ontario, were in a well known geological region [52]. There is no overburden, and highly resistive country rock apparently extends to 15–20 km. The presence of a conducting layer at about 20 km depth could be inferred from the data. This layer did not appear to be a continuous feature on a regional scale. A study in the Adirondacks [53], indicated a section comprising overburden, 15 km of highly resistive rock (12 000 Ω.m) a 5 km layer with a resistivity of 1000 Ω.m overlying a good conductor (resistivity <100 Ω.m) at 20 km depth.

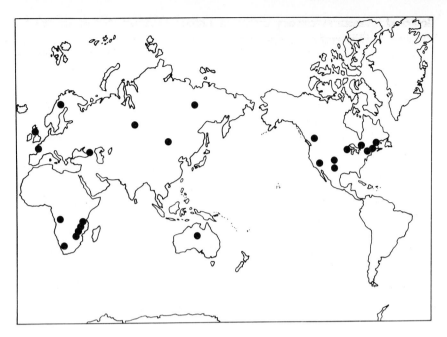

Fig. 4.9. Location of regions in which highly conducting layers have been detected in the lower crust. (Adapted from [51].)

In a study of the Fennoscandian region [54], the following model was proposed:
1   a crustal layer 6.5–16 km thick with a resistivity of 10^4 Ω.m;
2   a lower crustal layer with intermediate values of resistivity (215–400 Ω.m);
3   an upper mantle only slightly more conducting, with resistivity values of 70–95 Ω.m;
4   at a depth of 150–190 km, a highly conducting layer identified within the asthenosphere, with resistivity values of 2–8 Ω.m.

The thickness of the upper crust in this section is inferred to be 16 km from seismic evidence. The lower crustal layer has V_p velocities of 6.8–7.2 km s^{-1}, correlated with an amphibolite layer. There is only a slight change in resistivity across the Moho. Evidence for a highly conducting layer, at a depth of about 15–25 km, was also obtained in the Precambrian Shield areas in central Australia [51].

In summary, the geophysical data now appear to indicate the frequent presence of conducting layers at depths of about 15–20 km. The layers appear to be discontinuous on a regional scale. What are they due to? The most popular candidate is free water, presumably present as an intergranular film but many other suggestions have been advanced. The presence of water is apparently in conflict with the geological evidence from granulite facies rocks that the lower crust is dry. A basic problem with many of the hypotheses is that the conducting medium must be continuous. For example, a water film must be totally interconnected.

Dry rock of any composition has a resistivity $>10^9$ Ω.m. This value is lowered in near-surface rocks to values of about 10^5 Ω.m, due to the presence of water. Olhoeft [55] considers that the electrical properties of granite are dominated by *free* water, and by temperature. Hydrated hornblende schist (with structural water), for example, has electrical resistivity very close to that for dry granite. Bound water is thus not capable of explaining the high conductivity values observed. Explanations that the conducting layers represent buried slices of oceanic crust, amphibolites, serpentine, etc. are not viable, although they may provide a source for water during dehydration of these rocks.

One candidate which is often suggested is magnetite, which has a resisitivity of about 10^{-4} Ω.m [56]. If magnetite formed continuous films during alteration of olivine and pyroxene to serpentine, then it could explain the observations. Less than 0.1% of such a magnetite film could explain the data [51]. However, a continuous film of magnetite is completely unreasonable on geochemical grounds. If the low conductivities are due to free water, then a possible source of the water may be due to dehydration of the lower crust during upper amphibolite and granulite facies metamorphism. The water released travels upward until the combination of falling temperature and impermeable layers causes it to remain trapped as a highly conducting film, at depths of 15–20 km. Such a mechanism would produce discontinuous layers on a regional scale in agreement with the observations. The recent results of the COCORP investigations (Section 1.2) also indicate the possibility of large-scale underthrusting of wet oceanic crust or other water bearing sedimentary layers. The possibility of trapped free water in the deep crust thus appears to be much more likely than thought previously. Water originating from dehydration of the lower crust or from underthrust sheets may thus explain the enigma of these highly conducting layers in the lower crust. Clearly much work remains to be carried out on this problem to reconcile the petrological and geographical observations.

4.7 Proposed compositions for the lower crust

Various approaches are possible. One is to take the proposed model for the bulk crust, and calculate residual compositions following the extraction of granitic melts typical of the upper crust. In Table 4.3, such calculations are presented for the extraction of 10% minimum granitic melt from proposed total crustal compositions (column 6). Column 7 shows the residue following the extraction of 20% granite from the bulk crustal composition. A more realistic approach is to extract a granodioritic composition from the model total crust, assuming that the upper crust forms 25% of the total (column 8). Finally, in column 9 we present the composition resulting from the extraction of 25% of the upper crustal composition calculated in Chapter 2 (Table 2.15). These compositions indicate that the lower crust should contain compositions high in Al_2O_3, CaO, low in K_2O, with positive Eu anomalies (Eu/Eu* >1) and Sm/Nd ratios approaching chondritic values. Figure 4.10 shows the upper crustal REE pattern, and the predicted REE pattern for the lower crust.

Table 4.3. Possible lower crustal compositions resulting from the extraction of granitic and granodioritic melts.

	1	2	3	4	5	6	7	8	9
SiO_2	66.0	67.2	72.0	76.5	57.3	55.2	53.6	54.0	54.4
TiO_2	0.5	0.55	0.31	0.1	0.9	1.0	1.05	1.02	1.0
Al_2O_3	15.2	16.0	14.5	12.9	15.9	16.2	16.3	15.9	16.1
FeO	4.5	4.04	2.76	1.1	9.1	10.0	10.7	10.8	10.6
MgO	2.2	1.77	0.72	0.1	5.3	5.9	6.4	6.5	6.3
CaO	4.2	3.89	1.86	0.7	7.4	8.1	8.8	8.6	8.5
Na_2O	3.9	3.81	3.72	3.8	3.1	3.0	2.9	2.9	2.8
K_2O	3.4	2.78	4.11	4.7	1.1	0.7	0.35	0.54	0.33
Σ	99.9	100.0	100.0	99.9	100.1	100.1	100.1	100.3	100.0
Eu/Eu*	0.65	0.5	0.65	0.45	1.0	1.12	1.14	1.14	1.14

1. Upper crust (Table 2.15).
2. Average granodiorite (Table 3.1).
3. Average granite (Table 3.1).
4. Minimum melt composition [57].
5. Total crust (Table 3.5).
6. Residue following 10% extraction of minimum melt composition (Col. 4) from total crust (Col. 5).
7. Residue following 20% extraction of granite (Col. 3) from total crust (Col. 5).
8. Residue following 25% extraction of granodiorite (Col. 2) from total crust (Col. 5).
9. Residue following 25% extraction of upper crust (Col. 1) from total crust (Col. 5).

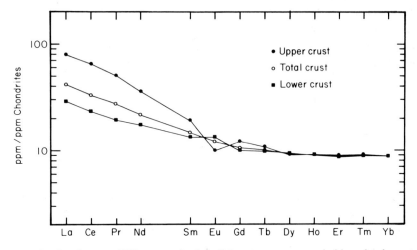

Fig. 4.10. Predicted average REE patterns for the bulk lower crust, compared with total and upper crustal patterns. Note the enrichment in Eu (Eu/Eu* = 1.1) compared with the upper crustal depletion in Eu (Eu/Eu* = 0.65) and the absence of an anomaly in the total crustal abundances. The lower crustal pattern also shows slight LREE enrichment.

These compositions provide some testable predictions. *Average* lower-crustal samples should have small positive Eu anomalies with Eu/Eu* values of about 1.1. The calculated Sm/Nd average ratio is 0.27, similar to chondritic values of 0.325, much higher than the upper-crustal average of 0.17. The estimated lower-crustal value is thus not far from chondritic or undepleted mantle values, if we allow for the uncertainties in the calculations. This has important implications for Sm–Nd systematics as old lower crust of this composition would not be readily distinguishable from younger mantle additions. The normative composition for the lower crust has about 54% plagioclase, consistent with the high Ca, Al, Sr and Eu contents. The composition of the lower crust during the Archean is different, as is discussed in Chapter 7, since intracrustal melting was at a minimum.

Numerous examples exist of lower crustal compositions which are high in Al, Ca, and Eu. A selection of these was given in Table 4.2. They include granulite xenoliths from Lesotho [39], Bournac, France [36], eclogites from Sauviat-sur-Vige, France [32], and the extensive suite from Hoggar, Algeria [37]. These compositions are typified by high Al_2O_3 and CaO, and low K_2O contents, and have Eu enrichment relative to chondritic normalized abundances of the other REE (Fig. 4.4). Sm/Nd ratios are higher than upper crustal or total crustal estimates.

Finally, we present, in Table 4.4, our estimate of the lower crustal composition. This has been calculated from the total crustal composition presented in Chapter 3 (Table 3.5) by extracting from that composition, the upper crustal values, given in Chapter 2. This lower crustal composition has the following properties. It is rich in normative plagioclase (54% Ab+An), is low in silica (54.4%), and high in alumina (16.1%). It thus resembles a high-alumina basalt. Potassium is very low (0.28%) and the K/Rb ratio (530) is

Table 4.4. Proposed bulk composition for the lower continental crust.

	%			NORM	
SiO_2	54.4		Q	3.7	
TiO_2	1.0		Or	2.0	
Al_2O_3	16.1		Ab	23.7	
FeO	10.6		An	30.4	
MgO	6.3		Di	9.8	
CaO	8.5		Hy	28.6	
Na_2O	2.8		Il	1.9	
K_2O	0.34				
Σ	100.0		Mg/Mg+Fe=0.51		

Li	11 ppm	Ni	135 ppm	In	50 ppb	Er	2.2 ppm
Be	1.0 ppm	Cu	90 ppm	Sn	1.5 ppm	Tm	0.32 ppm
B	8.3 ppm	Zn	83 ppm	Sb	0.2 ppm	Yb	2.2 ppm
Na	2.08%	Ga	18 ppm	Cs	0.1 ppm	Lu	0.29 ppm
Mg	3.80%	Ge	1.6 ppm	Ba	150 ppm	Hf	2.1 ppm
Al	8.52%	As	0.8 ppm	La	11 ppm	Ta	0.6 ppm
Si	25.42%	Se	0.05 ppm	Ce	23 ppm	W	0.7 ppm
K	0.28%	Rb	5.3 ppm	Pr	2.8 ppm	Re	0.5 ppb
Ca	6.07%	Sr	230 ppm	Nd	12.7 ppm	Ir	0.13 ppb
Sc	36 ppm	Y	19 ppm	Sm	3.17 ppm	Au	3.4 ppb
Ti	0.60%	Zr	70 ppm	Eu	1.17 ppm	Tl	230 ppb
V	285 ppm	Nb	6 ppm	Gd	3.13 ppm	Pb	4.0 ppm
Cr	235 ppm	Mo	0.8 ppm	Tb	0.59 ppm	Bi	38 ppb
Mn	1670 ppm	Pd	1 ppb	Dy	3.6 ppm	Th	1.06 ppm
Fe	8.24%	Ag	90 ppb	Ho	0.77 ppm	U	0.28 ppm
Co	35 ppm	Cd	98 ppb				

higher than the upper crustal values, as is the Rb/Cs ratio (53). The Rb/Sr ratio (0.023) is low compared to upper crustal values of 0.3 and is even lower than the whole Earth value of 0.031. While the Sm/Nd ratios of 0.25 is lower than the whole Earth value of 0.32, it is much higher than the value of 0.17 for the upper crust. Thorium (about 1 ppm) and U (about 0.3 ppm) are very low compared to upper crustal values.

4.8 Summary

1 The lower crust, comprising 75% of the total crust, remains an enigmatic region for which we lack large-scale natural sampling processes.
2 Granulite facies terrains, with pressure and temperature regimes indicative of lower crustal residence, exhibit great geochemical complexity. Commonly but not universally, they exhibit depletion in K, Rb, U and Th. Some terrains appear to have suffered dehydration only, others loss of alkalies, U and Th and others represent depleted residues following extraction of a granitic melt.
3 The present geochemical evidence, even in well studied regions such as the Lewisian, is often inadequate to decide between models which remove fluid phases and those which extract granitic melts. The bulk composition of the Lewisian amphibolites and granulites is not a satisfactory model for the overall composition of the lower crust.
4 Xenoliths of lower crustal origin generally appear to be more basic in composition than the granulite terrains, and frequently show enrichment in Eu. This suggests that the deeper portions of the lower crust are more basic than the composition of the exposed granulite terrains.
5 Isotope studies indicate that the lower crust is heterogeneous with respect to ε_{Nd} and ε_{Sr}. There is considerable fractionation of Sm–Nd and Rb–Sr between the upper and lower crusts.
6 Electrical conductivity measurements commonly indicate the presence of a highly conducting (1–100 Ω.m) layer at depths of 15–20 km. Petrological models indicate a dry lower crust. A possible resolution of the dilemma is that the conducting layers are due to trapped films of water derived by dehydration during granulite facies metamorphism.
7 Average lower crustal compositions are model dependent, but may be derived from the bulk crustal composition by removal of the upper crustal composition. Our proposed average is given in Table 4.4, for 62 elements. Key element values and ratios include: $SiO_2=54.4\%$; $Al_2O_3=16.1\%$; $K=0.28\%$; $K/Rb=530$; $Rb/Sr=0.023$; $La_N/Yb_N=3.8$; $Eu/Eu^*=1.14$; $Sm/Nd=0.25$; $U=0.28$ ppm; $Th=1.06$ ppm; $Cr=235$ ppm; $Ni=135$ ppm.

Notes and references

1 Tuttle, O.F. & Bowen, N.L. (1958) Origin of granite in the light of experimental studies in the system $NaAlSi_3O_8$–$KAlSi_2O_3$–SiO_2–H_2O. *GSA Mem.*, **74**; Wyllie, P.G. (1977) Crustal anatexis: an experimental view. *Tectonophysics*, **43**, 41.
2 Smith, R.B. (1978) Seismicity, crustal structure and intraplate tectonics in the interior of the Western Cordillera. *GSA Mem.*, **152**, 111.

3 Drummond, B.J. (1981) Crustal structure of the Precambrian terrains of northwest Australia from seismic refraction data. *BMR J.*, **6**, 123.
4 Finlayson, D.M. *et al.* (1979) Explosion seismic profiles and implications for crustal evolution in southeastern Australia. *BMR J.*, **4**, 243.
5 Windley, B.F. (1976) *The Early History of the Earth.* Wiley; (1977) *The Evolving Earth.* Wiley.
6 Kay, R.W. & Kay, S.M. (1981) The nature of the lower continental crust: Inferences from geophysics, surface geology and crustal xenoliths. *RGSP*, **19**, 271.
7 Fountain, D.M. & Salisbury, M.H. (1981) Exposed cross-sections through the continental crust: implications for crustal structure, petrology and evolution. *EPSL*, **56**, 263.
8 Percival, J.A. & Card, K.D. (1983) Archean crust as revealed in the Kapuskasing uplift, Superior Province, Canada. *Geology*, **11**, 323.
9 Fyfe, W.S. (1973) The granulite facies, partial melting and the Archean crust. *Phil. Trans. Roy. Soc.*, **A273**, 457.
10 Turner, F.J. (1968) *Metamorphic Petrology.* McGraw Hill.
11 Heier K.S. (1973) Geochemistry of granulite facies rocks and problems of their origin. *Phil. Trans. Roy. Soc.*, **A230**, 429.
12 Newton, R.C. *et al.* (1980) Carbonic metamorphism, granulites and crustal growth. *Nature*, **288**, 45.
13 An extensive discussion on 'Fluids in Metamorphism' is contained in *J. Geol. Soc. Lond.*, **140**, 529–663.
14 Winkler, H.G.F. (1976) *Petrogenesis of Metamorphic Rocks.* Springer-Verlag.
15 Janardhan, A.S. *et al.* (1979) Ancient crustal metamorphism at low P_{H_2O}. *Nature*, **278**, 511.
16 Touret, J. & Dietvorst, P. (1983). Fluid inclusions in high grade anatectic metamorphites. *J. Geol. Soc. Lond.*, **140**, 635.
17 Sighinolfi, G.P. *et al.* (1981) Geochemistry and petrology of the Jequie granulitic complex (Brazil). *CMP*, **78**, 263.
18 Weaver, B.L. & Tarney, J. (1981) Lewisian gneiss geochemistry and Archean crustal development models. *EPSL*, **55**, 172.
19 Weaver, B.L. & Tarney, J. (1984) Major and trace element composition of the continental lithosphere. *PCE*, **15**, Chapter 7.
20 Pride, C. & Muecke, G.K. (1980) Rare earth element geochemistry of the Scourian complex, NW Scotland—evidence for the granite-granulite link. *CMP*, **73**, 403.
21 Brown, G.C. & Fyfe, W.S. (1970) The production of granitic melts during ultrametamorphism. *CMP*, **28**, 310.
22 Hamilton, P.J. *et al.* (1979) Sm–Nd systematics of Lewisian gneisses: implications for the origin of granulites. *Nature*, **277**, 25.
23 Tarney, J. & Windley, B.F. (1977) Chemistry, thermal gradients and evolution of the lower continental crust. *J. Geol. Soc. Lond.*, **134**, 153.
24 Weaver, B.L. & Tarney, J. (1981) REE geochemistry of Lewisian granulite-facies gneisses, Northwest Scotland: Implications for the petrogenesis of the Archaean lower continental crust. *EPSL*, **51**, 279.
25 See Section 3.2.
26 Nesbitt, H.W. (1980) Genesis of the New Quebec and Adirondack granulites: evidence for their production by partial melting. *CMP*, **72**, 303.
27 Schmid, R. & Wood, B.J. (1976) Phase relationships in granulite metapelites from the Ivrea–Verbano zone (Northern Italy). *CMP*, **54**, 255.
28 Sighinolfi, G.P. & Sakai, T. (1977) Uranium and thorium in Archean granulite facies terrains of Bahia (Brazil). *Geochem. J.*, **11**, 33.
29 Allen, A.R. (1979) Metasomatism of a depleted granulite facies terrain in the Arunta Block, Central Australia. *CMP*, **71**, 85.
30 Spooner, C.M. & Fairbairn, H.W. (1970) Strontium 87/strontium 86 initial ratios in pyroxene granulite terrains. *JGR*, **75**, 6706.
31 Ben Othman, D. *et al.* (1984) Nd–Sr isotopic composition of granulites and constraints on the evolution of the lower continental crust. *Nature*, **307**, 510.
32 Bernard-Griffiths, J. & Jahn, B.M. (1981) REE geochemistry of eclogites and associated rocks from Sauviat-sur-Vige, Massif Central, France. *Lithos*, **14**, 263.
33 Drury, S.A. (1978) REE distributions in a high grade Archean gneiss complex in Scotland: Implications for the genesis of ancient sialic crust. *PCR*, **7**, 237.

34 Rogers, N.W. (1977) Granulite xenoliths from Lesotho kimberlites and the lower continental crust. *Nature,* **270,** 681.
35 Dupuy, C. et al. (1979) The lower continental crust of the Massif Central (Bournac, France). *PCE,* **11,** 401.
36 Dostal, J. et al. (1981) Geochemistry and petrology of meta-igneous granulitic xenoliths in Neogene volcanic rocks of the Massif Central, France—Implications for the lower crust. *EPSL,* **50,** 31.
37 Leyreloup, A. et al. (1982) Petrology and geochemistry of granulite xenoliths from Central Hoggar (Algeria)—Implications for the lower crust. *CMP,* **79,** 68.
38 Padovani, E. et al. (1982) Constraints on crustal hydration below the Colorado plateau from Vp measurements on crustal xenoliths. *Tectonophysics,* **84,** 313.
39 Rogers, N.W. & Hawkesworth, C.J. (1982) Proterozoic age and cumulate origin for granulite xenoliths, Lesotho. *Nature,* **399,** 409.
40 Arculus, R.J. et al. (1984) Eclogites and granulites in the lower continental crust: examples from eastern Australia and southwestern USA. *Proc. Eclogite Symp.* (in press).
41 Jackson, I. & Arculus, R.J. (1984) Laboratory wave velocity measurements on lower crustal xenoliths from Calcutteroo, South Australia. *Tectonophysics,* **101,** 185.
42 Manghani, M.H. et al. (1974) Compressional and shear wave velocities in granulite facies rocks and eclogites to 10 kbar. *JGR,* **79,** 5427.
43 Percival, J.A. et al. (1983) in A cross-section of the Archean Crust. *LPI Tech. Report 83–03.* LPI, Houston.
44 *Basaltic Volcanism on the Terrestrial Planets* (1981). Pergamon Press.
45 Upton, B.C.G. et al. (1983) The upper mantle and deep crust beneath the British Isles: Evidence from inclusions in volcanic rocks. *J. Geol. Soc. Lond.,* **140,** 105.
46 Smithson, S.B. et al. (1981) Mean crustal velocity: a critical parameter for interpreting crustal structure and crustal growth. *EPSL,* **53,** 323.
47 Pakiser, L.C. & Robinson, R. (1967) The composition of the continental crust as estimated from seismic observations. *AGU Mono.,* **10,** 620.
48 Jacobsen, S.B. & Wasserburg, G.J. (1978) Interpretation of Nd, Sr and Pb isotope data from Archean migmatites in Lofoten-Vesteralen, Norway. *EPSL,* **41,** 245.
49 McCulloch, M.T. et al. (1982) Isotopic and geochemical studies of nodules in kimberlite have implications for the lower continental crust. *Nature,* **300,** 166.
50 The SI unit for resistivity is the ohm meter (Ω.m). Conductivity is the reciprocal of resistivity and is expressed in siemens per metre (S m^{-1})
51 Constable, S. (1983) *Deep Resistivity Studies of the Australian Crust.* Ph.D. Thesis, Australian National University.
52 Duncan, P.M. (1978) *Electromagnetic Deep Crustal Sounding with a Controlled Pseudo-noise Source.* Ph.D. Thesis, University of Toronto.
53 Nekut, A. et al. (1977) Deep crustal electrical conductivity; evidence for water in the lower crust. *GRL,* **4,** 239.
54 Jones, A.G. (1982) On the electrical crust-mantle structure in Fennoscandia: Moho and the asthenosphere revealed? *Geophys. J.R. Astr. Soc.,* **68,** 371.
55 Olhoeft, G.R. (1981) Electrical properties of granite with implications for the lower crust. *JGR,* **86,** 931.
56 Parkhomenko, E. (1982) Electrical resistivity of minerals and rocks at high temperature. *RGSP,* **20,** 193.
57 White, A.J.R. & Chappell, B.W. (1977) Ultrametamorphism and granitoid genesis. *Tectonophysics,* **43,** 7.

5
Uniformity of Crustal Composition with Time

5.1 Measurement of secular variations

In the previous three chapters, we have examined methods of determining the composition of the continental crust and have arrived at estimates for the composition of present-day upper, total, and lower crusts. The next question to be addressed is how far back through time are such compositions valid? The obvious approach by direct geochemical examination of progressively older crustal blocks is inadequate. Problems of accessibility and representative sampling eventually become overwhelming. The use of the natural sampling processes is the only realistic alternative. The obvious drawback to any such method is that we only observe secular variations of the upper continental crustal compositions. Significant changes in upper crustal composition do not necessarily reflect changes in the total crust; the processes of intracrustal differentiation must also be considered (see Chapter 9). Accordingly, variations in the total crustal composition remain elusive, although examination of changes in heat flow through time does provide valuable information in this regard.

Weathering, transportation and diagenesis can all dramatically influence the composition of sedimentary rocks. In addition, the recycling of sedimentary rocks also may disguise long term trends [1, 2]. Consequently, great care must be taken in interpreting secular variations of sedimentary rock compositions. In Chapter 2, we outlined the rationale of using the abundance of certain trace elements in clastic sedimentary rocks as a measure of crustal composition. For the purposes of examining secular trends in upper crustal composition, we will use the distribution of the rare earth elements (REE), Th and Sc in fine-grained sedimentary rocks. To investigate changes in bulk crustal composition, we will examine the record of heat flow during geologic time.

5.2 The sedimentary record

5.2.1 Secular trends

In 1909, Daly [3] observed a statistically significant increase in Mg/Ca ratios in carbonate rocks with geologic age and thus began the study of the geochemical evolution of sedimentary rocks. Nanz [4] recognized several geochemical trends in argillaceous rocks of different ages and suggested that secular changes in grain size may be the dominating factor. Since that time, a large amount of data have been accumulated and compiled, mainly by Russian workers [e.g. 5, 6]. In this work, discussion is restricted to secular trends observed in the terrigenous clastic component of sedimentary rocks since these are most relevant to any evolution in the chemical composition of the upper continental crust. A number of possibly important trends have been noted for the oxidation states of some metals such as Fe and Mn [6, 7] and alkaline earth abundances [7, 8]. Such trends are likely to reflect evolution of the hydrosphere, atmosphere and biosphere, and may be coupled with changes in biochemical processes on the earth.

5.2.2 Major elements

Secular trends in the major element composition of various types of sedimentary rocks have been suggested by a number of workers on several continents. Garrels & Mackenzie [1, 9] examined data for shaly rocks from North America during post-Archean times. They noted that metal oxide to alumina oxide ratios (MeO/Al_2O_3) increased for CaO, SiO_2, Fe_2O_3, MgO and Na_2O and decreased for FeO and K_2O during post-Archean times. These authors argued that the changes in composition are those expected during post-depositional changes (i.e. diagenesis, burial metamorphism) and consequently concluded that the composition of the sedimentary rock mass had remained unchanged for the past 2.0 billion years [9]. Prior to that time, some types of chemical evolution could have taken place. Accordingly, these workers are of the opinion that the formation, preservation, destruction and composition of sedimentary rocks is controlled mainly by the recycling of the sedimentary mass. A weakness in this approach concerns the limited data base chosen for the various compositions for different periods of time. For example, for the Precambrian shale average, the data of Nanz [4] are used. However, these data are taken almost exclusively from the area around Lake Superior and represent a restricted area and a short time interval. Consequently, the question of how representative these data are must be raised [10].

In a very large-scale analytical programme, Ronov and his co-workers [5, 6] have attempted to examine the bulk composition of sedimentary rocks, and the various lithological components, throughout geological time. In general, a decrease in Al_2O_3, Na_2O, Fe, MgO and an increase in K_2O/Na_2O has been observed in younger rocks. The absolute abundance of K_2O increased in the early Palaeozoic and then decreased. The abundance of CaO showed the opposite trend. From these data, it was concluded that there is a gradual change from more basic to more granitic upper crustal compositions as well as changes in the patterns of weathering and evolution of the atmosphere and biosphere. These authors also drew attention to the discrepancy between average sedimentary rock compositions and average upper crustal compositions. The sedimentary rocks appeared to be considerably more basic in composition (also see [11]). It was suggested that this was due to the influence of an early mafic–ultramafic crust which had been recycled within the sedimentary mass throughout much of geologic time. Veizer [11] calculated that about 30% of typical oceanic basalt would have to be added to 70% of upper crustal composition to generate a composition consistent with that of the sedimentary data. Some of the discrepancy between the compositions of upper crustal and platform sediments may be related to changes in the sites of carbonate deposition [12].

Engel et al. [13] emphasized the episodic nature of changes in K_2O/Na_2O ratios in sedimentary and other crustal rocks. They noted a dramatic increase in K_2O/Na_2O ratios associated with the Archean–Proterozoic boundary and a sharp decrease during the Mesozoic. The increase at the Archean–Proterozoic boundary was related to a period of crustal growth and differentiation. The decrease during the Mesozoic was related to a period of particularly active continental rifting and plate movement. Alternatively, this may be a metastable stage, characteristic of young terrains, which disappears with later reworking and cratonization [14]. Intracrustal recycling, as described by Veizer and co-workers [2, 15], operates on time-scales of 10^8–10^9 years and accordingly, large time intervals must be considered when dealing with long term chemical evolution of the crust [2]. We will return to the question of sedimentary recycling at the end of this chapter. Clearly, however, the data of Engel et al. [13] contrast with those of Ronov, which indicate a more gradual change in K_2O/Na_2O ratios during the Precambrian and Palaeozoic. Ronov [5, 6] considered larger time intervals which results in a different perspective of chemical evolution [2].

More recently, Schwab [16] approached the problem in a slightly different manner. He calculated overall average major element compositions for the clastic portions of individual sedimentary sequences (e.g. Huronian Supergroup, Belt Series), since this would give the best information about source rock composition and minimize chemical effects related to local secular trends within individual basins and to diagenesis or metamorphism. These data were then grouped into broad age divisions of Archean (>2.5 Ae), early Proterozoic (2.5–1.7 Ae), late Proterozoic (1.7–0.7 Ae) and Phanerozoic (<0.7 Ae) and grand averages calculated accordingly. The results suggested a progressive decrease in Al_2O_3, Na_2O and MgO and progressive increase in SiO_2, K_2O and CaO with time. These trends were considered to reflect evolutionary changes in the composition of the upper continental crust.

McLennan [17] used the same approach but with an expanded data base, particularly for early Proterozoic sequences. A summary of these compilations is given in Table 5.1. The Archean sequences are primarily from late Archean greenstone belts but include the early Archean (c. 3.4 Ae) Fig Tree Group from South Africa. Insufficient data are available to compare early and late Archean sedimentary sequences in any systematic fashion. A statistical analysis of the data indicates that there are no differences in bulk major element composition of clastic sedimentary rocks during the post-Archean; this is consistent with uniform upper crustal composition throughout the post-Archean. However, the bulk composition of the Archean sedimentary rocks differ from their post-Archean counterparts in being depleted in Si, K and enriched in Na, Ca and Mg. These differences are consistent with the Archean upper crust being of significantly more mafic composition than the post-Archean upper crust. Thus, rather than a progressive change in sedimentary

Table 5.1. Average compositions of clastic sedimentary assemblages through geologic time [16, 17]*.

	Archean (n=11)	Early Proterozoic (n=6)	Late Proterozoic (n=6)	All Proterozoic† (n=12)	Phanerozoic (n=20‡)	All post-Archean (n=32§)
SiO_2	65.9±2.2	70.9±5.9	69.2±5.9	70.0±3.4	70.6±1.8	70.4±1.6
TiO_2	0.6±0.1	0.6±0.2	0.7±0.2	0.6±0.1	0.7±0.1	0.7±0.1
Al_2O_3	14.9±1.3	14.0±2.8	15.0±2.9	14.5±1.7	14.2±0.9	14.3±0.8
FeO	6.4±1.5	6.1±4.2	4.8±1.6	5.5±1.9	5.2±0.4	5.3±0.7
MgO	3.6±1.0	2.4±0.6	1.9±0.9	2.1±0.5	2.4±0.3	2.3±0.3
CaO	3.3±1.1	1.3±0.7	2.2±1.3	1.7±0.7	2.2±0.7	2.0±0.5
Na_2O	2.9±0.5	1.3±0.8	2.4±1.0	1.8±0.6	1.8±0.4	1.8±0.3
K_2O	2.2±0.3	3.0±0.9	3.5±1.0	3.5±0.6	2.8±0.3	3.0±0.3
Σ	99.8	99.6	99.7	99.4	99.9	99.8
MgO/Al_2O_3	0.24	0.17	0.13	0.14	0.17	0.16
CaO/Al_2O_3	0.22	0.09	0.15	0.12	0.15	0.14
Na_2O/Al_2O_3	0.19	0.09	0.16	0.12	0.13	0.13
K_2O/Al_2O_3	0.15	0.21	0.23	0.22	0.20	0.21
K_2O/Na_2O	0.76	2.3	1.5	1.8	1.6	1.7

*All concentrations in weight per cent, recalculated volatile-free. Uncertainties represent 95% confidence limits.
†Average of all Proterozoic clastic sedimentary assemblages considered in previous two columns.
‡Number of assemblages is 20 except for MgO (n=15), CaO (n=14) and TiO_2 (n=13).
§Number of assemblages is 32 except for MgO (n=27), CaO (n=26) and TiO_2 (n=25).

compositions, there appears to be a fairly sharp break at the Archean–Proterozoic boundary.

5.2.3 Trace elements

The trace element geochemistry of sedimentary rocks has received comparatively little attention. Consequently, the trace element data base for examining secular trends is considerably smaller than for the major elements. However, we will see that the interpretation of secular trends of the trace elements is relatively unambiguous in comparison with the major element data. In this approach, we are mainly (but not exclusively) concerned with elemental ratios because with such a small data base, absolute abundances can be significantly affected by factors such as quartz content.

Some workers have examined secular trends in the alkali (Rb) and alkaline earth trace elements (Sr) in sedimentary rocks [7, 8]. Trends have been noted but these are best related to changes in the chemistry of soils after the establishment of land plants or to chemical changes resulting from diagenesis. Accordingly, secular trends of such elements better reflect biological evolution or changes in sedimentary and/or diagenetic conditions rather than crustal evolution.

Ronov & Migdisov [6] first examined Th and U abundances in sedimentary rocks and continental crust with time, but there is some uncertainty in the quality of the data. For example, Th/U ratios of basement and sedimentary cover are quoted at 6–8 for the Archean and decrease through time to as low as 1–2 during the Palaeozoic. However, it is well established that most terrestrial igneous rocks, and the planet as a whole, have Th/U ratios of about 3.8. Most processes of sedimentation in oxidizing environments (dominant after about 2.3 Ae, if not before) tend to oxidize U^{4+} to the soluble form U^{6+} so increasing the Th/U ratios in most terrigeneous sedimentary rocks. Thus the low Th/U ratios quoted by Ronov & Migdisov [6] as averages are not realistic. We will return to the question of Th and U abundances in sedimentary rocks.

The REE abundances in sedimentary rocks are particularly well suited to studying the chemical evolution of the upper crust. Apart from the close correspondence between upper crustal REE patterns to sedimentary REE patterns (see Chapter 2), the trivalent REE exhibit a progressive decrease in ionic radius from La to Lu. Thus the light rare earths La–Sm (LREE) are more incompatible in typical igneous differentiation processes than the heavy rare earths Gd–Lu (HREE) and there is a general increase in LREE/HREE ratios from more mafic to more felsic compositions. Furthermore, the geochemical consequences of the two common oxidation states of europium may reveal information about the origin of the upper crust with time (Chapter 2).

Studies of REE in sedimentary rocks date from the early work of Minami [18]. The pioneering work of Haskin and his co-workers [19] firmly established the depletion of Eu in post-Archean rocks compared to their

Table 5.2. Trace element characteristics of fine-grained sedimentary rocks through geological time [20, 21].

Age (Ae)	>3.0	3.0–2.5	2.5–1.7	1.7–0.6	0.6–0.0	Archean	Post-Archean
La/Yb	12±2	11±4	19±3	13±2	14±1	11.5±2.5	15.5±1.3
Eu/Eu*	0.93±0.06	1.04±0.06	0.66±0.08	0.65±0.03	0.65±0.02	0.99±0.05	0.65±0.02
ΣREE	127±26	83±15	204±38	144±13	195±21	102±15	185±15
ΣLREE/ΣHREE	8.3±1.1	6.6±1.1	11±2	8.9±0.9	9.7±0.7	7.4±0.8	10±1
La	26±6	16±3	43±8	30±3	42±5	20±3	39±3
Th	8.3±1.7	5.6±1.3	14±2	13±1	15±2	6.7±1.1	14±1
Sc	19±2	18±3	16±3	13±2	17±2	19±2	16±1
La/Th	3.5±0.4	3.1±0.3	3.0±0.4	2.3±0.3	2.9±0.2	3.3±0.3	2.8±0.2
La/Sc	1.4±0.3	1.2±0.3	2.9±0.7	2.4±0.6	2.9±0.5	1.3±0.2	2.7±0.3
Th/Sc	0.46±0.09	0.40±0.12	1.1±0.3	1.0±0.1	1.0±0.1	0.43±0.07	1.0±0.1
U	2.1±0.4	1.5±0.3	3.5±0.7	3.2±0.8	3.0±0.3	1.7±0.3	3.2±0.3
Th/U	4.0±0.4	3.7±0.3	4.2±0.6	4.6±0.6	5.1±0.4	3.8±0.3	4.8±0.3
n	20†	25‡	19§	18‖	33¶	45	70

Uncertainties are quoted as 95% confidence interval.
† 19 for Sc, La/Sc; 17 for Th, U, La/Th, Th/Sc, Th/U.
‡ 12 for Sc, La/Sc, Th/Sc.
§ 18 for Sc, U, Th/Sc; 17 for La/Sc, Th/U.
‖ 17 for La/Yb, ΣREE; 16 for Eu/Eu*, U; 15 for Th/U.
¶ 32 for La/Yb, Eu/Eu*, ΣREE, U, Th/U, La/Sc.

Archean counterparts. Nance & Taylor [20] examined the REE patterns in Australian sedimentary rocks in a systematic manner; they confirmed the findings of Haskin and further suggested that post-Archean sedimentary rocks had higher LREE/HREE ratios and there was a general trend to higher ΣREE with time. Several studies have followed [21, 22].

In Table 5.2, we list a compilation of trace element data in fine-grained terrigenous sedimentary rocks for various ages. The age groupings were chosen to distribute sample numbers evenly and to correspond as closely as possible to those used for the major element analysis. Although the data base is small, the samples have been analysed for the full spectrum of major and trace elements and are considered to be representative. Only those elements and ratios pertinent to the present discussion are tabulated. The early Archean samples are from South Africa and Australia and the late Archean are from Australia only. Post-Archean samples are representative of widely separated basins, mainly from Australia, but also from New Zealand and Antarctica [20, 21].

The most striking evolutionary pattern of sedimentary trace element patterns is for Eu/Eu* (Fig. 5.1). Archean sedimentary rocks are not anomalous, or only slightly anomalous, with respect to Eu (i.e. Eu/Eu*=1.0) but post-Archean sedimentary rocks on average show a significant and constant depletion in Eu (Eu/Eu*=0.65). The break in composition corresponds to the Archean–Proterozoic boundary. The relatively large uncertainty in the early Proterozoic average is best related to rapid evolution from

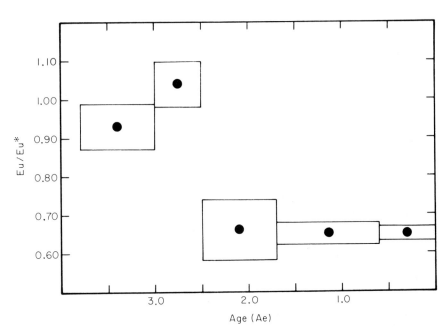

Fig. 5.1. Plot of Eu/Eu* against time for fine-grained sedimentary rocks [23]. Two features stand out. There is a dramatic shift in Eu/Eu* at the Archean–Proterozoic boundary (2.5 Ae) and Eu/Eu* remains constant throughout the post-Archean.

Archean-like to post-Archean-like patterns in some early Proterozoic sequences. This aspect is discussed in detail in Chapter 8. These data clearly indicate that sedimentary rocks display a constant depletion in Eu from the early Proterozoic to the present. However, this does not necessarily rule out other compositional changes since Eu-anomalies alone are not a reliable index of overall composition.

The ratio of LREE/HREE is a more reliable index of composition because the ratio typically changes with composition of the bulk rock. Some typical values for igneous rocks are listed in Table 5.3. The sedimentary data are plotted against time in Fig. 5.2. The patterns show considerably more scatter than for Eu/Eu*, but it is clear that there is a break at the Archean–Proterozoic boundary. Archean samples typically fall in the range LREE/HREE=6–9 and post-Archean samples typically have values in the range LREE/HREE=8–12. There is no obvious trend for the post-Archean samples. Nance & Taylor [20] indicated that there was some evidence of a secular increase in ΣREE with time. With the greatly expanded data base now available (Table 5.2), this trend is no longer apparent. Total REE in Archean samples are scattered but

Table 5.3. REE, Th and Sc abundances (in ppm) in the primitive mantle and various crustal reservoirs.

	% SiO$_2$	La	Th	Sc	LREE/HREE	La/Sc	Th/Sc
Primitive mantle	49.9	0.55	0.064	13	1.7	0.042	0.005
Ocean crust	49.5	3.7	0.22	38	1.4	0.10	0.006
Lower continental crust	54.4	11	1.06	36	3.9	0.31	0.03
Total continental crust	57.3	16	3.5	30	5.4	0.53	0.12
Average andesite	58.0	19	4.8	30	5.8	0.63	0.16
Upper continental crust	66.0	30	10.7	11	9.5	2.7	0.97

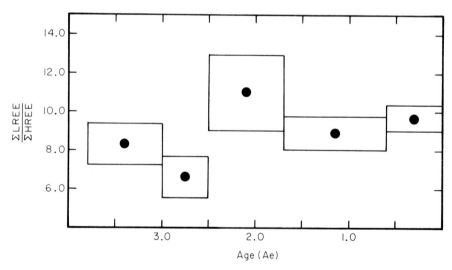

Fig. 5.2. Plot of ΣLREE/ΣHREE against time for fine-grained sedimentary rocks [23]. The break at the Archean–Proterozoic boundary is again apparent.

are significantly lower than for post-Archean samples. The post-Archean data are also scattered but show no systematic trends.

Of the elements under consideration, Th is the most incompatible and Sc is the most compatible. Thus it might be expected that a ratio of these elements would be the most sensitive index of overall chemical composition (i.e. mafic versus felsic; see Table 5.3). In Fig. 5.3, Th/Sc is plotted against age for

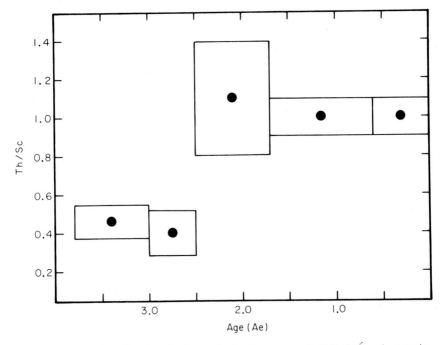

Fig. 5.3. Plot of Th/Sc against time for fine-grained sedimentary rocks [23]. A sharp increase in Th/Sc is recorded at the Archean–Proterozoic boundary. Th/Sc remains constant at about 1.0 throughout post-Archean time. Because Th is a highly incompatible element and Sc is quite compatible, this ratio should represent the most sensitive index of upper crustal compositions in sedimentary rocks.

sedimentary rocks. Two features are remarkable; one is the dramatic break in composition corresponding to the Archean–Proterozoic boundary, and the other is the constancy of Th/Sc ratios in sedimentary rocks during the post-Archean. Grouped together, the Archean fine-grained sedimentary rocks have Th/Sc=0.43±0.07 and post-Archean sedimentary rocks have Th/Sc 1.0±0.1. A similar pattern is seen for the ratio La/Sc (Fig. 5.4) although in this case the data are somewhat more scattered. Archean samples have La/Sc=1.3±0.2 whereas post-Archean samples have La/Sc ratios of 2.7±0.3.

Although some care must be taken in interpreting absolute abundances, it is clear that the change in Th/Sc and La/Sc ratios in sedimentary rocks at the Archean–Proterozoic boundary results from an enrichment in incompatible elements. This is seen best from an examination of the Th data (Fig. 5.5). There is a sharp increase in Th abundances in fine-grained sedimentary rocks at the Archean–Proterozoic boundary by a factor of about 1.5–3.0 with Th

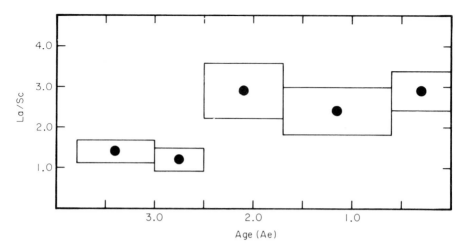

Fig. 5.4. Plot of La/Sc against time for fine-grained sedimentary rocks [23]. This index shows essentially the same features as Th/Sc ratios (Fig. 5.3).

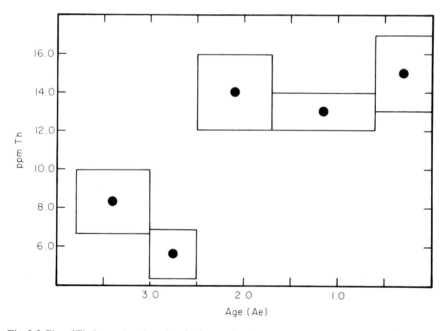

Fig. 5.5. Plot of Th (in ppm) against time for fine-grained sedimentary rocks [23]. The absolute abundances of this relatively immobile trace element also shows a sharp increase at the Archean–Proterozoic boundary. Within the uncertainties, Th abundances in fine-grained sedimentary rocks have remained constant during the post-Archean.

abundances remaining essentially constant during the post-Archean at about 14±1. La shows a similar increase (Table 5.2). On the other hand, Sc decreases in the post-Archean sedimentary sequences but the magnitude of the change is much smaller than for Th or La (Table 5.2). Even the comparatively soluble element U shows a sharp increase in abundance at the Archean–Proterozoic boundary (Fig. 5.6). McLennan & Taylor [24] suggested there

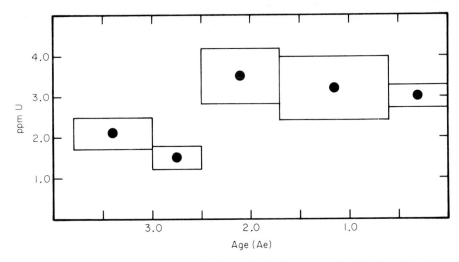

Fig. 5.6. Plot of U (in ppm) against time for fine-grained sedimentary rocks [23]. Even this rather mobile element shows the same pattern of evolution as Th. U abundances may decrease gradually during the post-Archean but the statistical validity of this suggestion is not well established.

may be a secular decrease in U abundances in post-Archean shales related to continual recycling of sedimentary material. This suggestion is not excluded using the present expanded data base, but the statistical validity is still uncertain. We will return to the question of recyling in the following section.

The conclusion from the study of the trace element data is that there is no compelling evidence to suggest any significant change in upper crustal abundances during post-Archean time. This conclusion rests mainly on the trends of Th/Sc, La/Sc and Eu/Eu* and is also consistent with absolute trace element abundances (e.g. Th, ΣREE) and other ratios (e.g. LREE/HREE) as well as with the major element data discussed in the previous section. On the other hand, there is ample evidence to suggest that the Archean exposed crust was considerably less differentiated. This conclusion is based on the more mafic major element composition of the Archean sedimentary sequences (Table 5.1) and the considerably lower Th/Sc, La/Sc, LREE/HREE ratios and lower absolute abundances of incompatible elements (e.g. Th, La, U) in fine-grained Archean sedimentary rocks.

5.3 Role of sedimentary recycling

The recycling of sedimentary and basement rocks through resedimentation and intracrustal processes such as remelting is well established. Recycling has been discussed and quantitatively modelled by several authors [e.g. 1, 15, 25]. What is the extent of this recycling and how does it influence the composition of sedimentary rocks and the upper continental crust itself?

The mass of the sedimentary record is of the order 2.7×10^{24} g. There is an exponential decrease of preserved sedimentary mass with geologic age. In Fig. 5.7, the evolution of the sediment mass with time is shown, based mainly on the work of Gregor [26]. Reliable data for much of the Precambrian is

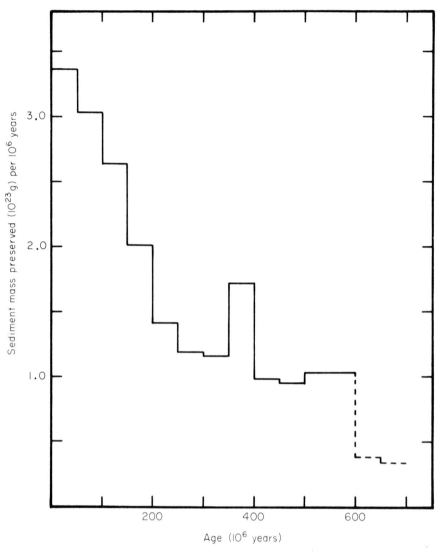

Fig. 5.7. Plot of preserved sedimentary mass per 10^6 a against age for 50 m.y. intervals. Data from [26] for 0–600 m.y. and from [27] for >600 m.y. There is an exponential decrease in the mass of preserved sediment with increasing age.

unavailable, but there is no reason to suppose any marked deviation from an exponential pattern at least back to the base of the Proterozoic [1, 2, 15]. Similarly, area–age relationships of continental basement rocks show an exponential decrease with age (see Table 1.2). Such trends are best explained if a portion of the sedimentary mass is recycled through the continental crust either by metamorphism and 'granitization' with subsequent re-sedimentation or by the more direct process of sedimentary recycling. The best estimate of the efficiency of recycling, based mainly on the Nd-isotopic composition of sedimentary rocks (see below), has been given by Veizer & Jansen [15] at 65%.

Thus 35% of a given sedimentary rock, on average, would be new sedimentary mass generated from young sources.

Do we see any evidence of sedimentary recycling in the trace element data? In a preliminary examination, it was suggested that the Th/U ratio has increased in fine-grained sedimentary rocks during the post-Archean in response to sedimentary recycling [24]. With each cycle of erosion and redeposition, some U^{4+} may be oxidized to the more soluble U^{6+} and removed, thus increasing the Th/U ratio. The ultimate sinks of such uranium could be in uranium deposits and altered oceanic crust. The larger data base compiled in Table 5.2 appears to confirm this secular trend in Th/U ratios in fine-grained sedimentary rocks, although the uncertainties remain large (Fig. 5.8).

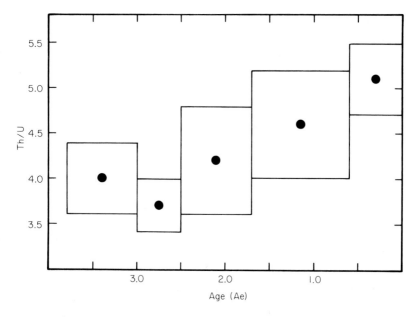

Fig. 5.8. Plot of Th/U against time for fine-grained sedimentary rocks [23]. The data suggest that there has been a secular increase in Th/U ratios during post-Archean time from about 4.2 to 5.1, although uncertainties are significant. The Archean data show generally low Th/U ratios, near the planetary value of 3.8.

If the sedimentary cycle is on average 65% cannibalistic, then significant changes in upper crustal compositions will only be revealed in the sedimentary record on long time-scales. Only substantial changes in upper crust compositions will be recorded. Transient phenomena will not appear [2]. The time required to reduce the continental sedimentary mass of given age by half, through recycling, is about 380 m.y. for continental sediments. Thus the rate of recycling of continental sediments is about 2.8×10^{15} g a^{-1} for a continental sedimentary mass of about 1.88×10^{24} g [2].

Veizer [2] has modelled the response of clastic sediment composition to changes in source rock composition, given varying degrees of recycling. For the recycling rates given above, the equivalent of about 3.5–4.0 sedimentary masses would be recycled during post-Archean time (2.5 billion years). If the composition of the upper crust were to change (e.g. from Source A to Source B in Fig. 5.9) we would see about 70% of Source B and 30% of Source A in the sedimentary record after about two billion years. This demonstrates the buffering capacity that sedimentary recycling has on the composition of sedimentary rocks.

How does this affect the interpretation of the post-Archean sedimentary record? In the previous section, we have established that the chemical composition of post-Archean sedimentary rocks has remained effectively constant for elements considered to reflect upper crustal abundances. To what extent can this be interpreted as indicating a constant upper crustal composition? In Fig. 5.10, we use the ratio of Th/Sc in sedimentary rocks to model the effects of an upper crust which is evolving towards a more differentiated composition. In the model shown in Fig. 5.10(a) the upper crust is assumed to

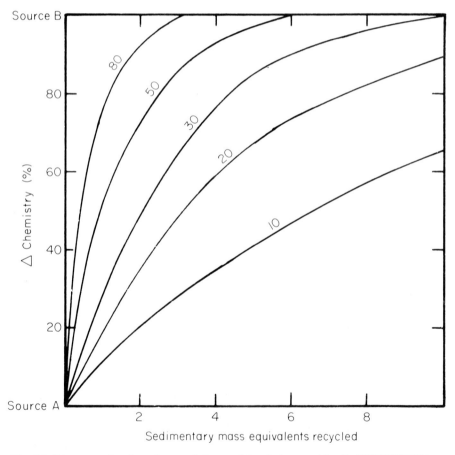

Fig. 5.9. Diagram to show how the rate of change of chemical composition (Δ CHEMISTRY) is controlled by the rate of sedimentary recycling. For typical recycling rates of continental sediments, the equivalent of about 3.5–4.0 sedimentary masses would be recycled during 2.5 billion years (x-axis). Lines indicate the degree of openness (in %) of the sedimentary system. On average, sedimentary recycling is about 35% open (65% cannibalistic) [15]. (Adapted from [2].)

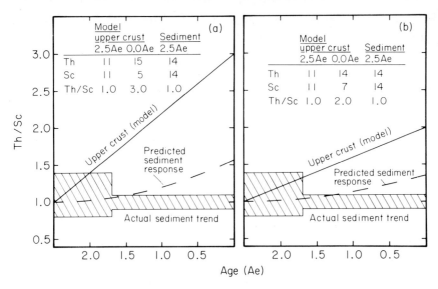

Fig. 5.10. Plot of Th/Sc against age to show the response of sediment compositions to changes in upper crustal composition for a sedimentary system which is 65% cannibalistic (35% open). Two models of post-Archean upper crustal evolution are shown: (a) Th/Sc evolving from 1.0–3.0; (b) Th/Sc evolving from 1.0–2.0. Absolute abundances are listed on the figure. Solid line shows the model of upper crustal evolution and dashed lines show the predicted sediment response. The hatched area is the measured evolution of Th/Sc in fine-grained sedimentary rocks (from Fig. 5.3). It can be seen that recycling has the effect of greatly buffering the response of sediment chemistry to changes in upper crustal composition. In spite of this, the sediment data appear to rule out significant changes in upper crustal composition during the post-Archean.

evolve, in a linear fashion, from Th/Sc=1.00 (the value preserved in post-Archean sedimentary rocks) to Th/Sc=3.00, using the absolute abundances listed on the figure. The sediment is assumed to have the same Th/Sc ratio as the upper crust at 2.5 Ae. This is a limiting case which minimizes relative changes in the sedimentary composition; if anything, the upper crust would be expected to have a higher Th/Sc ratio than the sedimentary mass at 2.5 Ae due to the recycling of Archean sedimentary rocks with low Th/Sc. The predicted evolution of the sedimentary composition for a system which is 65% recycled is plotted and it can be seen that changes in Th/Sc are greatly buffered. However, the constancy of Th/Sc ratios for the post-Archean sedimentary record appears to argue against changes in upper crust composition of the magnitude shown in Fig. 5.10(a). In Fig. 5.10(b), a more modest change in upper crustal Th/Sc ratio is modelled and we observe the predicted sediment response is still greater than the analytical constraints presently available. It is worth noting that it is only when ratios of elements with extreme differences in incompatibility are examined can subtle changes in upper crust composition be inferred from the sedimentary record; for many geochemical indices, predicted changes would fall within the uncertainty of the data [2]. A final comment is that the extreme changes in sedimentary composition at the Archean–Proterozoic boundary have implications for models of sedimentary recycling. We will return to this point in Chapter 8.

5.4 Nd model ages of sediments

The application of Nd model ages as a means of estimating crustal formation ages is now well established [28]. The basic assumption involved in this approach is that the major chemical fractionation of Sm and Nd occurs during differentiation of material from the mantle and incorporation into the continental crust. Further fractionation of Sm and Nd also is expected during intracrustal melting events but this is comparatively minor and generally follows crust formation within 100–200 m.y. For sediments, the Nd model ages provide unique information. They give the time of formation of the sources of the sediments rather than the age of deposition. (This depends on the assumption that Sm and Nd are not fractionated during sedimentation or diagenesis; see Chapter 2.) The Nd model ages (T_{DM}^{Nd}) [29] are calculated using depleted mantle parameters (see Appendix 4). In this case, we assume that the mantle source has evolved from $\varepsilon_{Nd}=0$ to $\varepsilon_{Nd}=+10$ (average value for MORB) over the past 3800 m.y. For samples with model ages younger that 3800 m.y., T_{DM}^{Nd} ages are older than those calculated using chondritic parameters (T_{CHUR}^{Nd}).

Table 5.4 summarizes the available Nd-model age data and in Fig. 5.11,

Table 5.4. Summary of Nd-model ages of sedimentary rocks.

Sequence	Abbreviation	Stratigraphic age (Ae)	Average T_{DM}^{Nd*}	n	Ref
Isua	I	3.75	3.6	3	[30]
Fig Tree–Moodies	F	3.4	≈3.5	?	[31]
Gorge Creek	G	3.4	3.6	1	[32]
Malene	M	3.05	3.0	2	[30]
Kalgoorlie	Ka	2.8	3.1	1	[28]
Whim Creek	W	2.7	3.2	1	[32]
Yellowknife	Y	2.7	≈2.7	?	[31]
Animike	A	2.0	≈2.5	?	[31]
Lewisian	L	2.0	2.4	2	[33]
Mt Isa	MI	1.7	2.3	1	[28]
Belt-Purcell	BP	1.3–0.9	≈1.8	?	[34]
Torridonian	T	1.0	2.2	4	[33]
Amadeus	Am	0.8	2.0	1	[28]
Kuiseb	K	0.75	2.4	1	[35]
Dalradian	D	0.6	2.3	7	[33]
NASC	NASC	0.5	2.0	1	[28]
Southern Uplands	SU	0.45	1.7	16	[33]
State Circle	SC	0.44	1.8	1	[28]
Canning Basin	C	0.3	1.9	1	[32]
Sierra Nevada	SN	0.25	1.5	2	[36]
Perth Basin	P	0.2	1.7	1	[32]
Baja Shale	B	0.1	1.1	1	[28]
Loess	Loess	0.01	1.5	8	[28, 37]
Marine Clay	MC	0.01	1.5	16	[28, 38]
British River Sediment	BR	0.00	≈1.5	?	[39]
San Gabriel	SG	0.00	2.3	1	[28]

*See Appendix 4.

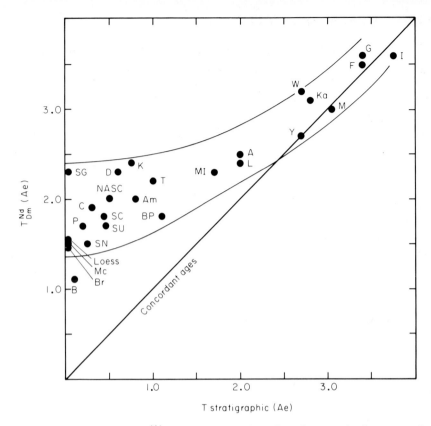

Fig. 5.11. Plot of Nd-model age (T_{DM}^{Nd}) against stratigraphic age for sediments and sedimentary rocks. Data and location abbreviations listed in Table 5.4. Nd-model ages become increasingly greater than stratigraphic ages as sedimentary sequences become younger. This trend is the result of sedimentary and crustal recycling [15]. The wide range of model ages at any given stratigraphic age is due to differing crustal formation ages and highly variable recycling rates [37].

the Nd-model ages (T_{DM}^{Nd}) of a number of sediments and sedimentary sequences are plotted against the stratigraphic age. An important observation is that the Nd-model age is commonly significantly older than the stratigraphic age, particularly for younger sediments [28]. Veizer & Jansen [15] used this observation to model sedimentary recycling rates. O'Nion's et al. [33] have noted that Archean and early Proterozoic sediments have T_{DM}^{Nd} ages fairly close to their stratigraphic age, suggesting they are primarily derived from new crustal material with an age close to the stratigraphic age of the sediments. Taylor et al. [37] examined the Nd-isotopic composition of Pleistocene loess deposits. On the basis of these and other very young sediments [28, 38] they concluded that sedimentary recycling rates were highly variable, with modern rates varying from at least 45–90%.

The Nd-model ages of sediments also provide some information regarding possible changes in upper crustal composition during the post-Archean. In Fig. 5.11, it is apparent that the T_{DM}^{Nd} model ages of post-Archean sediments cover a wide range from about 2.5 Ae to 1.0 Ae. These dates effectively represent

111

the average age of the upper crust which contributed to these sediments. Thus, the constancy of REE, Th and Sc distributions in post-Archean sediments, documented in previous sections, represent a wide range of provenance age. Accordingly, any additions to the upper crust during the post-Archean could not have differed significantly in composition to the upper crust itself.

5.5 Heat flow provinces and crustal uniformity

The question of the uniformity of bulk crustal composition with time can be examined also using the heat flow data (see Sections 2.6.5 and 3.2). According to Pollack [40, p. 187], 'a heat flow province is defined as that geographic area in which heat flow and heat production are linearly related. Each heat flow province has a characteristic reduced heat flow'. Data for 17 recognized provinces are listed in Table 5.5 [41]. Figure 5.12 shows the data plotted against the age of the last tectonothermal event to affect the area [42]. The high values for the younger regions reflect the effect of orogenic events. There is little difference from the mid-Palaeozoic to the base of the Proterozoic. We

Table 5.5. Heat flow and heat production for various heat flow provinces. (Modified from [41, 42].)

Province	Age of last tectonothermal event	q_s	q^*	q_A	b
Superior	Archean	—	28±1	—	13.6
Ukraine	Archean	37±8	25±2	12	7.1
Western Australia	Archean	39±8	26±8	13	4.5
Indian Shield	Archean	49±8	33±1	16	7.5
Baltic	Archean–Proterozoic	36±8	22±6	14	8.5
Central Australia	Proterozoic	83±21	27±6	56	11.1
Eastern USA	Proterozoic	57±17	33±4	24	7.5
Brazilian Coastal Shield	Proterozoic	56±15	28±7	28	13.1
Indian Shield	Proterozoic	71±11	38±2	33	14.8
Niger	Proterozoic	20±8	11±8	9	(7.5)
Zambia	Proterozoic	67±7	40±6	27	(7.5)
SE Appalachians	Palaeozoic	49±11	28±11	21	7.2
England and Wales	Palaeozoic	59±23	23±2	36	16.0
Indian Shield	Cenozoic–Mesozoic	71±11	38±2	33	14.8
Zambia	Cenozoic–Mesozoic	67±7	40±6	27	(7.5)
Basin and Range	Cenozoic	92±33	69±34	23	(10.0)
Sierra Nevada	Cenozoic	37±13	18±3	19	10.1
Bohemian Massif	Cenozoic	73±18	44±1	29	3.8
Eastern Australia	Cenozoic	72±27	57±22	15	(11.1)

Ages assigned to each province are age of last tectonothermal event. q_s is mean surface heat flow (mW m^{-2}) ± standard deviation; q^* is reduced heat flow (mW m^{-2}) ± uncertainty; q_A is estimated crustal heat production (mW m^{-2}) which is calculated by

$$q_A = q_s - q^*$$

b is depth parameter for upper crustal heat production (see Section 2.6.5); b values which are assumed are given in parentheses.

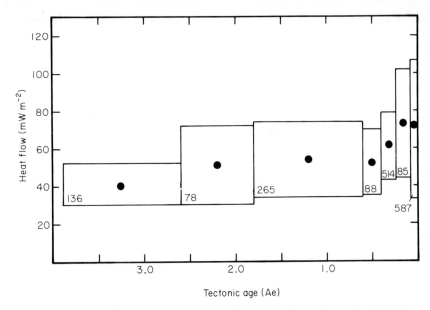

Fig. 5.12. Plot of continental heat flow against age of last tectonothermal event (modified from [42]). Boxes represent age range under consideration and one standard deviation of heat flow. Number of data used in each group are shown in each box. The higher values of young regions reflect orogenic processes and there is little difference from the mid-Palaeozoic to the base of the Proterozoic. Archean values are significantly lower and indicate the Archean crust had lower abundances of the heat-producing elements.

interpret this to mean that there is no significant change in crustal composition as reflected by the abundances of K, U and Th over this period. A significant drop in crustal heat production to 14 mW m^{-2} occurs for the Archean sites and we interpret this to indicate that the Archean crust had a significantly lower content of K, U, and Th (see Chapter 6).

The uniform heat flow data set extending from the mid-Palaeozoic to the base of the Proterozoic argues against the interpretation that the principal agent for reducing the overall heat flow is erosion of the uppermost layers, rich in K, U and Th [43]. Although such erosion and removal of a granitic layer will clearly reduce the heat flow, the effects are complex in detail. The cooling of the tectonothermal heat component is here thought to be more important. Much of the K, U and Th may be retained in the sediments. Although the blanketing effect of sediments may act to decrease the near-surface heat flow, the addition of the radioactive element component in the sediments will act in the opposite direction.

Watson [44] made the important observation that depths of erosion of Proterozoic terrains were in general no deeper than those in Caledonian terrains. Thus any effects of erosion on heat flow will occur within a few hundred million years of the orogenic events and have about the same time constant as the tectonothermal cooling. In this context, the difference between the Archean and Proterozoic provinces is not likely to be due principally to erosion, but rather to major compositional differences.

5.6 Summary

1 There are no secular changes in the chemical composition of sedimentary rocks (e.g. major elements, REE, Th, Sc) which can be related to changes in upper crustal composition during post-Archean time. This constancy of sedimentary composition is particularly significant for ratios of highly incompatible to compatible trace elements (Th/Sc, La/Sc, LREE/HREE).

2 Sedimentary recycling, at present, is about 65% efficient, on average, and buffers the response of sedimentary compositions to changes in the upper crust, making minor changes in upper crustal composition difficult to recognize.

3 The constancy of upper crustal composition during the post-Archean indicates that any additions to the upper crust must have had a composition indistinguishable from that of the upper crust itself, within the buffering capacity of sedimentary recycling.

4 In contrast, Archean sedimentary rocks have a different composition to those from the post-Archean, related to differences in the composition of the exposed crust. Our conclusion is that the bulk composition of the Archean exposed crust was less differentiated and more mafic.

5 Heat flow from mid-Palaeozoic to early Proterozoic age is uniform indicating that the crustal abundance of K, Th and U was similar throughout this time.

6 Archean heat flow provinces have much lower heat flow than their post-Archean counterparts. This cannot be readily related to erosional effects and is best explained by lower crustal abundances of K, Th and U in the Archean.

Notes and references

1 Garrels, R.M. & Mackenzie, F.T. (1971) *Evolution of Sedimentary Rocks.* Norton.
2 Veizer, J. (1984) Recycling on the evolving earth: geochemical record in sediments. *Proc. IGC* **11**, 325.
3 Daly, R.A. (1909) First calcareous fossils and the evolution of the limestones. *GSA Bull.*, **20**, 153.
4 Nanz, R.H. (1953) Chemical composition of Precambrian slates with notes on the geochemical evolution of lutites. *J. Geol.*, **61**, 51.
5 Ronov, A.B. (1964) Common tendencies in the chemical evolution of the earth's crust. *Geochem. Int.*, **1**, 713; (1968) Probable changes in the composition of sea water during the course of geological time. *Sedimentology*, **10**, 25; (1972) Evolution of rock composition and geochemical processes in the sedimentary shell of the Earth. *Sedimentology*, **19**, 157; Ronov, A.B. et al. (1974) Regularities of rare-earth element distribution in the sedimentary shell and in the crust of the earth. *Sedimentology*, **21**, 171; (1977) Regional metamorphism and sediment composition evolution. *Geochem. Int.*, **12**, 90.
6 Ronov, A.B. & Migdisov, A.A. (1971) Geochemical history of the crystalline basement and the sedimentary cover of the Russian and North American platform. *Sedimentology*, **16**, 137.
7 Veizer, J. (1973) Sedimentation in geologic history: recycling vs. evolution or recycling with evolution. *CMP*, **38**, 261; (1976) Evolution of ores of sedimentary affiliation through geologic history: relations to the general tendencies in evolution of the crust, hydrosphere, atmosphere and biosphere. *In:* Wolf, K.H. (ed.), *Handbook of Strata-bound and Strati-form Ore Deposits.* Vol. 3. Elsevier; (1978) Secular variations in the composition of sedimentary carbonate rocks, II. Fe, Mn, Mg, Si and minor constituents. *PCR*, **6**, 381; Veizer, J. & Garrett, D.E. (1978)

Secular variations in the composition of sedimentary carbonate rocks, I. Alkali metals. *PCR*, **6**, 367.
8 Reimer, T. (1972) The evolution of the rubidium and strontium content of shales. *N. Jb. Miner. Abh.*, **116**, 167.
9 Garrels, R.M. & Mackenzie, F.T. (1974) Chemical history of the oceans deduced from post-depositional changes in sedimentary rocks. *SEPM Spec. Pub.*, **20**, 193.
10 In this regard, it is of interest to note that in a later paper (Cameron, E.M. & Garrels, R.M. (1980) Geochemical compositions of some Precambrian shales from the Canadian Shield. *Chem. Geol.*, **28**, 181), significant variations (factors of 2–10) were observed in major element data for Proterozoic shales from differing sequences in the Canadian Shield.
11 Veizer, J. (1979) Secular variations in chemical composition of sediments: a review. *PCE*, **11**, 269.
12 Schwab, F.L. (1972) Geochemical history of the crystalline basement and the sedimentary cover of the Russian and North American platforms—a discussion. *Sedimentology*, **19**, 299.
13 Engel, A.E.J. et al. (1974) Crustal evolution and global tectonics: a petrogenic view. *GSA Bull.*, **85**, 843.
14 J. Veizer, pers. comm.
15 Veizer, J. & Jansen, S.L. (1979) Basement and sedimentary recycling and continental evolution. *J. Geol.*, **87**, 341.
16 Schwab, F.L. (1978) Secular trends in the composition of sedimentary rock assemblages—Archean through Phanerozoic time. *Geology*, **6**, 532.
17 McLennan, S.M. (1982) On the geochemical evolution of sedimentary rocks. *Chem. Geol.*, **37**, 335.
18 Minami, E. (1935) Gehalte an seltenen Erden in europäishen und japaneschen Tonschiefern. *Nach. Gess. Wiss. Goettingen*, **2**, Math-Physik KL. IV, 1, 155.
19 Haskin, L.A. & Gehl, M.A. (1962) The rare-earth distribution in sediments. *JGR*, **67**, 2537; Haskin, M.A. & Haskin, L.A. (1966) Rare earths in European shales: a redetermination. *Science*, **154**, 507; Haskin, L.A. et al. (1966) Rare earths in sediments. *JGR*, **71**, 6091; Wildeman, T.R. & Haskin, L.A. (1965) Rare earth elements in ocean sediments. *JGR*, **70**, 2905; (1973) Rare earths in Precambrian sediments. *GCA*, **37**, 419.
20 Nance, W.B. & Taylor, S.R. (1976) Rare earth element patterns and crustal evolution—I. Australian post-Archean sedimentary rocks. *GCA*, **40**, 1539; (1977) Rare earth element patterns and crustal evolution—II. Archean sedimentary rocks from Kalgoorlie, Australia. *GCA*, **41**, 225.
21 Nathan, S. (1976) Geochemistry of the Greenland Group (Early Ordovician) New Zealand. *NZJGG*, **19**, 683; Bavinton, O.A. & Taylor, S.R. (1980) Rare earth element geochemistry of Archean metasedimentary rocks from Kambalda, Western Australia. *GCA*, **44**, 639; McLennan, S.M. (1981) *Trace Element Geochemistry of Sedimentary Rocks: Implications for the Composition and Evolution of the Continental Crust*. Ph.D. Thesis, Australian National University; McLennan, S.M. et al. (1983) Geochemistry of Archean shales from the Pilbara Supergroup, Western Australia. *GCA*, **47**, 1211; (1983) Geochemical evolution of Archean shales from South Africa. I. The Swaziland and Pongola Supergroups. *PCR*, **22**, 93; McLennan, S.M. & Taylor, S.R. (1980) Rare earth elements in sedimentary rocks, granites and uranium deposits of the Pine Creek Geosyncline. *In:* Ferguson, J. & Goleby, A.S. (eds), Uranium in granites and uranium deposits of the Pine Creek Geosyncline. IAEA, 175.
22 McLennan, S.M. et al. (1979) Rare earth elements in Huronian (Lower Proterozoic) sedimentary rocks: composition and evolution of the post-Kenoran upper crust. *GCA*, **43**, 375; (1984) Geochemistry of Archean metasedimentary rocks from West Greenland. *GCA*, **48**, 1; Jenner, G.A. et al. (1981) Geochemistry of the Archean Yellowknife Supergroup. *GCA*, **45**, 1111.
23 Data taken from Table 5.2. Boxes represent time interval under consideration and 95% confidence interval of mean.
24 McLennan, S.M. & Taylor, S.R. (1980) Th and U in sedimentary rocks: crustal evolution and sedimentary recycling. *Nature*, **285**, 621.
25 Dacey, M.F. & Lerman, A. (1983) Sediment growth and aging as Markov Chains. *J. Geol.*, **91**, 573.
26 Gregor, C.B. (1985) The mass–age distribution of Phanerozoic sediments. *In:* Snelling, N.H. (ed.), *The Chronology of the Geological Record*. Geol. Soc. Lond. Spec. Pub.

27 Ronov, A.B. (1980) *Osadochnaya Obolochka Zemli*. Nauka (Moscow).
28 McCulloch, M.T. & Wasserburg, G.J. (1978) Sm–Nd and Rb–Sr chronology of continental crust formation. *Science*, **200**, 1003.
29 DePaolo, D.J. (1981) Neodymium isotopes in the Colorado Front Range and crust-mantle evolution in the Proterozoic. *Nature*, **291**, 193.
30 Hamilton, P.J. *et al.* (1983) Sm–Nd studies of Archaean metasediments and metavolcanics from West Greenland and their implications for the Earth's early history. *EPSL*, **62**, 263.
31 Miller, R.G. & O'Nions, R.K. (1983) The provenance of Precambrian clastic and chemical sediments (abst.). *2nd European Un. Geosci. Mtg., Strasbourg*, 131.
32 Allègre, C.J. & Rousseau, D. (1984) The growth of the continent through geological time studied by Nd isotopic analysis of shales. *EPSL*, **67**, 19.
33 O'Nions, R.K. *et al.* (1983) A Nd isotope investigation of sediments related to crustal development in the British Isles. *EPSL*, **63**, 229.
34 Frost, C.D. & O'Nions, R.K. (1983) Nd-isotopic study of a major Proterozoic sedimentary succession (abst.). *2nd European Un. Geosci. Mtg., Strasbourg*, 130.
35 Hawkesworth, C.J. *et al.* (1981) Old model Nd ages in Namibian Pan-African rocks. *Nature*, **289**, 278.
36 DePaolo, D.J. (1981) A neodymium and strontium isotopic study of the Mesozoic calc-alkaline granitic batholiths of the Sierra Nevada and Peninsular Ranges, California. *JGR*, **86**, 10470.
37 Taylor, S.R. *et al.* (1983) Geochemistry of loess, continental crustal composition and crustal model ages. *GCA*, **47**, 1897.
38 Goldstein, S.L. & O'Nions, R.K. (1981) Nd and Sr isotopic relationships in pelagic clays and ferromanganese deposits. *Nature*, **292**, 324.
39 Thoni, M. *et al.* (1983) Nd-isotopic study of British river sediments and particulates from the W. Atlantic (abst.). *2nd European Un. Geosci. Mtg., Strasbourg*, 132.
40 Pollack, H.N. (1980) The heat flow from the Earth: a review. *In:* Davies, P.A. & Runcorn, S.K. (eds), *Mechanisms of Continental Drift and Plate Tectonics*. Academic Press. 183.
41 Vitorello, I. & Pollack, H.N. (1980) On the variation of continental heat flow with age and the thermal evolution of continents. *JGR*, **85**, 983.
42 Morgan, P. (1984) The thermal structure and thermal evolution of the continental lithosphere. *PCE*, **15** (in press).
43 England, P.C. & Richardson, S.W. (1980) Erosion and the age dependence of continental heat flow. *Geophys. J. R. Astr. Soc.*, **62**, 421.
44 Watson, J.V. (1976) Vertical movements in Proterozoic structural provinces. *Phil. Trans. Roy. Soc.*, **A280**, 629.

6
Greywackes: Provenance and Tectonic Significance

6.1 Greywackes and the upper crust

The distinct compositions of fine-grained sedimentary rocks of Archean and post-Archean age, noted in the previous chapter, indicate a change in upper crustal composition between these two eras. The post-Archean trace element data were derived from a wide range of sedimentary and tectonic environments, but the bulk of the Archean trace element data were derived from the fine-grained portion of greywacke–mudstone turbidite [1] successions (see Chapter 7). These form the majority of Archean sedimentary sequences. Several questions occur. Does the composition of young greywacke–mudstone turbidite sequences reflect that of the upper crust for the key trace elements (e.g. REE, Th, Sc)? Are Archean greywackes comparable to younger examples? Do Archean clastic sedimentary rocks sample the Archean upper crust in a representative manner? In this chapter we will examine the petrography and geochemistry of Phanerozoic greywackes and modern deep sea sands (the present day analogue to greywackes) and compare them with Archean greywackes. The emphasis in the discussion will be on the greywackes rather than the mudstones, because sandstone petrography provides the best evidence of provenance.

Much of the early work on Archean greywackes emphasized the similarity of their petrography and major element geochemistry to typical Phanerozoic greywacke–shale turbidite sequences [2–4]. However, our understanding of Phanerozoic greywackes and modern deep-sea sands has advanced considerably during the past decade. Their compositions are highly variable and may be related to the tectonic environment of deposition [e.g. 5–9]. Consequently, it is inappropriate to compare Archean greywackes to 'average' Phanerozoic greywackes, and more detailed comparisons are warranted [10].

6.2 Deep-sea sands

Plate-tectonic theory coupled with the availability of deep-sea sediment samples has generated great interest in relating sand composition to tectonic setting. Deep-sea sand-mud sequences, deposited by turbidity currents, are clearly the modern analogue to ancient greywacke–shale turbidites [5]. One of the earliest findings was to confirm the belief that the matrix of greywackes was mainly a secondary phenomenon [e.g. 5, 11]. A more recent finding [9] has been that secondary diagenetic processes are not required to explain the high sodium content of ancient greywackes.

6.2.1 Tectonic settings

Modern sedimentary environments can be classified according to plate tectonic setting [e.g. 12]. A recent summary is given in Table 6.1 [9]. A fundamental distinction can be made between passive (or trailing-edge) settings and active margin settings. The first is related to continental rifting and separation. The active settings are more complex and can be subdivided into various categories, including collision zones, strike-slip settings and subduction settings (island-arc and continental-arc). Ancient deposits generally cannot be accommodated into such a refined classification scheme and several more simplified schemes have been proposed. Some of these are discussed below.

Table 6.1. Plate tectonic classification of sedimentary environments (after [9]).

I. PASSIVE OR SPREADING RELATED SETTINGS
  A. Intracratonic rifts—alluvial fans and lakes (e.g. East Africa).
  B. Failed rifts (aulacogens)—thick sequence of deposits from deep sea fan to fluvial (e.g. Benue Trough).
  C. Intercontinental rifts
    (1) Early—evaporites and clastics (e.g. Red Sea).
    (2) Late (trailing-edge)—early sediments along margins overlain by shelf or deltaic deposits, passing into oceanic crust overlain by turbidites and finally pelagics (e.g. Atlantic).

II. ACTIVE SETTINGS
  A. Continental collision-related
    (1) Remnant ocean basin—thick turbidite fan deposits eventually piled into imbricate thrust sheets (e.g. Bay of Bengal).
    (2) Late orogenic basin—dominantly fluvial with variety of other terrestrial and shallow marine deposits (e.g. sub-Himalayas).
    (3) Peripheral foreland basin—as in (2) above when on subducting plate.
  B. Strike-slip fault related settings—thick deep-marine to fluvial sediments in small basins (e.g. California).
  C. Subduction-related settings
    (1) Continental margin magmatic arcs (e.g. Andes).
      (a) Fore-arc (leading-edge)—turbidities to fluvial deposits.
      (b) Back-arc (ensialic back-arc, retro-arc)—thick, mostly terrestrial deposits.
    (2) Intraoceanic arcs (e.g. Japan, Aleutians).
      (a) Fore-arc—thin turbidite plus pelagic deposits.
      (b) Back-arc—clastics near the arc, passing into pelagics and possibly into terrigenous clastics again at the continental margin.

6.2.2 Petrography

Modern deep-sea sands show petrographical variations dependent on tectonic setting [e.g. 8]. In Table 6.2, the average petrographical features of modern deep-sea sands from various tectonic settings are listed [13]. The most useful parameters are the quartz content (Q) and the plagioclase/total feldspar (P/F) ratio. Sands deposited in trailing-edge basins tend to have quartz contents exceeding 50% and P/F ratios less than 0.4. Sands deposited in fore-arc basins typically have quartz contents less than 10% and P/F ratios typically about 0.9. Sediments in back-arc basins, and leading-edge basins (strike-slip

Table 6.2. Petrographic features of modern deep-sea sands [8].

	1	2	3	4	5
Q	2	13	12	20	56
F	12	30	43	39	24
R	86	57	45	41	20
P/F	0.90	0.64	0.72	0.58	0.31

1. Fore-arc basin (9 samples); e.g. North Pacific.
2. Back-arc basin (27 samples); e.g. Bering Sea.
3. Leading-edge with subduction (8 samples); e.g. Pacific Ocean, South America.
4. Leading-edge without subduction (strike-slip); (7 samples); e.g. Pacific Ocean, California.
5. Trailing-edge (29 samples); e.g. Atlantic Ocean.

margins, continental arcs) have intermediate values although those from strike-slip margins are distinctive since they contain a lower proportion of volcanic rock fragments [8].

6.2.3 Major element chemistry

The average analyses from various tectonic settings are listed in Table 6.3 [9]. Interbedded muds also have been examined, and the chemistry of these are listed in Table 6.4. In the sands, the most distinctive geochemical parameters are K_2O/Na_2O and $FeO+MgO$ [9]. Trailing edge sands typically have $FeO+MgO$ less than 5% and K_2O/Na_2O greater than or equal to 1.0. Fore-arc sands have $FeO+MgO$ greater than 8% and K_2O/Na_2O less than 0.5. Sands in back-arc basins and leading-edge basins tend to have intermediate values, consistent with their petrography. For the muds, the only unique parameter is the K_2O/Na_2O ratio. This is almost always greater than 1.0 for trailing-edge muds and typically less than or equal to 1.0 for the other tectonic environments.

Table 6.3. Average chemical composition of modern deep-sea sands [9].

	1	2	3	4	5
SiO_2	61.5	68.8	69.5	67.8	77.9
Al_2O_3	15.2	14.4	14.1	15.6	9.8
FeO	6.9	4.0	3.5	3.3	2.6
MgO	3.8	2.4	1.9	2.3	1.3
CaO	6.7	4.4	4.4	3.6	4.1
Na_2O	3.8	3.6	3.6	3.9	1.9
K_2O	1.4	2.0	2.6	2.9	2.0
Σ	99.3	99.6	99.6	99.4	99.6
K_2O/Na_2O	0.37	0.56	0.72	0.74	1.05
$FeO+MgO$	10.7	6.4	5.4	5.6	3.9

1. Fore-arc basin.
2. Back-arc basin.
3. Leading-edge with subduction.
4. Leading-edge without subduction (strike-slip).
5. Trailing-edge.

Table 6.4. Average chemical composition of muds associated with modern deep-sea sands [9].

	1	2	3	4	5
SiO_2	68.9	68.0	66.1	65.8	65.9
Al_2O_3	12.1	14.9	16.9	14.4	13.7
FeO	6.5	5.8	5.8	6.1	4.8
MgO	3.0	3.1	3.2	3.4	2.8
CaO	4.9	2.8	3.0	4.9	8.2
Na_2O	2.6	2.5	2.4	2.7	1.5
K_2O	1.5	2.3	2.5	2.0	2.6
Σ	99.5	99.4	99.9	99.3	99.5
K_2O/Na_2O	0.58	0.92	1.0	0.74	1.7
FeO+MgO	9.5	8.9	9.0	9.5	7.6

1. Fore-arc basin.
2. Back-arc basin.
3. Leading-edge with subduction.
4. Leading-edge without subduction (strike-slip).
5. Trailing-edge.

6.2.4 Trace element data

Trace element data, particularly for the REE, are scarce and this represents a major gap in our understanding of sedimentary geochemistry. In Table 6.5, REE data are presented for three deep-sea sediment samples taken from turbidites. They are plotted on chondrite-normalized curves in Fig. 6.1. The Woodlark Basin is a complex region but the sample shown (unpublished results from this laboratory) is volcaniclastic and best related to a fore-arc basin. The sample from the Cascadia channel of the western USA [14] was

Table 6.5 REE content of modern deep-sea turbidite sediments.

	1	2	3
La	8.7	22.8	39
Ce	17.0	52.6	81
Pr	2.51	5.4	—
Nd	10.7	22.4	—
Sm	2.30	4.66	6.6
Eu	0.72	1.28	1.6
Gd	2.29	—	—
Tb	0.40	0.76	1.0
Ho	0.51	0.95	—
Er	1.40	3.14	—
Yb	1.21	2.86	3.1
Lu	—	—	0.55
ΣREE	50.5	128.0	194.3
La_N/Yb_N	4.9	5.4	8.5
Eu/Eu*	0.96	0.76	0.75

1. Fore-arc basin—Woodlark Basin—mud (unpublished data).
2. Continental leading-edge (?)—Cascadia Channel [14].
3. Trailing-edge—West Atlantic—3 muds [15].

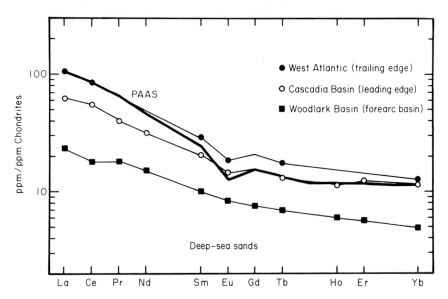

Fig. 6.1. Chondrite-normalized REE plot of modern deep-sea sediments from continental margins. Data from Table 6.5; PAAS plotted as heavy solid line for comparison. The samples from the leading-edge and trailing-edge environments have REE patterns similar to PAAS although Eu/Eu* is somewhat higher. The sample from the fore-arc environment (Woodlark Basin) is clearly different with lower ΣREE, La/Yb and no Eu-anomaly. Such a pattern is typical of calc-alkaline andesite material.

described as a terrigenous ocean sediment and is derived from a leading edge continental margin ('protocontinental arc') environment.

The third composition is an average of three clays from turbidite beds in the north-west Atlantic [15]. These samples are from a trailing-edge environment. The samples from the Atlantic basin (trailing-edge) and Cascadia channel (leading-edge) both show significant negative Eu-anomalies. These are not as great as that of PAAS (Eu/Eu*=0.75 and 0.76 respectively, compared to 0.66 for PAAS). The patterns are also somewhat less fractionated (lower La_N/Yb_N) than PAAS. Whether these differences are significant or simply reflect insufficient sampling awaits further work. The sample from the Woodlark Basin is clearly different with much lower ΣREE abundances, low La_N/Yb_N and most significantly, no Eu-anomaly. In these respects, it compares favourably to Archean sedimentary rocks (see below). The fore-arc setting, the volcaniclastic texture and the REE pattern reflect a predominantly andesitic source.

6.3 Phanerozoic greywackes

Interpretation of Phanerozoic greywacke terrains in terms of modern plate tectonic setting is contentious. This can be attributed to at least three reasons:
1 scarcity of modern deep-sea sand data for comparison;
2 considerable overlap in petrographical and geochemical features of many tectonic settings in deep-sea sands. This is highlighted by the more simplistic

tectonic classification schemes employed for greywacke petrography (see below);
3 development of up to 15–50% matrix, which is mainly secondary forming at the expense of labile rock fragments and feldspar.

A considerable amount of petrographical and geochemical data has been collected for Phanerozoic greywackes. Trace element analyses are less common, but there are enough data to make some comparisons with Archean sedimentary rocks in terms of tectonic setting and upper crustal composition.

6.3.1 Tectonic settings

The effort directed towards unravelling the tectonic settings of Phanerozoic turbidite sequences has resulted in a daunting nomenclature typical of that which haunts the study of sedimentary rocks. An early system of greywacke classification, introduced by Crook [5], has the advantage of a simple three-fold division designed specifically for greywackes and including both petrographical and major element geochemical data. Using the QFR system [13], Crook [5] divided greywackes into quartz-rich, quartz-intermediate and quartz-poor varieties, on the basis of SiO_2 content, K_2O/Na_2O ratios and proportion of quartz in the framework grains (Q) (Table 6.6). Quartz-rich greywackes were derived from trailing-edge (Atlantic-type) continental margins; quartz-intermediate varieties were derived from continental-arc (Andean-type) margins; quartz-poor greywackes were derived from fore-arc basins of island-arcs. Subsequent work has confirmed these findings (see [6]).

Table 6.6. Clasification of greywackes (after [5]).

	Quartz-poor	Quartz-intermediate	Quartz-rich
Q	<15	15–65	>65
SiO_2 (average) (%)	58	68–74	89
K_2O/Na_2O (average)	⩽1.0	<1.0	>1.0
Tectonic setting	Island-arcs	Leading-edge (with subduction)	Trailing-edge

Dickinson and his co-workers [7, 16, 17] have developed a more sophisticated petrographical scheme based on QFL components [13] which is applicable to sandstones in general (including greywackes). One of the more useful diagrams is displayed in Fig. 6.2. In this scheme, sandstones are derived from three major tectonic settings:
1 Continental Block—these sandstones come from continental interiors and uplifted cratonic blocks. This includes passive margin settings and plutonic belts from arc orogens as well as other wholly continental environments.
2 Recycled Orogens—this division comprises uplifted orogenic zones where recycled sedimentary debris is the dominant component in the sandstones. Included here are foreland basins flanking either arc or collision basins, collision zone basins such as Bengal Fan and recycled subduction complexes.
3 Magmatic Arcs

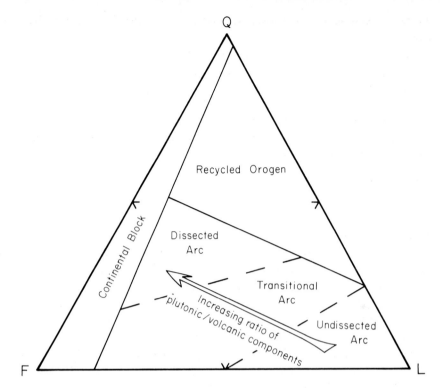

Fig. 6.2. *QFL* diagram for terrigenous sandstones showing inferred provenance type for various petrographical compositions. (Adapted from [17].)

(a) Undissected Arcs—comprising volcaniclastic (typically andesite) debris derived from active island-arcs and some continental arcs where only limited erosion has occurred. Depositional environments include trenches, fore-arc basins and marginal seas, among others.
(b) Dissected Arcs—representing more mature arcs where erosion has exposed plutonic root zones resulting in sandstones of mixed plutonic and volcanic origin. Depositional environments include fore-arc and back-arc basins, particularly at continental margin arcs.
(c) Transitional Arcs.

Practical usage of this classification system is often hampered by the lack of sufficient detail in many petrographic descriptions.

6.3.2 *Petrography*

Table 6.7 summarizes average petrographical analyses of a number of Phanerozoic greywacke sequences. The following characteristics may be noted:
1 There is a highly variable quartz content both in terms of absolute abundances and relative proportion of framework grains. This is one of the main parameters used for inferring tectonic setting (see Table 6.6 and Fig. 6.2).
2 Rock fragments are dominated by volcanic lithologies only in the quartz-

Table 6.7. Petrographic features of Phanerozoic greywackes.

	1	2	3	4	5	6	7	8	9	10
Quartz	0.4	1.8	6.1	9.0	—	27.6	25.0	31.9	45.8	60
Plagioclase	12.7	17.6	22.0	10.4	—	21.4	3.5	8.4	4.6	9
K-feldspar	—	tr.	2.0	7.1	—	2.1	6.0	2.9	—	<1
Rock fragments		(2.3)*	(7.3)*			(10)*	(24.4)*			
Felsic volcanic	—	—	11.7	1.8	—	—	—	—	—	—
Intermediate-mafic volcanic	63.6	50.8	28.5	23.3	—	—	—	—	2.0	—
Granitic	—	—	—	2.7	—	—	—	—	—	—
Sedimentary and metamorphic	—	—	—	20.4	—	21	—	8.9	4.9	≈10
Matrix	20.3	8.8	11.0	21.6	(15–20)	—†	31.3	38.7	36.7	≈20
Other	3.0	18.7	11.4	3.8	—	17.9	9.8	7.4	6.0	—
Framework grains										
Q	0.5	2.5	8	12	30	34	42	61	80	75
F	16.5	24.3	31	23	44	29	16	22	8	12
R	83.0	73.2	61	65	26	37	42	17	12	12
P/F	1.0	1.0	0.92	0.59	—	0.91	0.37	0.74	1.0	>9.0

* Undifferentiated rock fragments.
† Matrix not reported separately

1. Average of 31 volcanogenic quartz-poor greywackes, Baldwin Fm. (Devonian), Australia [18].
2. Average of 11 quartz-poor greywackes, North Range Group (Triassic), New Zealand [19].
3. Average of 9 quartz-poor greywackes, Taringatura Group (Triassic), New Zealand [19].
4. Average of 7 quartz-poor greywackes, Gazelle Fm. (Silurian), USA [20].
5 Average of 17 data sets (203 samples) of quartz-intermediate greywackes, Franciscan (Jurassic–Cretaceous), USA [21].
6. Average of 88 quartz-intermediate greywackes, Harz Mountain (Devonian–Carboniferous), Germany [22].
7. Average of 119 quartz-intermediate greywackes, Rensselaer Graywacke, USA [23].
8. Average of 5 quartz-intermediate sub-greywackes, Duzel Fm. (Ordovician), USA [20].
9. Average of 12 quartz-rich greywackes, Greenland Group (Ordovician), New Zealand [24].
10. Average of 14 quartz-rich greywackes, Mam Tor Beds (Carboniferous), England [25, 26].

poor greywackes, where mafic-intermediate varieties (mainly andesite) predominate.

3 The ratio of plagioclase/total feldspar (P/F) is consistently high (>0.9) in quartz-poor varieties; in the others it is highly variable and this parameter is not diagnostic of quartz-rich greywackes. Note that low P/F ratios were characteristic of modern trailing-edge basin sands (see above).

4 The matrix content is generally >10% and variable up to almost 50%. The generation of matrix is mainly a secondary phenomenon and forms at the expense of feldspar and unstable rock fragments [e.g. 5, 11, 16]. Thus an analysis of the framework grains (QFR) will tend to be biased towards higher Q values (see [10] and below for discussion).

6.3.3 Major element chemistry

Some of the available major element data for Phanerozoic greywackes are listed in Table 6.8 (a more extensive compilation may be found in [29]). For

Table 6.8. Average chemical composition of Phanerozoic greywackes.

	1	2	3	4	5	6	7	8	9	10	11
SiO_2	57.6	57.8	58.9	67.4	61.8	70.1	72.9	73.1	73.7	73.4	78.8
TiO_2	1.2	1.1	1.7	1.3	1.0	0.5	0.5	0.7	0.5	0.6	0.6
Al_2O_3	16.5	16.6	16.6	15.8	15.4	14.0	14.5	13.0	9.8	13.3	11.5
FeO	9.3	8.0	8.5	3.6	9.5	4.2	4.3	4.6	2.8	4.6	5.0
MgO	3.9	4.9	3.2	1.1	—	2.3	1.8	1.6	—	2.6	1.1
CaO	5.3	5.2	4.3	2.5	3.8	2.5	1.1	1.7	5.0	0.8	0.7
Na_2O	5.0	4.2	4.4	4.8	3.6	3.7	2.9	2.9	1.7	1.7	1.1
K_2O	0.7	1.8	1.3	2.6	1.1	1.8	1.5	2.3	1.4	2.8	0.9
Σ	99.5	99.6	98.9	99.1	96.2	99.1	99.5	99.9	94.9	99.8	99.7
K_2O/Na_2O	0.14	0.43	0.30	0.54	0.31	0.49	0.52	0.79	0.82	1.65	0.82
FeO+MgO	13.2	12.9	11.7	4.7	—	6.5	6.1	6.2	—	7.2	6.6

1. Average of 10 volcanogenic quartz-poor greywackes, Baldwin Fm. (Devonian), Australia [18].
2. Average of 29 volcanogenic quartz-poor greywackes, Goceano area (Hercynian), Sardinia [27].
3. Average of 7 greywackes, North Range Group (Triassic), New Zealand [19].
4. Average of 9 quartz-poor greywackes, Taringatura Group (Triassic), New Zealand [19].
5. Average of 9 greywackes, Gazelle Fm. (Silurian), USA [20].
6. Average of 21 quartz-intermediate greywackes, Franciscan (Jurassic–Cretaceous), USA [21].
7. Average of 25 greywackes, Harz Mountain (Devonian–Carboniferous), Germany [22].
8. Average of 119 quartz-intermediate greywackes, Rensselaer Graywacke, USA [23].
9. Average of 7 subgreywackes, Duzel Fm. (Ordovician), USA [20].
10. Average of 6 greywackes, Greenland Group (Ordovician), New Zealand [28].
11. Average of 14 quartz-rich greywackes, Mam Tor Beds (Carboniferous), England [26].

comparative purposes, the data have been recalculated on a volatile-free basis [5]. Mostly, these analyses correspond quite well with the subdivisions of Crook [5]. The volcanogenic quartz-poor greywackes from the Taringatura Group, New Zealand, have higher average SiO_2 (67.4%) and K_2O/N_2O (0.54) than expected for a Q-value of 8 (see Table 6.6). This can be related to the modest proportion of acid volcanic fragments in these sandstones [19]. This contrasts with the predominance of andesitic fragments in most other Phanerozoic volcanogenic greywackes, such as those derived from undissected magmatic arcs (see above). The quartz-rich greywackes from the Greenland Group have lower average SiO_2 (73.4%) than expected for a Q-value of 80. This is probably a result of the high matrix content for these samples.

The major element chemistry also compares well with the data from modern deep-sea sands, although there are some discrepancies. The quartz-poor greywackes show the lower SiO_2 content and K_2O/Na_2O ratios [30] and higher FeO+MgO contents expected for modern fore-arc basin sands. The exception again is the Taringatura Group; the distinctive composition for these rocks is related to the acid nature of the volcanic rock fragments. Conversely, the quartz-rich greywackes have high SiO_2 content and K_2O/Na_2O ratios expected for modern trailing-edge basin sands. However, FeO+MgO, which is generally low (but variable) in the modern trailing-edge sands (see Table 6.3) is highly variable and overlaps with the quartz-intermediate varieties. The chemistry of the quartz-intermediate varieties, which Crook [5] assigned to Andean margins, is intermediate with respect to SiO_2 and K_2O/Na_2O but are

Table 6.9. Chemical composition of selected Phanerozoic greywackes.

	Quartz-poor		Quartz-intermediate			Quartz-rich	
	M277	M285	P40136	MK64	T82/324	P39803	MK97
SiO_2	56.35	60.78	71.08	68.28	67.5	75.65	81.13
TiO_2	1.39	0.82	0.71	1.00	0.62	0.77	0.62
Al_2O_3	16.19	17.51	14.59	12.95	16.67	12.08	10.01
FeO	11.22	6.17	4.70	6.94	4.94	4.19	2.76
MnO	0.19	0.11	—	0.21	0.13	—	0.03
MgO	4.06	2.46	2.57	2.22	2.31	2.20	1.44
CaO	5.22	5.52	0.86	4.50	2.27	0.32	0.26
Na_2O	4.59	5.74	1.34	2.96	3.41	2.05	1.69
K_2O	0.55	0.74	3.99	0.83	1.91	2.57	1.93
P_2O_5	0.24	0.15	0.15	0.12	0.17	0.18	0.14
Σ	100.0	100.0	99.99	100.01	99.93	100.01	100.01
LOI	3.50	2.83	3.00	6.10	—	3.00	1.85
Cs	0.41	0.80	5.4	—	4.0	4.9	—
Ba	85	100	588	150	416	347	400
Rb	—	8.5	165	36	—	121	91
Sr	190	266	75	222	—	52	44
Pb	—	4.7	—	13	5.2	—	16
La	10	6.8	35.5	25	25.1	35.8	43
Ce	18	15	80.7	53	53.6	69.1	83
Pr	2.2	2.0	—	5.8	6.81	—	12
Nd	10	8.2	—	22	25.9	45	42
Sm	2.8	2.2	5.33	4.5	4.82	8.09	7.1
Eu	0.97	1.1	1.11	0.9	1.04	1.31	1.0
Gd	3.4	2.6	—	3.5	3.74	—	5.6
Tb	0.52	0.39	0.65	0.6	0.59	0.64	0.88
Dy	3.1	2.7	—	3.6	3.46	—	4.7
Ho	0.79	0.59	—	0.8	0.64	—	1.0
Er	2.2	1.9	—	2.2	1.72	—	2.9
Yb	2.3	1.8	1.62	2.0	1.68	2.85	2.9
Lu	—	—	0.34	—	—	0.48	—
ΣREE	56	45	176	124	130	186	207
La_N/Yb_N	2.9	2.6	14.8	8.4	10.1	8.5	10.0
Eu/Eu*	0.96	1.41	0.70	0.69	0.75	0.63	0.48
Y	19	15	—	22	21	—	32
Th	0.88	1.4	16.1	6.96	7.8	12.8	16.4
U	0.32	0.84	—	1.33	1.81	3.0	3.42
Zr	105	68	112	175	140	—	384
Hf	1.8	1.2	2.6	3.8	3.8	6.2	10.1
Nb	—	0.8	—	10	10.3	—	11
Th/U	2.8	1.7	—	5.2	4.3	4.3	4.8
La/Th	11.4	4.9	2.2	3.6	3.2	2.8	2.6
Cr	31	36	90	109	40	63	51
V	350	150	—	172	114	—	57
Sc	37	25	16.3	25	15	10.1	10
Ni	14	13	—	20	19	—	19
Co	31	17	13.3	11	10	10.5	13
Cu	190	35	—	9	31	—	11
Zn	—	—	—	95	—	—	53
Ga	23	20	—	15	19	—	13
La/Sc	0.27	0.27	2.2	1.0	1.7	3.5	4.3
Th/Sc	0.02	0.06	0.99	0.28	0.52	1.3	1.6
B	18	22	—	—	—	—	—

generally quite variable. These cannot be distinguished from back-arc or leading edge compositions (Table 6.3).

6.3.4 Trace-element data

Trace element data, including REE, are not abundant for Phanerozoic greywacke-shale suites. Virtually all of the analyses come from Australia and New Zealand [28, 31, 32; also see 20], but do cover the complete major element and petrographic compositional range. Representative analyses of various greywackes are listed in Table 6.9. These have been subdivided into quartz-poor, quartz-intermediate and quartz-rich varieties for consistency with the above discussion.

The quartz-poor greywackes are from the Devonian Baldwin Formation of

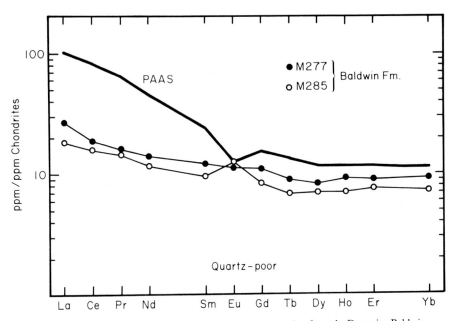

Fig. 6.3. Chondrite-normalized REE plot of quartz-poor greywackes from the Devonian Baldwin Formation. Data from Table 6.9; PAAS plotted as heavy solid line for comparison. The greywacke patterns are considerable lower in ΣREE, La/Yb and have no Eu-depletion. It is clear that calc-alkaline andesitic rocks were the primary source for these sedimentary rocks.

Table 6.9 *(footnotes)*

M277, M285: Baldwin Fm. (Devonian), Australia ([31]; unpublished data); probably deposited in fore-arc basin.

P40136: Robertson Bay Gp. (late Proterozoic–Cambrian), Antarctica [28]; possibly deposited at a continental-arc.

MK64: Waterbeach Fm., Hill End Trough (Upper Silurian–Lower Devonian), Australia [32]; possibly deposited at a continental-arc.

T82/324: Torlesse Group, Triassic, Ruataniwha Dam, Ohau River, New Zealand; S.34-7764; (unpublished data).

P39803: Greenland Gp. (Ordovician), New Zealand [28]; recycled provenance.

MK97: Bendigo Trough (Ordovician), Australia [32]; possibly deposited at passive margin.

the Tamworth Trough, Australia. These samples are volcanogenic (andesite) in origin and a fore-arc setting for deposition with an undissected magmatic arc provenance is strongly indicated [18, 31]. The REE patterns are plotted in Fig. 6.3 where they are compared to post-Archean average Australian shales (PAAS). The patterns are characterized by low ΣREE and La_N/Yb_N compared to PAAS. One sample has no Eu-anomaly and the other is enriched in Eu, related to plagioclase content [31]. These samples also have low Th/U ratios and generally low concentrations of all incompatible trace elements (e.g. Th, U, Cs, Rb, REE) resulting in low Th/Sc and La/Sc ratios. All of these observations including Cr and Ni values support derivation from an andesitic source in a fore-arc basin (see Chapter 3).

Quartz-rich examples come from the Ordovician Greenland Group of western New Zealand (the average composition of the Greenland Group greywacke and mudstone is given in Table 2.5) and the Ordovician Bendigo Trough of Victoria, Australia. Quartz-rich greywackes are typical of trailing-edge margins and such an environment has been suggested for the Bendigo Trough [32]. The tectonic setting of the Greenland Group, is not clear but a polycyclic source of plutonic-derived sedimentary material is indicated [28]. The REE patterns of representative samples from these successions are plotted in Fig. 6.4 and compared to PAAS. The REE patterns are LREE enriched, with $La_N/Yb_N >8$, and possess significant depletion in Eu, of comparable magnitude to those in PAAS. These greywackes are indistinguishable from typical post-Archean upper crust in terms of REE (as indicated by PAAS). They also have higher Th/U (>4) ratios than the quartz-poor greywackes and generally much higher levels of the incompatible elements (e.g. Th, U, REE). The Th/Sc, La/Sc and La/Th ratios are similar to those of the average post-Archean shale.

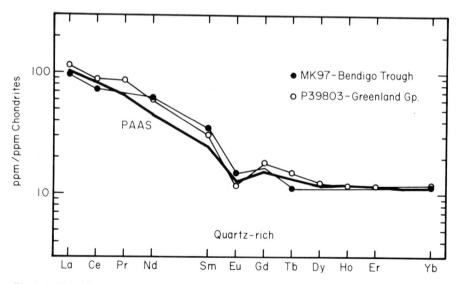

Fig. 6.4. Chondrite-normalized REE plot of selected quartz-rich greywackes from New Zealand and Australia. Data from Table 6.9; PAAS plotted as heavy solid line for comparison. These greywackes have REE patterns essentially identical to PAAS.

Quartz-intermediate greywackes may be derived from a variety of sources including recycled orogenic, continental block and dissected magmatic arcs. The overall similarity of petrography and major element chemistry for sands deposited at Andean-margins, strike-slip margins and back-arc basins make it difficult to distinguish these environments [8, 9]. The representative samples in Table 6.9 are taken from the late Precambrian–early Cambrian Robertson Bay Group, Antarctica, the Silurian–Devonian Waterbeach Formation of the Hill End Trough, NSW, Australia, and the Triassic Torlesse Group of New Zealand. REE patterns are plotted on Fig. 6.5 where they are compared to

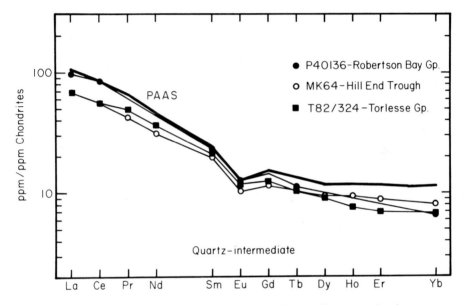

Fig. 6.5. Chondrite-normalized REE plot of selected quartz-intermediate greywackes from Antarctica, Australia and New Zealand. Data from Table 6.9; PAAS plotted as heavy solid line for comparison. The greywacke patterns are essentially parallel to PAAS but have slightly lower ΣREE and slightly higher Eu/Eu* (though still possessing a substantial negative Eu-anomaly).

PAAS. The patterns are quite similar with La_N/Yb_N ranging from 8 to 15 and Eu/Eu* ranging from 0.69 to 0.75, and compare favourably with PAAS. Whether higher Eu/Eu* is a general characteristic of these varieties of greywackes is not clear, due to lack of data, although preliminary work in Canberra suggests it may be. Other trace element levels are scattered although incompatible element levels (Th, U, REE, etc.) are generally much higher than in quartz-poor varieties, resulting in higher Th/Sc and La/Sc ratios.

6.4 Archean greywackes

There were two major periods of Archean greenstone belt development at 3.6–3.2 Ae and 2.8–2.6 Ae [4, 33, 34]. It has also been suggested that important sedimentary and tectonic differences could exist between early and late Archean greenstone belts [35, 36]. These distinctions are discussed in greater detail in the following chapter. Thick sequences of greywacke-mudstone turbidite strata are preserved in the upper portions of the early Archean greenstone belt successions (e.g. Swaziland Supergroup, Pilbara Supergroup) and at the top of volcanic-sedimentary cycles in younger greenstone belts (e.g. Yilgarn Block, Slave Province, Superior Province). In the following discussion, we group early and late Archean greywackes together because there are insufficient data to warrant separation.

A recurring observation by students of Archean sediments is the paucity of evidence for shallow marine environments, indicative of a lack of wide continental shelves. Rapid facies changes are observed, from terrestrial deposits, such as braided stream or alluvial fan deposits to deeper water turbidites [37–43]. These environments have been named Continental or Non-Marine Association and Resedimented Association [38, 44]. At the top of

Table 6.10. Petrographic features of Archean greywackes.

	1	2	3	4	5	6	7	8	9	10
Quartz	8.5	13.8	23	24.9	26.1	27.8	29.5	41.4	20.4	63.5
Plagioclase	24.7	18.9	—	12.6	1.8	3.7	16.2	10.7	7.2	7.7
K-feldspar	0.6	—	16‡	—	8.1	15.8	0.3	3.9	0.3	2.1
Rock fragments					(3.6)*	(4.2)*			(<5.0)*	
Felsic volcanic	40.2	6.2	11.2	7.7	—	—	14.2	—	—	—
Intermediate mafic volcanic	2.0	tr.	2.4	3.5	—	—	tr.	—	—	—
Granitic	1.4	tr.	0.3	1.3	—	—	tr.	—	—	—
Sedimentary and metamorphic	—	tr.	0.1	6.0	16.3	5.9	—	(25.0)†	—	—
Matrix	12.2	59.1	48.0	37.0	35.4	37.0	37.2	(19.9)†	66.1	22.1
Other	10.4	2.1	—	7.0	8.7	5.6	1.3	—	1.0	4.7
Framework components										
Q	11	35	43	44	47	48	49	51	62	87
F	33	49	30	23	18	34	27	18	22	13
R	56	16	26	33	36	18	24	31	15	0
P/F	0.98	—	—	1.00	0.18	0.19	0.98	0.73	0.96	0.79
Provenance	FI	FM(G)	GMF	FMG	GFM	GFM	G(SFM)	SGM	SG	G or S

* Undifferentiated rock fragments.
† Matrix and some varieties of sedimentary rock fragments indistinguishable; divided equally between two classifications.
‡ Total feldspar.

1. Average of 10 greywackes, Vermilion District, Minnesota, USA [46].
2. Average of 14 greywackes, Gamitagama Lake, Superior Province [47].
3. Average of 10 greywackes, Abrams Group, Superior Province, Canada [38].
4. Average of 9 greywackes, Burwash Fm., Slave Province, Canada [48].
5. Average of 84 greywackes, Sheba Fm., Barberton Mountain Land, South Africa [49, 50].
6. Average of 16 greywackes, Belvue Road Fm., Barberton Mountain Land, South Africa [49, 50].
7. Average of 6 greywackes, Minnitaki Group, Superior province, Canada [37].
8. Average of 10 greywackes, Chitaldrug Schist Belt, India [51].
9. Average of 23 greywackes, South Pass Greenstone Belt, Wind River Mountains, Wyoming, USA [3]; recalculated to exclude undifferentiated quartz and feldspar.
10. Average of 7 greywackes, North Spirit Lake, Superior Province, Canada [52].

Provenance key: F=felsic volcanics; M=mafic volcanics; I=intermediate volcanics; G=granite–granitic gneiss; S=recycled sedimentary; listed in decreasing importance with minor components in parentheses.

the Swaziland Supergroup, in the Moodies Group, stable continental shelf or shelf rise margin deposits are well developed [42, 45], indicating perhaps the initiation of cratonic conditions. In addition, the lower volcanic sequences of early Archean greenstone belts also display evidence for shallow water deposition of sediments, possibly in association with volcanic islands [e.g. 36].

6.4.1   Petrography

In Table 6.10, some typical petrographic analyses of Archean greywackes are listed. The following observations are significant (see reviews in [4, 10, 47]):
1   Archean greywackes (and other sandstones) have highly variable compositions; purely volcanogenic varieties occur, such as in the Vermilion district [46] and Gamitagama Lake greenstone belt, Superior Province [47] and the source of such sediments is typically felsic in composition.
2   The abundance and proportions of various rock fragments are variable. Felsic volcanic fragments typically dominate with mafic and intermediate volcanic lithologies being considerably less abundant. Granitic and granitic gneiss fragments are not volumetrically abundant but are observed in many areas; where present, they are typically of tonalite–trondhjemite (–granodiorite) composition. Sedimentary (mainly argillite and chert) and metamorphic rock fragments are also observed in some cases.
3   The ratio of plagioclase/total feldspar is typically very high (>0.7), although the Fig Tree and Moodies Groups are important exceptions. The dominance of plagioclase feldspar cannot be taken as evidence for a volcanic origin because early Archean granitic rocks are primarily sodium-rich varieties (tonalite–trondhjemite).
4   Many Archean greywackes contain a high proportion of sand-sized quartz ($>20\%$), with high quartz/feldspar ratios (>1). In some cases an acid volcanic origin can be inferred [e.g. 47], but it is generally agreed that acid volcanic rocks cannot be the dominant source of abundant sand-sized quartz in extensive sandstone bodies [37, 52–54] so that a granitic, granitic gneiss or recycled sedimentary provenance is indicated.

6.4.2   Major element chemistry

The average major element composition of several Archean greywackes are listed in Table 6.11. Archean greywackes are notable for their low Al_2O_3/Na_2O ratios (<6) compared to most other sandstones [53] indicative of chemical immaturity and derivation from a relatively unweathered source. Data from the North Spirit Lake area are an exception to this; these samples are derived from a recycled sedimentary or highly weathered granitic provenance [52]. Archean greywackes also contain high contents of Fe and Mg [4, 10]. There is no difference in major element chemistry between those derived from a volcanogenic source and those derived from a mixed or granitic source. The major element composition of Archean greywackes is close to the present-day composition of the upper continental crust (i.e. granodiorite), but they lack the characteristic signature of depletion in Eu (see below). It would be

Table 6.11. Average chemical composition of Archean greywackes.

	1	2	3	4	5	6	7	8
SiO_2	64.3	71.6	68.1	70.7	64.7	63.4	65.6	82.8
TiO_2	0.5	0.6	0.7	0.6	0.6	0.4	0.6	0.3
Al_2O_3	15.6	14.5	15.7	10.9	14.0	15.0	15.8	8.1
FeO	5.3	4.2	5.3	6.7	6.4	4.7	6.0	2.9
MgO	3.6	2.3	2.8	4.8	4.8	3.7	3.2	1.2
CaO	4.2	1.8	1.8	2.1	3.4	5.9	2.3	1.8
Na_2O	2.9	3.2	3.2	1.9	3.1	4.2	3.8	0.5
K_2O	2.5	1.7	2.0	1.7	2.4	2.0	2.5	0.9
Σ	98.9	99.9	99.6	99.4	99.4	99.3	100.1	98.5
K_2O/Na_2O	0.86	0.53	0.63	0.89	0.77	0.48	0.66	1.80
FeO+MgO	8.9	6.5	11.1	11.5	11.2	8.4	9.2	4.1

1. Average of 4 greywackes, Vermilion District, Minnesota, USA [55, 56].
2. Average of 4 greywackes, Gamitagama Lake, Superior Province, Canada [47].
3. Average of 3 greywackes, Burwash Fm., Slave Province, Canada [48].
4. Average of 17 greywackes (4 only for MgO), Sheba Fm., Barberton Mountain Land, South Africa [49].
5. Average of 7 greywackes (3 only for MgO), Belvue Road Fm., Barberton Mountain Land, South Africa [49].
6. Average of 10 greywackes, Chitaldrug Schist Belt, India [51].
7. Average of 23 greywackes, South Pass greenstone belt, Wind River Mountains, Wyoming, USA [3].
8. Average of 2 greywackes, North Spirit Lake, Superior Province, Canada [52].

incorrect to equate greywacke compositions directly to upper crustal compositions because argillaceous rocks comprise a significant part of the greywacke–mudstone turbidite sequences and their composition is not similar to the greywackes (Table 6.12). The mudstones are generally lower in Si, Ca, Na and

Table 6.12. Average chemical composition of Archean mudstones from greywacke–mudstone turbidite sequences.

	1	2	3	4	5	6
SiO_2	56.8	56.2	63.9	61.8	65.4	60.8
TiO_2	0.9	1.0	0.8	0.6	0.4	0.7
Al_2O_3	21.3	21.6	20.1	14.3	22.1	24.1
FeO	8.3	8.6	6.6	11.8	4.0	5.5
MgO	4.9	5.0	2.4	6.7	0.5	3.5
CaO	2.0	1.3	0.4	1.0	2.0	0.01
Na_2O	2.9	2.3	2.8	1.1	2.4	0.6
K_2O	2.8	3.7	2.6	2.5	3.6	4.8
Σ	99.9	99.7	99.6	99.8	100.4	100.0
K_2O/Na_2O	1.0	1.6	0.93	2.3	1.5	8.0
FeO+MgO	13.2	13.6	9.0	18.5	4.5	9.0

1. Vermilion District, Knife Lake [55].
2. Average of 3 mudstones, Burwash Fm., Slave Province, Canada [48].
3. Minnitaki Group, Superior Province, Canada [37].
4. Average of 5 mudstones (3 only for MgO), Fig Tree Group, Barberton Mountain Land [49, 57].
5. North Spirit Lake, Superior Province, Canada [52].
6. Average of 10 mudstones, Gorge Creek Group, Pilbara Block [58].

high in Ti, Al, Fe, K, Al_2O_3/Na_2O, K_2O/Na_2O when compared to the greywackes.

6.4.3 Trace element data

Most analyses for trace elements in Archean sedimentary rocks have been carried out on mudstones (which are discussed in detail in the following chapter). In Table 6.13 several typical Archean greywacke analyses are given and REE patterns are shown in Fig. 6.6. The suggested source terrains include mixed felsic–mafic (Kalgoorlie, Knife Lake); mixed felsic–mafic volcanic and granitic (Yellowknife); and dominantly granitic and recycled sedimentary (Fig Tree, Wyoming). In spite of the differing sources, the REE patterns are remarkably similar with no or very slight depletion in europium.

6.4.4 Provenance

The petrographic and geochemical evidence points to a wide diversity in source rock lithologies for Archean greywacke–mudstone turbidite sequences. The most important are felsic volcanics (including porphyries), granite and granitic gneiss, mafic volcanic and recycled sedimentary rocks, with minor intermediate volcanics and other metamorphic rocks. In some cases the provenance is dominated by one lithology such as at Knife Lake (felsic volcanic) or North Spirit Lake (granitic or recycled sedimentary), although in most cases, a bimodal mafic–felsic mixture of source lithologies is indicated. The lack of intermediate compositions is significant and is discussed below.

6.5 Comparisons

The Archean greywacke–shale turbidite sequences show both similarities and differences with modern deep-sea sands and Phanerozoic turbidite sequences. In the following sections we will compare these and discuss restrictions on using Archean sedimentary rocks as typical or average upper crustal samples.

6.5.1 Petrography

In terms of their petrographic character, Archean greywackes (Section 6.4.1) cover the range of quartz-poor to quartz-rich, but the majority are quartz-intermediate varieties. Compared to modern deep-sea sands, they resemble those deposited at trailing-edge (e.g. Atlantic type) margins in terms of the framework mineralogy. This may be misleading because matrix forms at the expense of feldspar and rock fragments and results in higher apparent framework quartz abundances. An approximate correction can be made [10] using:

$$Q_O = Q_A - MQ_A \qquad (1)$$

where Q_O is the original framework quartz content, Q_A is the apparent (measured) framework quartz content and M is the proportion of matrix. This assumes all of the matrix is derived from the breakdown of feldspar and rock fragments. The original framework quartz content for Archean greywackes is

Table 6.13. Chemical composition of selected Archean greywackes.

	DD9	YK2	KH44	C28	G21
SiO_2	65.8	67.79	69.76	67.5	69.35
TiO_2	0.52	0.56	0.52	0.42	0.59
Al_2O_3	15.9	15.44	13.79	11.8	14.98
FeO	5.36	5.94	7.79	4.72	4.52
MnO	0.07	0.07	0.02	0.14	0.05
MgO	3.56	2.54	2.91	4.9	1.89
CaO	2.87	1.90	1.27	5.54	2.13
Na_2O	3.65	4.26	1.78	2.75	4.26
K_2O	2.17	1.40	2.11	2.22	2.24
P_2O_5	0.11	0.09	0.05	—	—
Σ	100.0	99.99	100.00	99.99	100.01
LOI	2.88	1.62	6.23	—	—
Cs	—	—	1.3	—	—
Ba	566	418	790	489	—
Rb	—	50	52	81	73
Sr	457	357	93	318	324
La	—	18	17	22	25
Ce	32.6	41	33	45	41
Pr	—	4.7	4.4	—	5.8
Nd	14.8	19	17	17.2	25
Sm	2.68	3.9	3.1	3.1	4.4
Eu	0.785	1.1	1.1	0.80	1.28
Gd	2.24	3.0	3.2	2.6	4.1
Tb	—	—	0.49	0.4	0.54
Dy	1.74	2.4	3.1	2.3	—
Ho	—	—	0.64	0.49	0.54
Er	0.913	1.2	1.8	0.93	1.5
Tm	—	—	—	—	0.24
Yb	0.845	1.0	1.6	1.27	1.4
Lu	0.140	—	—	0.19	0.28
ΣREE	77.2	97	84	101	114
La_N/Yb_N	≈13	12.2	7.2	11.7	12.1
Eu/Eu*	0.98	0.98	1.07	0.86	0.92
Y	—	17	12	—	14.8
Th	—	9.6	6.3	—	—
U	—	—	1.6	—	—
Zr	—	130	113	153	171
Hf	—	—	2.8	—	—
Nb	—	7	6	—	—
Th/U	—	—	3.9	—	—
La/Th	—	1.9	2.7	—	—
Cr	—	144	110	—	—
V	—	116	72	—	—
Sc	—	—	16	10	—
Ni	—	59	95	234	64
Co	—	—	30	—	—
Cu	—	46	680	—	—
Zn	—	72	—	—	—
Ga	—	20	19	—	—
La/Sc	—	—	1.1	2.2	—
Th/Sc	—	—	0.39	—	—
B	—	—	38	—	—

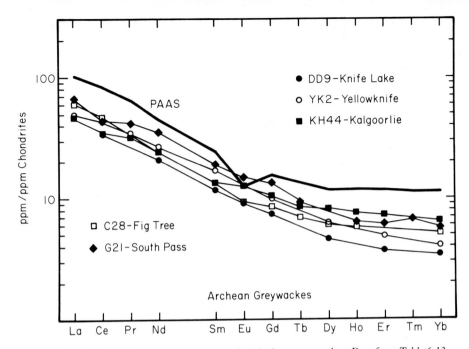

Fig. 6.6. Chondrite-normalized REE plot of selected Archean greywackes. Data from Table 6.13; PAAS plotted as heavy solid line for comparison. The greywacke patterns do not resemble PAAS. The patterns are all generally similar, having no substantial Eu-anomalies and fairly steep HREE patterns, even though they are derived from several different source terrains. Of the post-Archean greywackes, they resemble only the quartz-poor varieties in lacking negative Eu-anomalies, but differ even from these in the steepness of the REE patterns.

thus less than that measured in the rock by amounts varying between 0 and 40%, and these lower framework quartz values would suggest back-arc basin, leading-edge basin (e.g. Andean margins strike-slip margins) or dissected island-arc environments [8, 16].

In Archean greywackes, the plagioclase/total feldspar ratio is typically high. This is also characteristic of Phanerozoic quartz-poor greywackes and modern sands deposited in fore-arc basins but such analogies are inappropriate for Archean greywackes. The high plagioclase content in Phanerozoic quartz-poor greywackes and modern fore-arc sands reflect a dominant volcanogenic component of mainly andesitic origin. In other tectonic environments the K-rich granitic rock sources dilute plagioclase with K-feldspar, which results in a lower P/F ratio. During the early Archean in contrast, Na-rich granitic rocks (tonalite–trondhjemite suites) dominate over K-rich granitic rocks. K-rich varieties become abundant only in the late Archean, towards the end of greenstone belt development (see Chapter 9).

Table 6.13 (*footnotes*)

DD9: Knife Lake Group, Superior Province [56].
YK2: Walsh Fm., Yellowknife Supergroup, Slave Province [59].
KH44: Kalgoorlie District, Yilgarn Block ([31] and unpublished data).
C28: Belvue Road Fm., Fig Tree Group, South Africa; MgO not reported for this sample and formation average used [49, 60].
G21: South Pass greenstone belt, Wind River, Wyoming [3, 60].

There are also differences in the Archean greywacke rock fragment data compared to those of Phanerozoic greywackes. In the Archean, volcanic rock fragments constitute a significant and commonly a major proportion of the total rock fragment suite. Felsic volcanic rock fragments dominate. This contrasts with Phanerozoic greywackes, in which the volcanic fragments are almost exclusively andesitic (Fig. 6.7), reflecting the importance of arc environments as the source areas. In the quartz-poor Archean greywacke suite from the volcanogenic Vermilion District (Table 6.10), felsic volcanic fragments exceed mafic-intermediate volcanic fragments by a factor of about 20. Similarly, in the volcanogenic greywackes from the Gamitagama Lake greenstone belt, felsic volcanic fragments greatly outnumber intermediate volcanic fragments [47]. If Archean volcanogenic quartz-poor sedimentary rocks were deposited in island-arc settings, such arcs had very different compositions to modern examples.

6.5.2 Major element chemistry

There are significant differences between the major element composition of Archean greywackes and Phanerozoic greywackes and modern deep-sea sands.

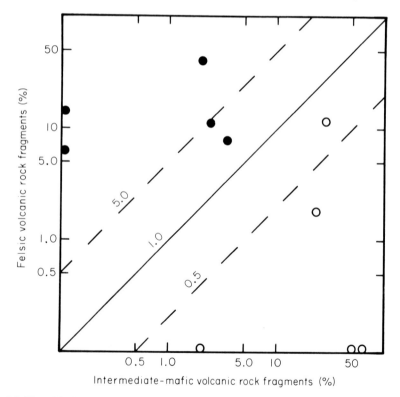

Fig. 6.7. Plot of felsic volcanic rock fragments versus intermediate-mafic volcanic rock fragments for Archean and Phanerozoic greywacke suites. Filled circles are Archean; open circles are Phanerozoic. Data from Tables 6.7 and 6.10. Phanerozoic greywackes are characterized by andesitic volcanic rock fragments whereas Archean greywackes are characterized by felsic volcanic rock fragments. (Adapted from [10].)

The Archean greywackes generally have intermediate abundances of K_2O and Na_2O and intermediate K_2O/Na_2O ratios which correspond to Phanerozoic quartz-intermediate greywackes (Fig. 6.8). In modern environments, such levels would indicate back-arc basin or leading-edge basin (Andean or strike-slip margins) environments. Some care is needed in simple tectonic interpretations based on K and Na abundances. Variations in Na_2O and K_2O in sandstones are controlled by feldspar distribution [53] and in modern deep-sea sands, the abundance of plagioclase of volcanogenic origin is a key index of tectonic setting [9]. As pointed out in the previous section, early Archean granitic rocks were generally Na-rich and accordingly the proportion of plagioclase and thus K_2O/Na_2O ratios on their own are not diagnostic of a volcanic origin. Another factor is the albitization (replacement by Na) of the Ca and K constituents of feldspars during diagenesis of sandstones [61].

Other differences also exist between Phanerozoic quartz-intermediate greywackes (Table 6.8) and Archean greywackes (Table 6.11). The Archean greywackes generally are about 5% lower in SiO_2 than the Phanerozoic

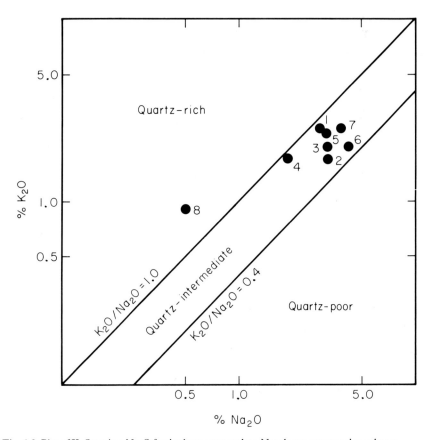

Fig. 6.8. Plot of K_2O against Na_2O for Archean greywackes. Numbers correspond to column headings in Table 6.11. Divisions used for classification apply to Phanerozoic greywackes and are approximate only [5]. In terms of these parameters, most Archean greywacke suites, including the volcanogenic greywackes from the Vermilion district, are similar to Phanerozoic quartz-intermediate greywackes. (Adapted from [10]).

quartz-intermediate varieties. In addition, the FeO+MgO content, which Maynard et al. [9] considered particularly diagnostic of tectonic environment (see Table 6.3), is much higher in Archean quartz-intermediate greywackes (generally >8% and commonly >10%) than in their Phanerozoic quartz-intermediate counterparts (<7%). These features suggest that mafic volcanics were important source rocks [10]. The lack of mafic volcanic rock fragments may be a result of degradation during matrix formation, due to their high calcium content and fine grain size.

A final observation is that even the Archean quartz-poor volcanogenic greywacke suite, from the Vermilion District, Minnesota (Table 6.11, column 1), has a major element composition indistinguishable from that of the other Archean greywacke suites. The major element composition of volcanogenic greywackes from the Gamitagoma Lake greenstone belt is similar to other quartz-intermediate Archean greywackes, reflecting the felsic nature of the volcanic sources (see previous section).

6.5.3 Trace element data

REE abundances of Archean sedimentary rocks resemble those Phanerozoic greywacke–shale turbidites and modern deep-sea turbidites deposited in fore-arc basins of immature island-arcs (i.e. quartz-poor greywackes; undissected magmatic arcs). Even this similarity is superficial because Archean sedimentary rocks show variable but higher La/Yb ratios. The steepest Archean sedimentary REE patterns show HREE depletion suggesting that garnet was an important fractionating phase in the igneous precursor (see Chapter 7). Steep REE patterns of this type are rare in modern orogenic andesitic rocks [62].

Of more fundamental significance is the fact that the petrography and major element chemistry of Archean greywackes are not comparable to Phanerozoic quartz-poor greywackes or recent sands deposited in fore-arc basins. The Archean suite compares more with quartz-intermediate and quartz-rich Phanerozoic greywackes and modern sands deposited at the edge of continents (trailing-edge, leading-edges, back-arc basins). If comparisons of provenance or tectonic setting between Archean and younger turbidites are to be made, it is appropriate to consider quartz-intermediate and quartz-rich varieties. As demonstrated above, the REE patterns of these sedimentary rocks reflect those of the present upper continental crust, particularly for quartz-rich greywackes (trailing-edge margins). The slightly higher Eu/Eu^* observed in quartz-intermediate varieties (leading-edge margins, back-arc basins) probably reflects a component of volcanogenic (andesitic) debris which would dilute the upper crustal REE signature. However, during the post-Archean, this does not appear to be a major factor and the upper crustal signature dominates [63].

6.5.4 Early and late Archean greywackes

The only early Archean greywacke–mudstone turbidite suite discussed previously was the Fig Tree Group. The greywackes from this sequence contain considerable amounts of K-feldspar, probably of granitic origin [42, 49]. The overlying sandstones from the Moodies Group are also rich in K-feldspar [42]. This difference is less apparent in the major element data. K_2O/Na_2O ratios from the Fig Tree Group, while high, are indistinguishable from those of many late Archean greywackes (Table 6.11; Fig. 6.8). It is hazardous to make generalizations based on one location, but it is interesting to note that extremely high K_2O values (up to 7.1%) were observed in shales from the early Archean Gorge Creek Group from the Pilbara Block ([58]; Table 6.12). If this also reflects the presence of K-feldspar in the source regions, then it is possible that the granitic source rocks of early Archean greywacke were more K-rich than the tonalite–trondhjemite suites which are the dominant granitic source for late Archean greywackes. The fact that the early Archean greywackes and mudstones do not possess significant negative Eu-anomalies (see Chapter 7) indicates the granitic source rocks were either not depleted in Eu, or that their contribution to the sedimentary rock REE pattern was minor.

6.5.5. Archean turbidites as upper crustal samples

We conclude from the above discussion that Archean greywacke–mudstone turbidite sequences do provide a useful sampling of the Archean upper crust. These sediments are derived from several source lithologies. Dominantly volcanogenic sources are found, particularly in late Archean sequences. In contrast to Phanerozoic volcanogenic greywackes which are derived mainly from andesitic sources, Archean volcanogenic greywackes are derived from felsic sources (mainly dacites). In this context, Archean felsic volcanics are geochemically indistinguishable from the Archean tonalite–trondhjemite suites (see Chapters 7 and 9). It is not easy to decide, on chemical grounds, between plutonic and volcanic provenances for Archean sedimentary rocks. This contrasts with the situation for the post-Archean crust, where volcanogenic provenances, typically andesitic, are readily distinguished from granitic provenances.

6.6 Summary

1 The petrographical and chemical compositions of Phanerozoic greywacke–mudstone turbidites and modern deep-sea sand-mud turbidites reflect the tectonic setting.
2 Only the volcanogenic greywackes and sands derived from immature island-arcs have REE patterns which are dissimilar to typical post-Archean upper crust. The REE patterns of these sediments are similar to their andesitic source rocks. Other greywackes and sands have REE patterns similar to the upper crust or intermediate between those of andesite and the upper crust.
3 The petrography, major element chemistry and provenance of Archean greywacke–mudstone turbidite sequences do not resemble those of Phanerozoic greywackes or modern deep-sea sands.
4 The provenance of Archean greywackes includes many lithologies such as felsic volcanic rocks, granite–granitic gneiss (of mainly tonalite–trondhjemite

composition), mafic volcanic rocks, recycled sedimentary rocks and minor intermediate volcanic rocks and metamorphic rocks.

5 Dominantly volcanogenic greywackes are common in the Archean, particularly in late Archean greenstone belts, but differ from volcanogenic Phanerozoic greywackes and deep-sea sands in being derived from felsic volcanic rather than andesitic debris. These felsic volcanic sources are chemically similar to the tonalite–trondhjemite granitic rocks and have a similar origin.

6 Archean greywacke–mudstone turbidites provide a reasonably representative sampling of the exposed part of the Archean upper crust.

Notes and references

1 The term 'greywacke' is probably the most controversial name in all petrological nomenclature. Introduced as 'grauewacke' by Werner in 1787 (see Crook, K.A.W. (1970) *Encycl. Brit.*) to describe Palaeozoic rocks in the Harz Mountains, it came into disrepute by 1818 ('Geologists differ much respecting what is, and what is not, Gray Wacce' (Mawe, J. (1818). *Catalog of Minerals*, London)), the same year it was introduced to the English language (Jameson, R. (1818) *In:* translation of M. Cuvier, *Theory of the Earth*. Kirk & Mercein). By 1854, Murchison asked, 'when will my valued friends, the mineralogists and geologists of Germany, abandon a word which has led to such endless confusion?' (quoted in Blatt, H. *et al.* (1980) *Origin of Sedimentary Rocks*. Prentice-Hall). A good introduction to the problem can be found in Dott, R.H. (1964) (*JSP*, **34**, 625). While being fully aware of the problems attending the use of the word greywacke, we have retained it (as have most students of sedimentary rocks!) because of its deep entrenchment. Our meaning is adapted from that of G. H. Packham ((1954) *AJS*, **252**, 466). Thus greywackes are taken to be arenaceous sedimentary rocks deposited by turbidity currents and possessing the characteristic sedimentary structures of the classic flysch profiles (see Crook (ref. [5]) for discussion). This definition is particularly well suited for this book as turbidites are the hallmark of Archean sedimentation.

2 Pettijohn, F.J. (1943) Archean sedimentation. *GSA Bull.*, **54**, 925; (1972) The Archean of the Canadian Shield: a résumé. *GSA Mem.*, **135**, 131.

3 Condie, K.C. (1967) Composition of the ancient North American crust. *Science*, **155**, 1013; (1967) Geochemistry of early Precambrian greywackes from Wyoming. *GCA*, **31**, 2135.

4 Condie, K.C. (1981) *Archean Greenstone Belts*. Elsevier.

5 Crook, K.A.W. (1974) Lithologenesis and geotectonics: the significance of compositional variation in flysch arenites (graywackes). *SEPM Spec. Pub.*, **19**, 304.

6 Schwab, F.L. (1975) Framework mineralogy and chemical composition of continental margin-type sandstone. *Geology*, **3**, 487.

7 Dickinson, W.R. & Valloni, R. (1980) Plate settings and provenance of sands in modern ocean basins. *Geology*, **8**, 82.

8 Valloni, R. & Maynard, J.B. (1981) Detrital modes of recent deep-sea sands and their relation to tectonic setting: a first approximation. *Sedimentology*, **28**, 75.

9 Maynard, J.B. *et al.* (1982) Composition of modern deep-sea sands from arc-related basins. *Geol. Soc. Lond. Spec. Pub.*, **10**, 551.

10 McLennan, S.M. (1984) Petrological characteristics of Archean greywackes. *JSP*, **54**, 889.

11 Whetten, J.T. & Hawkins, J.W. (1970) Diagenetic origin of graywacke matrix minerals. *Sedimentology*, **15**, 347.

12 Reading, H.G. (1982) Sedimentary basins and global tectonics. *Proc. Geol. Ass.*, **93**, 321.

13 There is considerable debate over the question of petrographical classification of sandstones in general and greywackes in particular. The two most important systems adopted may be termed the QFR and QFL systems. In both systems, classification considers only the framework quartz (Q), feldspar (F, with plagioclase, P; K-feldspar, K) and rock fragment (R or L) grains with matrix, heavy minerals and cement being ignored. In QFR, all polymineralic grains (including chert and quartz arenite) are counted as rock fragments (R). Polycrystalline and monocrystalline quartz grains (excluding chert and quartz arenite) are

counted as Q (see [5]). In the QFL system, only aphanitic lithic rock fragments are counted separately (L) with separate minerals being counted individually as quartz (Q) or feldspar (F) in coarser grained rock fragments. In this system, chert and quartz arenite are included in the Q pole. Several other subdivisions can also be employed in this system (see Dickinson, W.R. (1970) *JSP*, **40**, 695; Dickinson, W.R. & Suczek, C.A. (1979) *AAPG Bull.*, **63**, 2164; for details). In this book, we have adopted the QFR system even though there are many arguments in favour of the QFL system. We have done this for strictly practical reasons. Much of the published data, particularly for Archean greywackes, is not reported in a manner which allows for classification in the QFL system.

14 Wildeman, T.R. & Haskin, L.A. (1965) Rare earth elements in ocean sediments. *JGR*, **70**, 2905.
15 Addy, S.K. (1979) Rare earth element pattens in manganese nodules and micronodules from the northwest Atlantic. *GCA*, **43**, 1105.
16 Dickinson, W.R. (1970) Interpreting detrital modes of graywacke and arkose. *JSP*, **40**, 695; Graham, S.A. *et al.* (1976) Common provenance of lithic grains in Carboniferous sandstones from Ouachita Mountains and Black Warrior basin. *JSP*, **46**, 620; Dickinson, W.R. & Suczek, C.A. (1979) Plate tectonics and sandstone compositions. *AAPG Bull.*, **63**, 2164.
17 Dickinson, W.R. *et al.* (1983) Provenance of North American Phanerozoic sandstones in relation to tectonic setting. *GSA Bull.*, **94**, 222.
18 Chappell, B.W. (1968) Volcanic greywackes from the upper Devonian Baldwin Formation, Tamworth-Barraba district, New South Wales. *J. Geol. Soc. Austr.*, **15**, 87.
19 Boles, J.R. (1974) Structure, stratigraphy, and petrology of mainly Triassic rocks, Hokonui Hills, Southland, New Zealand. *NZJGG*, **17**, 337.
20 Condie, K.C. & Snansieng, S. (1971) Petrology and geochemistry of the Duzel (Ordovician) and Gazelle (Silurian) formations, Northern California. *JSP*, **41**, 741.
21 Dickinson, W.R. *et al.* (1982) Provenance of Franciscan greywackes in coastal California. *GSA Bull.*, **93**, 95.
22 Huckenholz, H.G. (1963) Mineral composition and texture in greywackes from the Harz Mountains (Germany) and in arkoses from Auvergne (France). *JSP*, **33**, 914.
23 Ondrick, C.W. & Griffiths, J.C. (1969) Frequency distribution of elements in Rensselaer Graywacke, New York. *GSA Bull.*, **80**, 509.
24 Laird, M.G. (1972) Sedimentology of the Greenland Group in the Paparoa Range, West Coast, South Island. *NZJGG*, **15**, 372.
25 Allen, J.R.L. (1960) The Mam Tor Sandstone: a 'turbidite' facies of the Namurian deltas of Derbyshire, England. *JSP*, **30**, 193.
26 Spears, D.A. & Amin, M.A. (1981) A mineralogical and geochemical study of turbidite sandstones and interbedded shales, Mam Tor, Derbyshire, U.K. *Clay Minerals*, **16**, 333.
27 Ricci, C.A. & Sabatini, G. (1976) An example of sedimentary differentiation in volcano-sedimentary series: the high chromium meta-graywacke of central Sardinia (Italy). *N. Jb. Miner.*, **7**, 307.
28 Nathan, S. (1976) Geochemistry of the Greenland Group (early Ordovician), New Zealand. *NZJGG*, **19**, 683.
29 Bhatia, M.R. (1983). Plate tectonics and geochemical composition of sandstones. *J. Geol.*, **91**, 611.
30 Some caution is warranted in interpreting differing K_2O/Na_2O ratios. Maynard *et al.* [9] observed higher K_2O/Na_2Oa ratios in both ancient greywackes and ancient shales. Early Palaeozoic sediments contain high K_2O contents, related to the ubiquitous presence of detrital K-feldspar. The cause of this high K_2O abundance in early Palaeozoic clastic sedimentary rocks is yet to be investigated.
31 Nance, W.B. & Taylor, S.R. (1977) Rare earth element patterns and crustal evolution—II. Archean sedimentary rocks from Kalgoorlie, Australia. *GCA*, **41**, 225.
32 Bhatia, M.R. (1981) *Petrology, Geochemistry and Tectonic Setting of some Flysch Deposits*. Ph.D. Thesis, Australian National University.
33 Windley, B.F. (1977) *The Evolving Continents*. Wiley.
34 Proponents of this idea have yet to address the position of the early Archean volcanic-sedimentary sequences preserved in high grade terrains (such as Isua) in the classification, if indeed such belts can be considered 'greenstone belts'.
35 Lowe, D.R. (1980) Archean sedimentation. *Ann. Rev. Earth Planet. Sci.*, **8**, 145.

36 Lowe, D.R. (1982) Comparative sedimentology of the principal volcanic sequences of Archean greenstone belts in South Africa, Western Australia and Canada: implications for crustal evolution. *PCR*, **17**, 1.
37 Walker, R.G. & Pettijohn, F.J. (1971) Archaean sedimentation: analysis of the Minnitaki Basin, Northwestern Ontario, Canada. *GSA Bull.*, **82**, 2099.
38 Turner, C.C. & Walker, R.G. (1973) Sedimentology, stratigraphy and crustal evolution of the Archean greenstone belt near Sioux Lookout, Ontario. *CJES*, **10**, 817.
39 Hyde, R.S. & Walker, R.G. (1977) Sedimentary environments and the evolution of the Archean greenstone belt in the Kirkland Lake area, Ontario. *GSC Paper*, **77-1A**, 185.
40 Walker, R.G. (1978) A critical appraisal of Archean basin-craton complexes. *CJES*, **15**, 1213.
41 Eriksson, K.A. (1980) Hydrodynamic and paleogeographic interpretation of turbidite deposits from the Archean Fig Tree Group of the Barberton Mountain Land, South Africa. *GSA Bull.*, **91**, 21.
42 Eriksson, K.A. (1980) Transitional sedimentation styles in the Moodies and Fig Tree Groups, Barberton Mountain Land, South Africa: evidence favouring an Archean continental margin. *PCR*, **12**, 141.
43 Eriksson, K.A. (1981) Archean platform-to-trough sedimentation, East Pilbara Block, Australia. *Geol. Soc. Austr. Spec. Pub.*, **7**, 235.
44 Hyde, R.S. (1980) Sedimentary facies in the Archean Timiskaming Group and their tectonic implications, Abitibi greenstone belt, northeastern Ontario, Canada. *PCR*, **12**, 161.
45 Eriksson, K.A. (1979) Marginal marine depositional processes from the Archaean Moodies Group, Barberton Mountain Land, South Africa: evidence and significance. *PCR*, **8**, 153.
46 Ojakangas, R.W. (1972) Archean volcanogenic graywackes of the Vermilion District, northeastern Minnesota. *GSA Bull.*, **83**, 429.
47 Ayres, L.D. (1983) Bimodal volcanism in Archean greenstone belts exemplified by greywacke composition, Lake Superior Park, Ontario. *CJES*, **20**, 1168.
48 Henderson, J.B. (1975) Sedimentology of the Archean Yellow-knife Supergroup at Yellow-knife, District of Mackenzie. *GSC Bull.*, 246.
49 Condie, K.C. et al. (1970) Petrology and geochemistry of early Precambrian graywackes from the Fig Tree Group, South Africa. *GSA Bull.*, **81**, 2759.
50 Reimer, T.O. (1972) Diagenetic reactions in early Precambrian graywackes of the Barberton Mountain Land (South Africa). *Sed. Geol.*, **7**, 263.
51 Naqvi, S.M. & Hussain, S.M. (1972) Petrochemistry of early Precambrian metasediments from the central part of the Chitaldrug schist belt, Mysore, India. *Chem. Geol.*, **10**, 109.
52 Donaldson, J.A. & Jackson, G.D. (1965) Archaean sedimentary rocks of the North Spirit Lake area, northwestern Ontario. *CJES*, **2**, 622.
53 Pettijohn, F.J. et al. (1972) *Sand and Sandstone*. Springer-Verlag.
54 Blatt, H. et al. (1980) *Origin of Sedimentary Rocks*. 2nd Ed. Prentice-Hall.
55 Grout, F.F. (1933) Contact metamorphism of the slates of Minnesota by granite and gabbro magmas. *GSA Bull.*, **44**, 989.
56 Arth, J.G. & Hanson, G.N. (1975) Geochemistry and origin of the early Precambrian crust of north-eastern Minnesota. *GCA*, **38**, 325.
57 McLennan, S.M. et al. (1983) Geochemical evolution of Archean shales from South Africa. I. The Swaziland and Pongola Supergroups. *PCR*, **22**, 93.
58 McLennan, S.M. et al. (1983) Geochemistry of Archean shales from the Pilbara Supergroup, Western Australia. *GCA*, **47**, 1211.
59 Jenner, G.A. et al. (1981) Geochemistry of the Archean Yellowknife Supergroup. *GCA*, **45**, 1111.
60 Wildeman, T.R. & Condie, K.C. (1973) Rare earths in Archean graywackes from Wyoming and from the Fig Tree Group, South Africa. *GCA*, **37**, 439.
61 Land, L.S. & Milliken, K.L. (1981) Feldspar diagenesis in the Frio Formation, Biazoria County, Texas Gulf Coast. *Geology*, **9**, 314; Ogunyomi, O. et al. (1981) Albite of secondary origin in Charny Sandstones, Quebec: a re-evaluation. *JSP*, **51**, 597.
62 Gill, J. (1981) *Orogenic Andesites and Plate Tectonics*. Springer-Verlag.
63 PAAS has $Eu/Eu^* = 0.66$; typical orogenic andesites have $Eu/Eu^* = 1.00$ (see Chapter 3) and quartz-intermediate greywackes from Table 6.9 have $Eu/Eu^* = 0.69$–0.75, much closer to PAAS.

7

The Archean Crust

7.1 The problems

The composition of the post-Archean crust has been addressed in Chapters 2 through 5, and its composition has been shown to be effectively uniform, with no compelling evidence of secular changes in composition in the upper crust over a period of about 2500 m.y. (Chapter 5). The consequences of this observation will be addressed in Chapter 9 on models for crustal evolution. In this chapter, we explore the relationship of the composition of Archean sedimentary rocks to that of the Archean upper crust and attempt to constrain the bulk composition of the preserved total Archean crust. The enquiry is a little more difficult than for Proterozoic and Phanerozoic rocks, on account of the restricted occurrences of the remnants of Archean crustal segments. A further question to be addressed is whether there are any geochemical differences between the crust exposed during the deposition of early Archean and late Archean greenstone belts. Amongst other problems, the thickness of the Archean crust and the inferences to be drawn from the presence of rocks which have undergone granulite facies metamorphism, must be examined.

7.2 The Archean record

Remnants of Archean rocks are exposed on all continents (Fig. 7.1). Three major lithological associations are evident:
1 'low-grade terrains' composed of the granite–greenstone association. There is a consensus that there were at least two major periods of greenstone belt development at 3.6–3.2 Ae and 2.8–2.6 Ae [1, 2, 3];
2 'high-grade terrains' including the early Archean 'grey gneiss' association and the generally younger 'paragneiss' belts;
3 'cratonic terrains' which include the late Archean sedimentary successions of the Kaapvaal Province, South Africa.

We are particularly concerned with understanding the geochemistry of potential source lithologies of Archean sedimentary rocks. The occurrence and characteristics of these rocks have been described in great detail in several recent books [1, 4–10] and only brief comments relevant to the theme of this book will be discussed here.

7.3 Low-grade terrains

Low-grade Archean terrains are typically composed of greenstone belts, preserved between circular to elongate granite–granitic gneiss batholiths. The

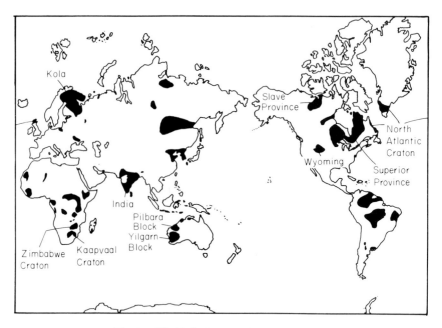

Fig. 7.1. World distribution of Archean terrains.

nature of Archean granitic rocks will be discussed in Chapter 9 and that of the greenstone belts will be emphasized here. Archean greenstone belts are composed of thick sequences of volcanic and sedimentary rocks preserved in typically elongate basins, generally 10–15 km in width and 100–300 km in length. Preserved or inferred thicknesses are usually in excess of 10–20 km although considerable thickening due to tectonic processes may have occurred [11, 12].

Early Archean greenstone belts are particularly well preserved in the Pilbara Block of Western Australia and in the Barberton Mountain Land, South Africa. Late Archean greenstone belts are found in most shield areas with those of the Yilgarn Block and Superior, Slave and Zimbabwe Provinces being particularly well studied. In older greenstone belts, a single thick volcanic sequence is overlain by a major clastic sedimentary succession whereas in younger greenstone belts, there commonly is more than one cycle of volcanic rocks overlain by clastic sediments [13]. In Fig. 7.2, schematic stratigraphic columns of two early Archean and two late Archean greenstone belt successions are shown.

7.3.1.   Volcanic rocks

The students of the sedimentary sequences have emphasized differences between early and late Archean greenstone belts. Whether there are contrasts in the volcanic associations is less clear. Anhaeusser [3] cautioned that no two greenstone belts are identical. For example, in greenstone belts of the Superior Province, Canada, the ratio of volcanic rocks to sedimentary rocks is about 4:1

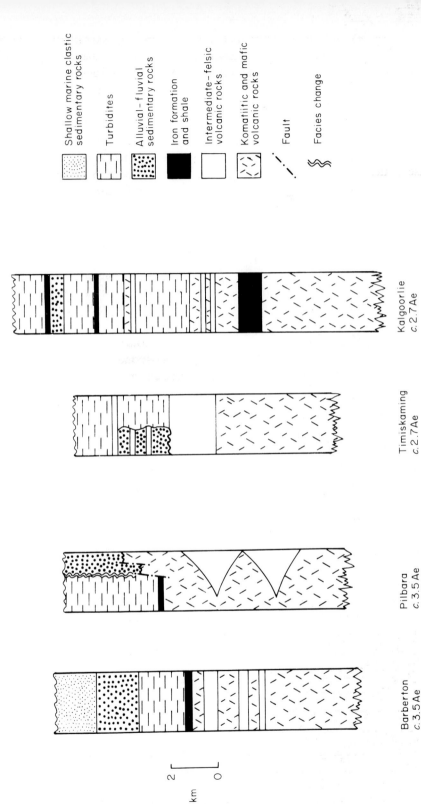

Fig. 7.2. Schematic stratigraphic columns for representative early and late Archean greenstone belt successions. (Adapted from [1] and [14].)

[15] whereas in the Slave Province the ratio is about 1:4 [16]. The relative abundance of major volcanic lithologies for some selected greenstone belt terrains are shown in Table 7.1. Some of the common features of the volcanic stratigraphy can be summarized as follows:

1 Volcanic rocks show an overall upward trend of dominantly mafic–ultramafic to mixed mafic–felsic composition. At the base, bimodal mafic–felsic volcanism, commonly with tholeiitic and/or komatiitic major-element chemistry predominates. The overlying rocks show mafic to felsic cycles. In some cases, the bimodal nature of the volcanism is only apparent from the occurrence of felsic volcaniclastic sedimentary rocks [22].

2 Ultrabasic volcanics (e.g. komatiites), if present, are more abundant in the basal levels. Andesitic rocks are more common in the upper levels.

3 The occurrence of volcanic rocks of intermediate composition is highly variable. Often they are missing from the volcanic sequences (Table 7.1). They never appear to be the dominant volcanic lithology. There is some evidence that they are found in abundance only in late Archean greenstone belts.

Table 7.1. Lithological proportions of volcanic rocks in some Archean greenstone belts [17–21].

	Australia		Canada		Southern Africa	
	Pilbara Block	Yilgarn Block	Superior Province	Slave Province	Barberton	Zimbabwe Craton
Ultramafic–mafic	4.3	20	0.7	—	24.3	9.8
Mafic	59.3	62	54.0	49	72.0	60.2
Andesite	7.8	5	31.4	35*	—	22.5
Felsic	28.6	13	13.9	16	3.7	7.5

*Comprises about 80% basaltic andesite.

4 There is an increase upwards in the ratio of volcaniclastic to volcanic material reflecting the overall mafic to felsic trend of the volcanic rocks.

5 There is an upward trend from volcanic to sedimentary rocks.

The geochemistry and isotopic characteristics of Archean volcanic rocks have been reviewed by a number of workers [1, 23, 24, 25]. Significant element mobility during hydrothermal alteration and to a lesser extent, low grade metamorphism affects the distribution of alkali and alkaline earth elements, many transition metals, silica, water and the oxidation state. In an important study, MacGeehan & MacLean [26] showed that hydrothermal alteration of a bimodal basalt–rhyolite tholeiitic suite resulted in transformation to a suite with apparent calc-alkaline chemistry. Because of their generally immobile nature, the REE have become particularly important in deciphering the original composition of Archean volcanics.

In Fig. 7.3 REE patterns for several typical Archean peridotitic and basaltic komatiites are shown. REE patterns are generally flat, commonly with slight LREE depletion which is less than that seen in typical MORB. This may indicate a somewhat less depleted upper mantle than the present day MORB source. Such differences may indicate a secular evolution of the upper mantle composition related to continuing crustal extraction. In a given

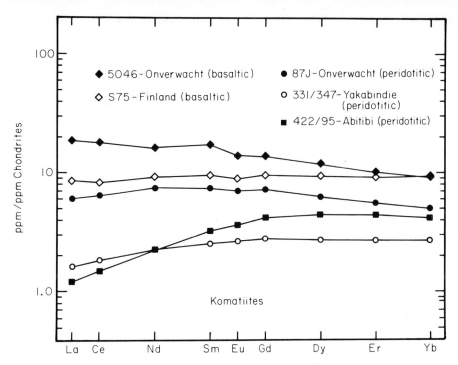

Fig. 7.3. Chondrite-normalized REE patterns for typical Archean peridotitic and basaltic komatiites [27]. The patterns are flat or show slight LREE depletion, consistent with derivation from a depleted mantle source.

greenstone belt, total REE abundances correlate inversely with MgO content, with basaltic komatiites (MgO=9–18%) at 6–20 times chondrite and peridotitic komatiites (MgO>28%) at 2–4 times chondrite. Slight Eu-anomalies, commonly observed, may not be related to plagioclase fractionation but rather to alteration [24].

In the absence of komatiitic basalts, low-K tholeiites normally predominate. Some typical REE patterns are shown in Fig. 7.4. Less commonly, and typically only at the top of greenstone belt successions, basaltic rocks with more fractionated REE patterns ($La_N/Yb_N=4$) and higher abundances may be found [1].

The petrography and geochemistry of Archean andesites was reviewed recently by Condie [1, 19]. On the basis of REE geochemistry, he has divided them into 3 groups (Fig. 7.5):
1 Type I—with slight REE enrichment similar to modern calc-alkaline andesites.
2 Type II—with greater LREE enrichment than Type I and generally similar to modern high-K andesites.
3 Type III—exhibiting flat REE patterns with Eu-depletion, most comparable to some oceanic island-arc andesites. This type appears to be restricted to the Abitibi belt, Canada.

Enough differences exist between Archean and modern andesitic rocks to

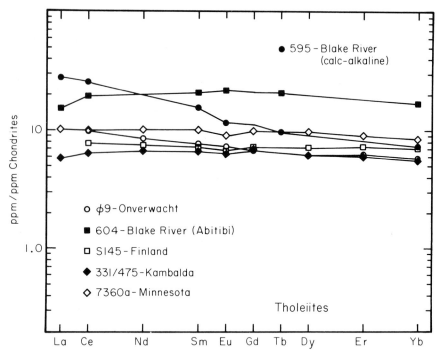

Fig. 7.4. Chondrite-normalized REE patterns for Archean tholeiitic basalts [27]. Note the generally flat patterns, generally without substantial enrichment or depletion in LREE. These represent typical patterns for the basaltic end-member of the Archean bimodal igneous suite. The steeper pattern is from an Archean 'calc-alkaline' volcanic rock with 53% SiO_2, from the top of a greenstone belt sequence.

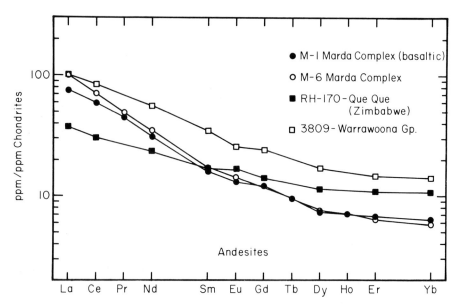

Fig. 7.5. Chondrite-normalized REE patterns for Archean andesites [27]. See text for discussion.

warrant extreme caution in making direct analogies. For example, some Archean andesites show severe HREE depletion, indicative of garnet in the source regions. Such patterns are virtually unknown in modern examples [28].

Felsic volcanics occur as part of bimodal volcanic suites and as an end member of calc-alkaline mafic to felsic suites [1]. Pyroclastic varieties are most

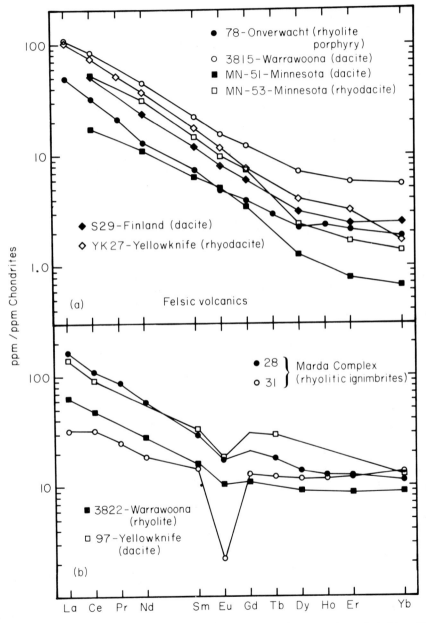

Fig. 7.6. (a) Chondrite-normalized REE patterns for Archean felsic igneous rocks, typical of the acidic end-member of the Archean bimodal basaltic–felsic suite of igneous rocks [27]. Note the severe depletion in HREE and the absence of Eu-anomalies. (b) Rare examples of Archean felsic volcanic rocks showing depletion in Eu [27].

common. Such rocks typically show very fractionated REE patterns with severe HREE depletion [1, 24]; typical examples are shown in Fig. 7.6(a). The HREE depletion cannot be derived from fractional crystallization of basalt and indicates a separate origin through partial melting of basaltic rocks at mantle depths where garnet is a stable residual phase [24].

Less commonly, Archean felsic volcanics lack HREE depletion and possess negative Eu-anomalies, as in the Marda Complex of Western Australia. In this example, the felsic volcanics are probably related to the more mafic rocks and derived through higher degrees of partial melting at crustal depths [29].

7.3.2   Sr and Nd isotopes in Archean volcanics

Initial $^{87}Sr/^{86}Sr$ ratios in Archean volcanic sequences are generally low but scattered between 0.700 and 0.702 [24]. The majority of the data are consistent with derivation from an undepleted upper mantle with Rb/Sr in the range 0.026–0.034. Significant variations of $(^{87}Sr/^{86}Sr)_I$ in Archean volcanic rocks of similar age are thought to be related to any of three processes including:
(a) mantle isotopic heterogeneity
(b) metamorphic resetting
(c) sea water alteration

Nd-isotopic data provide more persuasive evidence of Archean mantle heterogeneity since isotopic resetting is considerably less likely than with the

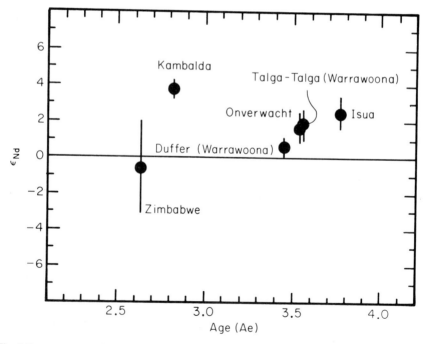

Fig. 7.7. ε_{Nd} versus geological age diagram for Archean volcanic rocks. Some sequences lie close to the chondritic evolution line ($\varepsilon_{Nd}=0$), implying derivation from undepleted mantle sources. Many other sequences have positive ε_{Nd} values, implying derivation from mantle sources depleted in LREE on long-time scales. Based on compilation given in [30].

Rb–Sr system. Some of the available data are plotted on an ε_{Nd} versus age diagram in Fig. 7.7. The initial $^{143}Nd/^{144}Nd$ ratios (ε_{Nd}) of a number of sequences fall on or near $\varepsilon_{Nd}=0$, suggesting derivation from an undepleted (i.e. chondritic REE) mantle source. However, many sequences display low $(^{143}Nd/^{144}Nd)_I$ with ε_{Nd} ranging up to about +3.5 [30] suggesting derivation from a long-lived LREE-depleted mantle source. Such isotopic heterogeneities date back to at least 3.75 Ae with the Isua volcanics [31, 32].

The isotopic evidence from Archean volcanic rocks is consistent with the presence of significant regions in the Archean mantle with long-lived depletions in LREE and Rb relative to Sr. Such heterogeneities indicate the extraction of continental material as far back as 3.75 Ae.

7.3.3 Sedimentary rocks

Sedimentary rocks are preserved in two major settings in Archean greenstone belts [13]:
1 volcaniclastic and chemical sediments interbedded with volcanic rocks in the lower volcanic portion of greenstone belts;
2 thick clastic sedimentary deposits above the volcanic sequences of greenstone belts.

Lowe [2] has also pointed out possible differences in the sedimentary character of early Archean greenstone belts (such as those from the Pilbara Block and Barberton Mountain Land) and late Archean greenstone belts (for example, those from the Superior and Slave Provinces and Yilgarn Block). Sediments deposited in early Archean greenstone belts (Onverwacht Group, Barberton Mountain Land; Warrawoona Group, Pilbara) include silicified volcaniclastic sediments, cherts, silicified carbonates and evaporites, and local development of stromatolites. Significant amounts of terrigenous sediments and iron-formations are missing. Much of this material appears to have been deposited in shallow water without significant tectonic activity. The sediments show no sign of continental (sialic) material. The distinction between the volcanic and sedimentary sequences is less clear in late Archean greenstone belts because more than one volcanic–sedimentary cycle often is preserved. In late Archean greenstone belts, sedimentary rocks interbedded with the volcanic rocks are usually unsilicified, comprising volcaniclastic and terrigenous clastic sediments and iron-formations. These rocks probably were deposited in fairly deep water during periods of tectonic activity with abundant evidence for a sialic basement.

The thick sedimentary sequences of Archean greenstone belts are found at the top of volcanic cycles and are typically 1000–5000 m in thickness. An outstanding feature of all these sequences, regardless of age, is the widespread evidence for a terrigenous component in the clastic sediments and the rapid facies changes from fluvial–alluvial sandstones to deep water turbidites. Development of substantial shelf environments occurs (e.g. Moodies Group), but is not common [13, 33, 34].

Early Archean sedimentary sequences from the Barberton Mountain Land

and the Pilbara have been studied by Eriksson [33, 35–41]. The volcanic sequences are overlain by a laterally persistent iron-formation–shale unit, interpreted to represent basin deepening, since the underlying volcanic associations are thought to be of shallow water origin. The iron-formation, in turn, is overlain by thick sequences of aerially extensive dominantly terrigenous clastic turbidite deposits. At Barberton, the Fig Tree Group (turbidites) grades vertically into fluvial–alluvial and shallow marine deposits of the Moodies Group. Volcaniclastic sediments are abundant at the base of the Fig Tree, but become less common with stratigraphic height. In the Pilbara, there is an abrupt lateral facies change between turbidites and fluvial–alluvial sediments, with no evidence of a shallow marine facies. These sediments are now considered to have been deposited at a tectonically active continental margin.

Late Archean sedimentary sequences may be preserved between major volcanic cycles as well as at the top of greenstone belt stratigraphies. The characteristic feature of late Archean sedimentation is the development of thick turbidite greywacke–shale successions, with lesser amounts of arkoses, conglomerates and chemical sediments. Rapid facies changes from terrestrial fluvial and alluvial fan deposits to the deeper water turbidites are common and there is little sedimentological evidence to suggest the development of extensive continental shelves. Iron formations accumulated during periods of non-sedimentation of clastic debris. The turbidites range from almost entirely volcaniclastic [22, 42], with felsic volcanic sources dominating, to almost entirely terrigenous [e.g. 43], with sialic and recycled sources dominant. The petrographic and geochemical evidence relating to the origin of early and late Archean greywacke–shale turbidite sequences were described in the previous chapter.

7.3.4 The basement of greenstone belts

A major controversy, concerning greenstone belt development, is about the nature of the basement rocks on which the volcanic–sedimentary successions were deposited. Two contrasting theories are in vogue:
1 the supracrustal rocks were deposited on a basement of granitic and gneissic material;
2 they were deposited on basaltic or ultramafic basement (probably oceanic crust).

It is again instructive to consider early and late Archean greenstone belts separately. The basement to the early Archean greenstone terrains has been thought to be either entirely sialic [e.g. 35, 44, 45] or entirely mafic [46, 47]. There is no evidence for a significant terrigenous (sialic) contribution to the sediments preserved in the lower volcanic sequences of early Archean greenstone belts; this has been cited as evidence that sialic basement was absent [2]. Other geological evidence, such as the predominance of shallow water deposition for volcanic sequences and detailed examination of greenstone–granitic gneiss relationships could indicate sialic basement. In the

Barberton Mountain Land, the Ancient Gneiss Complex has been proposed as the sialic basement on which the Swaziland Supergroup was deposited [33, 44]. However, recent Sm–Nd isotopic data on the lower Onverwacht Group [48] and the Ancient Gneiss Complex [49] indicate ages of 3530±50 Ae and a maximum of 3450±40 Ae, respectively. Thus, the Ancient Gneiss Complex is probably a little younger than the lowermost part of the Swaziland Supergroup. In contrast, sialic material played an important role in the source of sediments from the upper part of the Swaziland and Pilbara Supergroups [40]. For the late Archean greenstone belts, the evidence is more certain. Geological evidence for unconformities, isotopic age data, and sedimentological data, suggests that these greenstone belts were deposited on or at least in very close proximity to older sialic basement [50, 51, 52].

7.4 High-grade terrains

The metamorphosed gneissic terrains of the Archean have received less attention because of their structural complexity and high metamorphic grades (commonly granulite facies). The best studied areas include the West Greenland and Labrador gneisses of the North Atlantic Craton [53, 54], Lewisian–Scourian terrains of north-west Scotland [55, 56], the Limpopo belt of southern Africa [10], the Western Gneiss Terrain of the Yilgarn Block [50, 57] and the paragneiss belts of the Superior Province, Canada [58]. Structural, geochemical and petrological evidence point to depths of burial corresponding to at least 20–50 km and considerable crustal thickening due to horizontal tectonic movement [59].

Table 7.2. Simplified Archean stratigraphy of the North Atlantic Craton [after 54, 60].

Greenland	Age	Labrador
Intrusion of post-tectonic K-rich granite (Qôrqut granite)	c. 2.53 Ae	Intrusion of post-tectonic, K-rich granites (Igukshuk granite)
Deformation and granulite-facies metamorphism (minor granitic intrusions)	c. 2.9–2.7 Ae	High-grade metamorphism and formation of Kiyuktok gneisses
Thrusting, syntectonic intrusion of tonalites, trondhjemites, granodiorites (Nuk gneisses), polyphase deformation	c. 3.1–2.7 Ae	Intrusion of tonalite–granodiorite sheets and plutons (Ikarut, Kammarsuit gneisses)
Intrusion of layered gabbroic and anorthositic bodies		Intrusion of layered gabbroic and anorthositic bodies
Deposition of Malene volcanics and sediments	>3.05 Ae	Deposition of Upernavik volcanics and sediments
Intrusion of Ameralik mafic dykes	c. 3.4–3.2 Ae	Intrusion of Saglek mafic dykes
Intrusion of Fe-diorites, quartz monzonites; deformation and metamorphism up to granulite facies	c. 3.6–3.5 Ae	Intrusion of Fe-diorites; quartz monxonites; deformation and metamorphism up to granulite facies
Intrusion of tonalite–trondhjemite–granodiorite parents of the Amîtsoq grey gneisses	c. 3.75–3.6 Ae	Intrusion of tonalite–trondhjemite–granodiorite parents of the Uivak gneisses
Deposition of Isua and Akilia volcanics and sediments	c. 3.75 Ae	Deposition of Nulliak volcanics and sediments

The dominant lithologies in Archean high-grade terrains are quarto-feldspathic tonalite–trondhjemite–granodiorite gneisses and metamorphosed volcanic–sedimentary rocks composed of metavolcanic amphibolites (including dykes and sills), and metasedimentary rocks (including marbles and quartzites). Layered complexes of peridotite–gabbro–anorthosite and gabbro–anorthosite are also common. Lithological relationships are complex and uncertain, but in West Greenland and Labrador a detailed stratigraphic framework has been established (Table 7.2). Metavolcanic and metasedimentary rocks (Akilia–Nulliak assemblages) comprise the oldest preserved material.

The Archean bimodal suite of high-grade terrains comprises mafic volcanics (now amphibolites) and sodium-rich igneous rocks, including tonalite–trondhjemite granitic gneisses and dacites [61, 62]. Intermediate compositions are scarce and the calc-alkaline trends observed do not correspond to those found at young continental margins [61]. The geochemistry and isotopic composition of the granitic rocks will be discussed in Chapter 9. The volcanic and sedimentary rocks have received much less attention. Where original volcanic lithologies can be recognized [63, 64], trace element patterns are similar to volcanic rocks of equivalent major element composition in greenstone belts.

7.4.1   Relationships between high- and low-grade terrains

Archean high-grade and low-grade terrains of similar age are commonly juxtaposed, separated by faults or thrusts, and in some cases by steep metamorphic gradients. This has resulted in a wealth of geological debate. A popular view that the differences in metamorphic grade are related to differential uplift of essentially similar material can be discounted because the sedimentary facies are distinctly different. Low-grade terrains are characterized by greywackes whereas high-grade terrains typically contain quartzites, marbles and pelitic schists as the dominant sedimentary lithologies. This important difference in sedimentary facies effectively eliminates models which suggest that high-grade terrains are deeply eroded equivalents to greenstone belts. The best explanation is that these terrains represent distinct tectonic environments which underwent differing thermal histories [65, 66].

7.5   Cratonic terrains

Such sequences are well preserved in the late Archean of the Kaapvaal Craton of southern Africa. These include the Pongola Supergroup (*c.* 3.1 Ae), Dominion Group (*c.* 2.8 Ae) and Witwatersrand Supergroup (*c.* 2.7–2.6 Ae). These sequences closely resemble the flat-lying early Proterozoic sedimentary rocks common in many parts of the world. It is one conclusion of our study that such sequences are deposited following the major episodic evolution of the upper crust during the period 3.2–2.5 Ae. In the next chapter, we discuss the evidence from the REE patterns in these sediments. From this we infer that the intrusion of K-rich granites into the upper crust in southern Africa, effectively completing the major evolution of the crust in that area, occurred before the deposition of the Pongola Supergroup.

7.6 Tectonic models

Tectonic models to explain the distribution of Archean rocks are many and appeal to processes ranging from catastrophic to strictly uniformitarian. It is naive to expect that the development of Archean greenstone belts can be explained by a single tectonic model. In this respect, the cautionary comments of Anhaeusser [3] are noteworthy. He pointed out that the widely divergent models of Archean crustal evolution indicate that many tectonic settings have been preserved or, alternatively, that there are insufficient data to constrain speculation.

Some of the most elegant models of greenstone belt development are the least plausible. For example, one model suggested that Archean greenstone belts were analogous to lunar mare basins, with early magmatism being triggered by the impact [67]. Apart from being based on the false assumption that the mare basalts filling lunar multi-ringed basins are related to the formation of the basin itself (see Taylor [68] for a discussion), most greenstone belts post-date the decline in the massive meteorite flux at about 3.8–3.9 Ga. None of the expected shock metamorphic features have been found.

Condie [1, 69] has reviewed the tectonic setting of greenstone belts: the models fall generally into six classifications:
1 impact models (see above);
2 density-inversion models;
3 oceanic crust models;
4 hot-spot models;
5 convergent plate boundary models (as either fore-arc, back-arc or continental margin basins);
6 continental rifting models.

To some of these can be added the variation of plate tectonic or non-plate tectonic mechanisms.

Most models have relied heavily on evidence from structure, metamorphism and igneous petrology. Tectonic associations based on igneous petrology and geochemistry can give the wrong answer. For example, the Archean Marda Complex of Western Australia [29, 70] was not formed in an island-arc setting despite the similarity in chemistry with modern high-K calc-alkaline rocks. Until quite recently, evidence from sedimentary studies has been largely ignored although such evidence has been very useful in unravelling the tectonic history of Phanerozoic regions [e.g. 71]. We consider models which integrate sedimentological data with other geological and petrological evidence to be the most useful.

In the late Archean Slave Province, the greenstone belts developed as graben-like structures in response to intracratonic rifting of sialic continental blocks [72, 73]. A similar model has been proposed for the Yilgarn Block [50].

A thorough synthesis of evidence relating to development of the late Archean southern Abitibi Belt, Canada, suggests it developed as part of an immature island-arc [74]. A major problem with the island-arc model for greenstone belts is the general lack of andesitic rocks. Detailed studies of the early Archean Fig Tree, Moodies and Gorge Creek Groups of South Africa and the Pilbara Block reveal that these rocks are apparently synorogenic and were deposited at an evolving continental margin with little evidence for sediment contribution from an island-arc [33, 35–41]. Less attention has been given to volcanic and sedimentary rocks preserved in high-grade terrains. At Isua, the sedimentary rocks appear to have been deposited at a site remote from sialic crust [75] whereas part of the younger Malene supracrustals appear to lie on Amîtsoq Gneiss basement and were in part derived from it.

The origin of high-grade terrains has received less attention. Some models suggest they form at the site of upwelling mantle convection currents [1, 69]; others that they form at the site of mantle downwelling [76]. Possibly such associations are similar to exposed root zones of continental-arc margins such as those in California and the Andes [65].

7.7 Crustal thickness

The large depletion in europium in late Archaean K-rich granites (Chapter 9), and which is characteristic of the post-Archaean upper crust (Chapter 2), is explained here as due to partial melting in the lower crust, leaving Ca-feldspar as a residual phase. The thickness of the late Archaean crust is thus a critical factor since partial melting is unlikely to occur in a thin crust. Present concepts are highly polarized, estimates ranging from very thin (<15 km; [77, 78]) to thick (25–30 km; [79]). The evidence from metamorphic and experimental studies [79, 80, 81, 82] strongly suggests that by about 2.9 Ae and possibly earlier, geothermal gradients had decreased sufficiently to permit crustal thicknesses of up to 40 km or more.

Evidence from laboratory studies have generally indicated a relatively thick Archean crust, at least in the late Archean. The development of charnockites at high temperatures, but low pressures, is not supported by the experimental data [81]. Minimum thicknesses of at least 30 km for the late Archean crust are demanded by geobarometry measurements in charnockites from Southern India and data from many localities indicate crustal thicknesses of 20–40 km [82, 83, 84]. Data from nearly 40 locations [83] do not support interpretations that Archean metamorphism was the intermediate P-T type [79] or of a low-pressure type [85]. Archean localities such as the Scourian gneisses of Harris and Lewis retain evidence of metamorphic pressures of 6–12 kbar or depths of 20–40 km [86].

The geological and geochemical evidence is thus consistent with the presence of thick crust in the late Archean. This permits the formation of K-rich granites by intracrustal melting, in accordance with the isotopic and experimental evidence (see Chapter 9). The record of the sedimentary rocks themselves and the presence of 'granitic' detritus from distant sources at

Kambalda, for example, indicate that substantial areas of crust were above sea-level. Temperature gradients inferred in the Archean continental crust from metamorphic conditions do not indicate higher thermal gradients [87]. Accordingly, the additional heat loss due to higher Archean heat production probably occurred through the ocean floors, implying accelerated plate production and consumption [88, 89].

There is no consensus on the thickness of the Archean oceanic crust. Recently, it has been argued that the Archean oceanic crust may have been dominated by komatiitic rocks [88, 89]. This model has led both to suggestions that the oceanic crust was substantially thinner [88, 90] and slightly thicker [89] than modern ocean crust.

7.8 Regional geochemistry

In earlier chapters, we used sedimentary rock trace element data, particularly

Table 7.3. Provenance characteristics of selected Archean sedimentary sequences [91].

Sequence	Approx. age (Ae)	Suggested provenance
1. Gorge Creek Group (Western Australia)	3.4	Not documented in detail. Sandstone petrography indicates complex source of granitic rocks, felsic–mafic volcanics, chert, iron formation and recycled quartz arenite.
2. Fig Tree Group (South Africa)	3.4	Varies with geography and stratigraphic height. Dominant lithologies include felsic and mafic (to ultramafic) volcanics, granitic rocks, chert and metasedimentary rocks. The proportion of granitic detritus increases significantly towards the top.
3. Moodies Group (South Africa)	3.4	Similar to the upper part of the Fig Tree Group but with a greater proportion of granitic material.
4. Kalgoorlie (Yilgarn Block) (Western Australia)	2.8–2.7	Dominantly from variable mixtures of felsic–mafic volcanics. Associated Na-rich granitic rocks could also be an important component.
5. Kambalda (Yilgarn Block) (Western Australia)	2.8–2.7	Mixture of mafic volcanics and Na-rich granitic rocks. Role of felsic volcanics not fully assessed
6. Yellowknife Supergroup (Canada)	2.7–2.6	Widespread turbidites (Burwash–Walsh Formations) derived from felsic volcanics granite–granite gneiss and mafic volcanics. Restricted alluvial facies (Jackson Lake Formation) derived mainly from felsic volcanics with variable mafic volcanic component.
7. South Pass Greenstone Belt	2.7–2.6	Recycled quartz-rich metasedimentary rocks and granitic rocks.
8. Isua supracrustals–Akilia association (West Greenland)	3.8	Not thoroughly studied. Dominantly felsic volcanic with variable admixture of mafic rocks
9. Malene Supracrustals (West Greenland)	>3.05	Not well documented. Probably complex in detail. Field evidence suggests at least partly laid down on Amîtsoq gneiss basement.

for the REE, Th and Sc, to place constraints on upper crustal compositions during the post-Archean. In this section, we will examine the Archean sedimentary record and attempt to relate it to the nature and composition of the Archean crust. This question was recently reviewed by McLennan & Taylor [91]. Some individual Archean sequences, where detailed trace element work has been carried out, will be examined first, followed by a discussion of the general features of Archean clastic sedimentary rocks. Table 7.3 lists the inferred provenance characteristics of the Archean sedimentary sequences which have been studied. In Tables 7.4–7.6, typical geochemical analyses of the areas under consideration are listed.

Table 7.4. Chemical composition of selected shale samples from early Archean greenstone belts.

	Pg2	Pg6	C-3	79NC118	79NC124	79NC131
SiO_2	59.94	63.37	62.7	63.64	64.02	60.77
TiO_2	0.48	0.61	0.51	0.56	0.41	0.47
Al_2O_3	26.91	20.91	11.5	14.11	14.32	16.46
FeO	4.03	6.38	14.5	10.35	6.47	8.03
MnO	0.012	0.027	0.18	0.089	0.150	0.16
MgO	2.85	4.50	—	7.63	5.35	5.55
CaO	0.011	0.012	2.43	0.52	2.07	0.88
Na_2O	0.42	0.33	0.31	1.18	2.23	1.66
K_2O	5.34	3.86	0.81	1.92	4.98	6.01
Σ	99.99	100.00	92.9	100.00	100.00	99.99
LOI	4.96	4.91	—	7.52	7.18	8.61
Cs	5.54	3.47	—	3.44	4.03	6.55
Ba	1095	998	516	246	547	458
Rb	—	—	28	—	—	—
Sr	41.1	43.6	85	—	—	—
Pb	21.4	18.8	—	2.62	9.67	5.86
La	32.5	36.2	4.6	16.3	16.7	26.7
Ce	65.7	71.5	10.9	37.5	35.7	54.8
Pr	8.33	8.42	—	4.20	3.77	6.38
Nd	29.9	29.9	5.5	16.8	14.7	24.3
Sm	5.62	6.00	1.39	3.28	2.73	4.64
Eu	1.43	1.45	0.44	0.88	0.73	1.30
Gd	3.51	4.13	1.5	2.86	2.14	3.94
Tb	0.61	0.70	0.27	0.48	0.37	0.61
Dy	3.52	4.03	1.73	3.19	2.18	3.53
Ho	0.81	0.90	0.40	0.78	0.50	0.79
Er	2.41	2.70	1.2	2.07	1.57	2.22
Yb	2.51	1.95	1.09	2.00	1.38	2.13
ΣREE	157.6	168.5	30.7	90.9	82.9	132.0
La_N/Yb_N	8.8	12.5	2.9	5.5	8.2	8.5
Eu/Eu^*	0.98	0.89	0.93	0.88	0.92	0.93
Y	28.5	31.3	—	21.3	20.7	23.1
Th	9.98	10.1	—	4.21	5.18	7.21
U	2.01	2.23	—	1.20	1.54	2.31
Zr	404	221	34	121	131	175
Hf	4.79	4.29	—	2.67	3.10	5.52
Sn	30.8	9.37	—	3.35	4.17	5.04
Nb	14.7	13.1	—	9.83	10.1	10.7
Mo	1.23	1.02	—	1.26	0.99	1.46
W	3.94	3.07	—	1.30	1.13	3.81

Table 7.4 (continued)

	Pg2	Pg6	C-3	79NC118	79NC124	79NC131
Th/U	5.0	4.5	—	3.5	3.4	3.1
La/Th	3.3	3.6	—	3.9	3.2	3.7
Cr	625	788	—	854	577	855
V	125	169	—	144	91	109
Sc	15	19	28	22	15	15
Ni	239	500	958	555	287	366
Co	21	36	—	38	36	36
Cu	67	81	—	78	3.3	25
Ga	30	12	—	16	13	14
La/Sc	2.2	1.9	0.16	0.74	1.1	1.8
Th/Sc	0.67	0.53	—	0.19	0.35	0.48
Bi	0.29	0.24	—	0.04	0.13	0.06
B	83	77	—	33	31	87

Pg2, Pg6: Gorge Creek Group, Pilbara Block [92].
C3, 79NC118: Fig Tree Group, Barberton Mountain Land [93, 94, 95].
79NC124, 79NC131: Moodies Group, Barberton Mountain Land [93].

Table 7.5. Chemical composition of selected sedimentary rocks from late Archean greenstone belts.

	KH38	KH47	KH21	8781	481	YK17	YK28	YK1	YK7	WC1
SiO_2	64.16	66.76	69.60	59.88	60.14	69.72	65.55	67.52	63.80	61.68
TiO_2	0.47	0.82	0.83	0.30	0.50	1.07	0.44	0.70	0.75	0.75
Al_2O_3	13.10	19.16	23.08	7.08	13.19	18.04	10.94	16.85	18.09	25.03
FeO	17.03	1.41	0.73	15.92	10.82	9.26	8.44	6.05	7.64	5.07
MnO	0.02	0.01	0.01	0.17	0.04	0.04	0.16	0.05	0.06	0.068
MgO	1.45	0.63	0.49	1.91	4.85	0.49	3.83	3.01	3.58	2.81
CaO	0.59	1.46	0.06	6.85	1.66	0.11	7.34	0.54	1.01	0.54
Na_2O	0.52	2.65	0.67	0.81	0.31	0.24	2.14	3.30	2.62	1.68
K_2O	2.57	7.08	4.47	1.41	2.82	0.99	1.06	1.94	2.36	2.37
P_2O_5	0.08	0.02	0.05	0.06	0.14	0.04	0.10	0.04	0.07	—
S	—	—	—	5.62	5.52	—	—	—	—	—
Σ	99.99	100.00	99.99	100.01	99.99	100.00	100.00	100.00	99.98	100.00
LOI	14.64	7.00	3.31	7.17	4.93	2.53	11.29	3.07	3.08	4.42
Cs	4.7	1.5	8.9	—	—	—	—	—	—	2.45
Ba	540	900	1200	180	719	172	393	595	689	647
Rb	61	105	—	40	39	37	27	55	70	—
Sr	51	108	130	26	332	49	259	120	166	—
Pb	44	85	—	18	21	—	—	—	—	24.8
La	23	11	36	3.23	15.5	17	20	21	17	38.0
Ce	45	23	75	8.28	29.9	32	38	50	40	76.1
Pr	5.5	2.2	8.9	0.98	3.33	3.7	3.9	6.5	5.1	8.98
Nd	21	9.1	31	4.46	13.4	15	17	25	20	35.0
Sm	4.0	2.1	4.9	1.26	2.87	3.2	3.0	5.1	4.4	6.26
Eu	1.3	0.80	1.2	0.47	0.92	0.9	0.9	1.6	1.6	1.55
Gd	3.6	2.0	3.7	1.55	2.97	2.6	2.3	4.1	3.9	4.84
Tb	0.54	0.31	0.49	0.29	0.49	—	—	—	—	0.79
Dy	3.5	1.9	2.5	1.71	2.93	2.7	1.6	3.1	3.0	4.59
Ho	0.82	0.40	0.39	0.37	0.59	—	—	—	—	1.07
Er	2.5	1.3	0.87	1.06	1.71	1.9	1.0	1.5	1.6	3.05
Yb	2.5	1.3	0.44	1.10	1.66	1.8	0.4	1.4	1.2	2.73
ΣREE	110	54	162	25.1	76.7	82	89	121	99	183.8

Table 7.5 (continued)

	KH38	KH47	KH21	8781	481	YK17	YK28	YK1	YK7	WC1
La_N/Yb_N	6.2	5.7	55.3	2.0	6.3	6.4	33.8	10.1	9.6	9.4
Eu/Eu*	1.05	1.19	0.86	1.03	0.96	0.95	1.05	1.07	1.18	0.86
Y	22	11	11	7.35	19.1	30	13	20	20	39.4
Th	6.9	4.1	16	1.26	5.11	12.3	5.8	6.0	6.5	12.4
U	1.9	0.97	2.0	0.37	1.56	—	—	—	—	3.40
Zr	147	133	180	52	124	153	81	119	121	271
Hf	4.1	3.3	3.7	1.37	3.33	—	—	—	—	5.39
Sn	2.9	11	1.3	—	—	—	—	—	—	9.06
Nb	10.3	10.5	7.2	2.2	3.0	8	10	14	8	15.2
Mo	—	—	—	—	—	—	—	—	—	0.63
W	1.1	3.3	0.63	—	—	—	—	—	—	2.56
Th/U	3.6	4.2	8.0	3.4	3.3	—	—	—	—	3.6
La/Th	3.3	2.7	2.3	2.6	3.0	1.4	3.4	3.5	2.6	3.1
Cr	69	210	370	113	362	260	96	173	179	353
V	65	98	185	62	114	229	124	156	183	209
Sc	24	17	22	—	—	—	—	—	—	24
Ni	56	11	*	73	268	65	70	69	69	162
Co	96	5.1	*	61	61	—	—	—	—	17
Cu	400	81	3.9	445	265	16	25	49	54	66
Zn	—	—	—	700	1400	82	55	89	89	—
Ga	23	18	27	10	16	19	15	20	23	25
La/Sc	0.96	0.65	1.6	—	—	—	—	—	—	1.6
Th/Sc	0.29	0.24	0.73	—	—	—	—	—	—	0.52
Bi	1.7	1.4	0.22	—	—	—	—	—	—	0.15
B	62	4	110	—	—	—	—	—	—	45

KH38, KH47: Group I shales, Kalgoorlie, Yilgarn Block ([96, 97] unpublished data).
KH21: Group II shale, Kalgoorlie, Yilgarn Block ([96, 97] unpublished data).
8781: Sediment from Footwall Basalt, Kambalda, Yilgarn Block [98].
481: Internal sediment, Kambalda, Yilgarn Block [98].
YK17: Jackson Lake Fm. Group A lithic wacke, Yellowknife Supergroup, Slave Province [99].
YK28: Jackson Lake Fm. Group B lithic wacke, Yellowknife Supergroup, Slave Province [99].
YK1, YK7: Burwash Fm. greywacke–shale, Yellowknife Supergroup, Slave Province [99].
WC1: Whim Creek Group shale, Pilbara Block [92].

Table 7.6. Chemical composition of selected sedimentary rocks from the Archean high-grade terrains of West Greenland [100].

	Akilia		Isua		Malene			
	221127	152769	248484G	248484A	221136	201429	201424	221137
SiO_2	56.65	52.28	53.81	58.83	55.70	52.97	78.68	83.82
TiO_2	0.53	1.09	0.68	0.58	0.54	1.68	0.31	0.14
Al_2O_3	12.56	21.41	18.49	16.62	12.79	25.06	11.07	9.16
FeO	16.97	9.71	16.64	13.99	17.37	7.40	2.79	1.11
MnO	0.18	0.03	0.08	0.09	0.16	0.12	0.02	0.02
MgO	6.19	5.29	2.63	2.43	6.35	3.35	1.19	0.75
CaO	3.94	2.46	2.75	2.26	4.16	3.65	1.34	1.29
Na_2O	1.24	4.88	2.19	2.45	1.27	4.02	2.98	1.91
K_2O	1.74	2.85	2.74	2.74	1.66	1.75	1.62	1.80
Σ	100.00	100.00	100.01	99.99	100.00	100.00	100.00	100.00
LOI	0.85	0.99	0.99	0.91	0.77	1.06	0.80	0.73

Table 7.6 (continued)

	Akilia		Isua		Malene			
	221127	152769	248484G	248484A	221136	201429	201424	221137
Cs	2.38	125	7.62	25.9	1.68	8.62	1.71	2.19
Ba	114	487	177	201	87.7	452	450	416
Pb	5.08	14.8	7.06	7.13	4.37	17.8	12.2	15.3
La	2.55	19.1	19.6	31.8	2.10	14.9	10.2	7.02
Ce	6.59	44.7	44.2	66.8	6.15	36.2	23.3	13.7
Pr	1.19	4.63	5.43	7.18	1.07	4.11	2.56	1.64
Nd	6.71	18.3	22.0	28.6	6.15	17.3	10.6	6.66
Sm	2.80	3.46	4.84	5.25	2.47	4.39	1.81	1.53
Eu	1.13	1.07	1.20	1.38	0.98	1.84	0.55	0.45
Gd	3.75	3.58	4.22	3.81	3.58	3.94	1.23	1.18
Tb	0.76	0.56	0.71	0.63	0.73	0.69	0.19	0.16
Dy	5.05	3.72	4.01	3.66	5.08	3.86	1.07	0.84
Ho	1.13	0.82	0.87	0.80	1.18	0.90	0.22	0.16
Er	3.00	2.40	2.36	2.16	3.41	2.58	0.59	0.38
Yb	2.72	2.55	2.20	1.86	3.11	2.46	0.60	0.34
ΣREE	38.2	105.6	96.1	154.5	37.0	93.9	53.1	34.2
La_N/Yb_N	0.63	5.1	6.0	11.6	0.46	4.1	11.5	14.0
Eu/Eu*	1.07	0.93	0.81	0.94	1.01	1.35	1.13	1.02
Y	27.5	24.9	31.4	23.9	34.3	21.2	4.42	2.77
Th	0.75	3.97	3.49	4.53	0.36	3.32	2.80	1.84
U	0.25	0.82	0.70	0.93	0.23	0.94	0.78	0.46
Zr	48.4	125	167	163	28.2	93.1	40.2	29.3
Hf	3.26	3.30	4.25	4.72	3.77	4.32	2.52	2.22
Sn	1.68	1.72	3.16	3.25	1.33	4.28	1.89	1.02
Nb	4.84	8.29	5.98	6.09	4.95	8.50	3.51	2.69
Mo	0.95	1.12	0.39	0.22	0.74	1.12	1.12	*
W	*	0.49	0.65	0.40	*	1.02	0.19	*
Th/U	3.0	4.8	5.0	4.9	1.6	3.5	3.6	4.0
La/Th	3.4	4.8	5.6	7.0	5.8	4.5	3.6	3.8
Cr	315	325	63	69	220	195	89	33
V	76	260	54	54	65	268	46	21
Sc	19	29	13	13	19	30	6.5	3.3
Ni	162	207	56	52	155	148	32	15
Co	24	50	17	15	20	46	10	*
Cu	14	84	25	21	13	71	11	3.0
Ga	12	12	14	15	10	15	*	*
La/Sc	0.13	0.66	1.5	2.4	0.11	0.50	1.6	2.1
Th/Sc	0.04	0.14	0.27	0.35	0.02	0.11	0.43	0.56
Bi	*	0.07	0.06	0.05	*	0.08	0.03	*
B	14	15	58	37	11	375	7	38

221127, 152769: Mica schist and garnet–mica schist, Akilia association.
248484G, 248484A: Metapelites, Isua volcanic–sedimentary belt.
221136, 201429: Mica schist and metasedimentary gneiss from low silica group, Malene supracrustals.
221424, 221137: Semi-pelitic gneisses from high silica group, Malene supracrustals.

In the following sections, we will emphasize two major themes. Firstly, the trace elements which best reflect provenance, notably REE, have a different distribution pattern in Archean sedimentary rocks compared to that of post-Archean sedimentary rocks. Secondly, these patterns commonly reflect a mixture of complex source lithologies (including ultramafic–mafic volcanics, felsic volcanics, granitic rocks and recycled sedimentary and metamorphic debris). In some cases exclusively local sources are apparent and this provides important information concerning the dominant lithologies contributing to Archean sediments. In most examples a mixture of source rocks is required, indicating a reasonably wide sampling of the exposed crust.

7.8.1 Early Archean greenstone belts

(a) *Pilbara Block, Western Australia.* The Pilbara Block covers about 60 000 square kilometres in Western Australia (Fig. 7.8). Low grade volcanic and sedimentary supracrustal rocks of the Pilbara Supergroup form arcuate greenstone belts around domal granitic batholiths. The generalized stratigraphy is shown in Fig. 7.2. Mafic-felsic volcanics dominate the basal Warrawoona Group. This passes up into mainly terrigenous sedimentary rocks of the Gorge Creek Group. The Whim Creek Group, a sequence of volcanic and sedimentary rocks, unconformably overlies the Gorge Creek Group and is of late Archean age (see below).

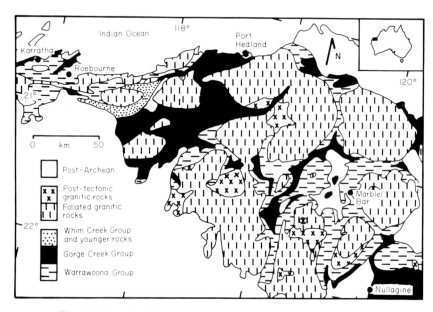

Fig. 7.8. Geological sketch map of the Pilbara Block, Western Australia.

The base of the Pilbara Supergroup is marked by tectonic and intrusive contacts with the granitic batholiths. Volcanic rocks of the Warrawoona Group have been dated by Sm–Nd and zircon U–Pb methods which yield ages of 3.56 and 3.45 Ae, respectively [101]. The Gorge Creek Group is older than 2.9 Ae [102] and the general conformity with the underlying Warrawoona Group suggest it is not much younger than about 3.4 Ae [39, 40]. The Whim Creek Group is older than the Mount Bruce Supergroup of the Hamersley Basin and is probably older than about 2.75 Ae [103].

The Gorge Creek Group has a platform (alluvial) to trough (turbidite) facies relationship with no evidence of shallow marine shelf facies, consistent with deposition at a continental margin [39–41]. The geochemistry of shales from the Gorge Creek (Table 7.4) [92] is particularly interesting because of their high K_2O content (1.9–7.1%) and high K_2O/Na_2O ratios (2–19). This feature may be unique to sedimentary rocks deposited in early Archean greenstone belts (see Section 6.5.4). The Gorge Creek samples are also characterized by high large ion lithophile (LIL) element (e.g. Th, U, REE, Cs) abundances which correlate with K_2O and muscovite abundances. Again, the levels are similar to post-Archean rather than other Archean shales. The REE (Fig. 7.9) show high abundances and are characterized by LREE enrichment ($La_N/Yb_N>7.5$) at a level normally seen in post-Archean shales (e.g. PAAS $La_N/Yb_N=9.2$). Regardless of these features, the Pilbara Supergroup shales show no or only very slight Eu-depletion ($Eu/Eu^*=0.82$–0.99) which is typical of Archean sedimentary rocks. The abundances of Cr and Ni in the Gorge Creek Group shales are also anomalously high with average Cr values of 600 ppm and average Ni contents of 350 ppm.

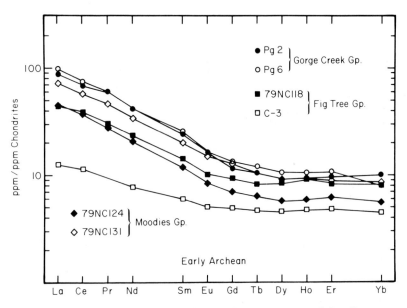

Fig. 7.9. Chondrite-normalized REE patterns for early Archean greenstone belt sedimentary sequences. Data for Archean shales from the Gorge Creek Group, Western Australia, and from the Fig Tree and Moodies Groups, South Africa. (Data from Table 7.4.) Note the absence of Eu-anomalies.

The provenance of Gorge Creek Group shales has been modelled from trace element abundances [92]. The source was composed of about 70–80% felsic igneous rocks (felsic volcanics and/or granitic rocks) and 20–30% mafic volcanics [92]. Much of the associated sandstones are arkosic [39] indicating that granitic rocks must have been a significant contributor. The general lack of Eu-depletion in the sediments indicate that igneous rocks of intracrustal origin, with negative Eu-anomalies, must have been a minor component of the provenance.

(b) *Barberton Mountain Land, South Africa.* The Swaziland Supergroup, Barberton Mountain Land, South Africa, is a classical example of an Archean greenstone belt [10]. A generalized stratigraphic column is shown in Fig. 7.2. Volcanic rocks of the Onverwacht Group yield a Sm–Nd isochron age of 3.54 Ae [48]. The minimum age of the Fig Tree and Moodies Groups is less well established but is probably in excess of 3.3 Ae [10].

The Swaziland Supergroup comprises three groups. The lower Onverwacht Group contains a sequence of up to 15 km of ultramafic–mafic to felsic volcanic rocks with intercalated volcanogenic and chemical sediments in the upper parts. The Onverwacht grades up into the Fig Tree Group which comprises 2000 m of dominantly terrigenous clastic sedimentary rocks. The Fig Tree Group can be divided into two facies [38]. The southern facies consists of 10–80 m of iron-formation and tuffs which grade up into arenaceous–conglomeratic sediments derived from the underlying Onverwacht Group, with a minor granitic contribution. The northern facies is subdivided into three formations which include, from the base, the Sheba (c. 700–1000 m), Belvue Road (c. 600–1000 m) and Schoongezicht (200–600 m) Formations. The Sheba Formation is a sequence of greywacke with minor shale and iron-formation. The Belvue Road Formation includes interbedded greywacke–siltstone with chert, iron-formation and local pyroclastics. The uppermost Schoongezicht Formation consists of pyroclastics, volcanics, conglomerate and greywacke. The Fig Tree Group has been interpreted as a turbidite sequence with the Sheba and Belvue Road Formations representing fan-basin plain and slope deposits and the Schoongezicht Formations representing feeder channels [38].

The petrography of greywackes from the Sheba and Belvue Road Formations clearly indicates an upward increase in granitic rock fragments and decrease in volcanic fragments [33, 94]. An important feature of the Fig Tree and Moodies Groups (see below) is that K-feldspar is an important component, indicative of a major granodiorite–granite source. In most Archean sandstones, plagioclase greatly exceeds K-feldspar (see Chapter 6).

The Fig Tree sediments become coarser as they pass up into the Moodies Group. The Moodies comprises some 1500–3600 m of more mature sandstones, siltstones, shales and conglomerates. These rocks represent various sedimentary facies deposited in alluvial and marginal marine environments. The framework grains of sandstones from the Moodies Group are almost exclusively monocrystalline quartz, K-feldspar and chert with lesser polycrystalline quartz and rock fragments. This indicates a decreasing role of volcanic sources and a continuing continental source [33].

A tectonic model, involving continental rifting, to explain the nature of the sedimentary rocks of the Fig Tree and Moodies Groups has been developed [33, 35–41], based on the suggestion that the associated Ancient Gneiss Complex formed a basement to the Swaziland Supergroup [44]. There is no direct geological evidence of this [104] and recent Sm–Nd dating suggests the Ancient Gneiss Complex is of similar or probably slightly younger age as the Onverwacht volcanics [49]. A more recent tectonic model suggests the Fig Tree and Moodies Groups may represent foredeep flysch and molasse deposits which were derived from an uplifting continental margin [104]. These sedimentary rocks contain no evidence for a contribution from an island-arc.

Trace element data are available for greywackes from the Sheba and Belvue Road Formations of the Fig Tree Group [94, 95] and for shales from the Fig Tree and Moodies Groups [93–95]. The greywackes were discussed in Chapter 6. Representative shale analyses are given in Table 7.4. There is a wide range in REE abundances for the greywackes and shales (Figs 6.6 and 7.9). There is also a general enrichment of Eu, in comparison with post-Archean sedimentary rocks, although for the greywackes it is not as large as in most other Archean sedimentary rocks. There are upward trends in the trace element geochemistry of Fig Tree greywackes [94, 95]. Most important among these is a decrease in Ni and an increase in LREE and La_N/Yb_N, consistent with an increasing importance of granitic source rocks with stratigraphic height.

There are differences in composition of shales from the Fig Tree and Moodies Group [93]. The Moodies had generally lower concentrations of ferromagnesian elements and higher abundances of incompatible trace elements, such as LREE, Zr and Th, resulting in higher ratios of La_N/Yb_N, Th/Sc, La/Sc and Th/Co (Table 7.4). This indicates an increased amount of granitic source material for the Moodies, as suggested by the petrographic evidence [33, 94]. In summary, the Fig Tree was derived from equal proportions of mafic and felsic igneous rocks whereas the Moodies required a much larger component of parental felsic rocks [93].

High levels of Cr and Ni occur in Fig Tree shales with Cr >800 ppm and Ni >400 ppm [93, 105]. The Moodies Group shales are also enriched in Cr and Ni (Cr >550 ppm, Ni >250 ppm). Danchin [105] suggested an ultramafic source to explain these levels, but the REE patterns and other incompatible trace element abundances are not consistent with such a source. It is most likely that mafic–ultramafic rocks contributed to these sedimentary rocks but that Cr and Ni abundances were further enriched by secondary processes [93].

7.8.2 Late Archean greenstone belts

(a) *Yilgarn Block, Western Australia.* The Yilgarn Block is situated in south-west Australia and occupies more than 600 000 km². It is mainly composed of late Archean granite–greenstone terrains with smaller areas of early Archean gneissic terrains in the west. The greenstone belt successions consist of ultramafic–mafic, felsic and clastic associations. Mafic rocks which are mainly low-K tholeiites make up about 50% of the successions with ultramafic rocks being more common at basal levels. Felsic volcanics occur in the upper levels. Thick sequences of clastic sedimentary rocks are found at the top of the sequences. A generalized stratigraphic column is shown in Fig. 7.2. These rocks are about 2.6–2.8 Ae in age.

The sedimentary rocks from the Kalgoorlie area were deposited mainly as turbidites and petrographic evidence indicates derivation from a mixture of Na-rich felsic porphyries, mafic volcanics and recycled sedimentary material (quartzite, chert) ([106]; also see [107]). Na-rich granites which are found in the area and may pre-date the sedimentary sequence also may have contributed. The fine-grained sedimentary rocks of the Kambalda area differ from those at Kalgoorlie in that they occur exclusively as lenticular horizons within the lower ultramafic–mafic volcanic sequence at Kambalda [108]. They originate from a mixture of locally derived mafic–ultramafic volcanics, and distal felsic volcanic–granitic detritus (e.g. Fig. 7.10). The distribution of some elements (e.g. S, Ni, Cu, K, Na) is also affected by fumarolic and hydrothermal processes.

Kalgoorlie sedimentary rocks were divided into two groups on the basis of REE [96] with Group I having $La_N/Yb_N \leq 6$, $Eu/Eu^* \geq 1.00$ and Group II having $La_N/Yb_N > 10$, $Eu/Eu^* \geq 1.00$ (see Table 7.5, Fig. 7.11). Group I samples were taken from sediments closely associated with mafic volcanic rocks and were modelled as being derived from either calc-alkaline island-arc volcanics or a mixture of mafic and felsic rocks, with the latter model favoured. Group II samples came from the top of the volcanic cycles, more closely associated with felsic volcanics and porphyries. The HREE-depleted patterns of Group II were best explained by having a large proportion of felsic

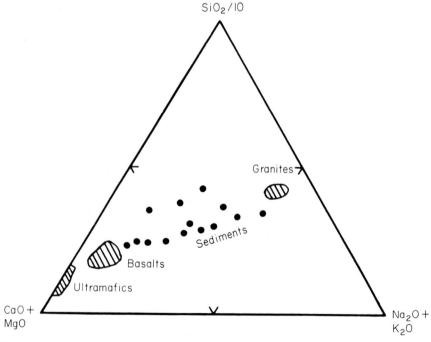

Fig. 7.10. Comparison between the composition for major elements of the Kambalda sedimentary rocks with those of granites, basalts and ultramafic rocks from the Kambalda district. The sedimentary compositions show evidence for a strong granitic component, although they occur principally interbedded in the basaltic and ultramafic sequences. (Adapted from [98, 108].)

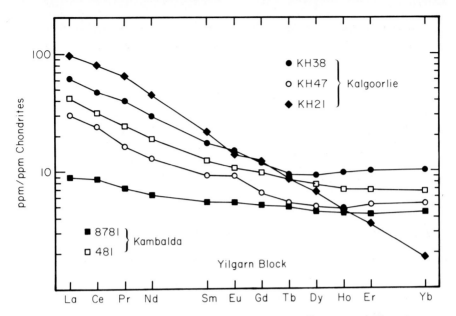

Fig. 7.11. Chondrite-normalized REE patterns for late Archean sedimentary rocks from the Yilgarn Block. (Data from Table 7.5.) Note the variable La/Yb ratios, interpreted as evidence for mixing, and the absence of Eu-anomalies.

volcanic and Na-rich granitic material in the source. Positive Eu-anomalies, found in some samples from both groups, could be attributed to local derivation, for example, from trondhjemites.

The Kambalda sedimentary rocks (Table 7.5) have quite uniform REE-characteristics (Fig. 7.11) with $\Sigma REE = 67 \pm 12$ ppm; $La_N/Yb_N = 4.6 \pm 0.7$ and $Eu/Eu^* = 1.01 \pm 0.07$ for 14 samples [98]. The most significant observation was that the local ultramafic–mafic volcanics could not be the sole source but that a more distant felsic source also contributed detritus.

(b) *Yellowknife Supergroup, Slave Province, Canada.* The Yellowknife Supergroup comprises the Archean supracrustal rocks of the Slave Province, NWT, Canada (Fig. 7.12). This sequence differs from most Archean terrains, containing about 80% sedimentary rocks and 20% volcanic rocks. These proportions are the reverse of those in most Archean greenstone belts. The isotopic data suggest a depositional age of about 2.67 Ae [109]. The Yellowknife Supergroup is of special interest because, as at Point Lake and Benjamin Lake, there is good evidence that the supracrustal rocks unconformably overlie granitic basement [52]. This is supported by isotopic evidence [72].

Generalized stratigraphic sections are given in Fig. 7.12. Relationships between the west and east sides of Yellowknife Bay are obscure but the Jackson Lake and Burwash Formations appear to be of equivalent facies [109]. Lithic sandstones of the Jackson Lake Formation are regarded as a fluvial deposit of largely felsic volcanic origin. Interbedded greywacke and mudstones

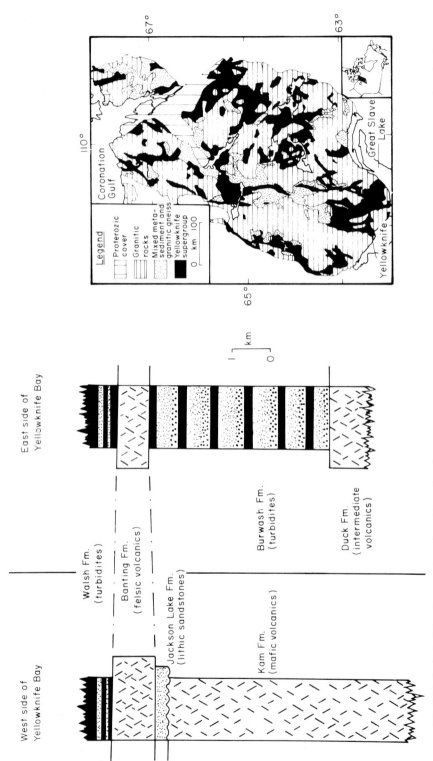

Fig. 7.12. Geological sketch map and stratigraphic sections for the Archean supracrustal rocks of the Yellowknife Supergroup, Canada (adapted from [109]). The stratigraphic sections are based on detailed work in the vicinity of the town of Yellowknife.

of the Burwash and Walsh Formations are turbidites. Petrographic evidence indicates these rocks were derived from a mixed source of felsic–mafic volcanic, plutonic (mainly sodic granitic rocks), minor recycled metamorphic material and recycled intraformational sedimentary material. The Yellowknife Supergroup was apparently deposited in a series of fault-bounded basins within a granitic to tonalitic basement. The entire sequence may have been deposited over a period as short as 10–15 m.y. [72, 109].

Trace element geochemistry of the sedimentary rocks of the Yellowknife Supergroup show that lithic wackes of the Jackson Lake Formation were divisible into two groups mainly on the basis of REE [99] with Group A having $La_N/Yb_N = 5-8$ and Group B having $La_N/Yb_N = 11-55$ (Table 7.5, Fig. 7.13). All samples are characterized by no Eu-anomaly or slight Eu-enrichment. Trace element modelling of Group B indicates derivation from

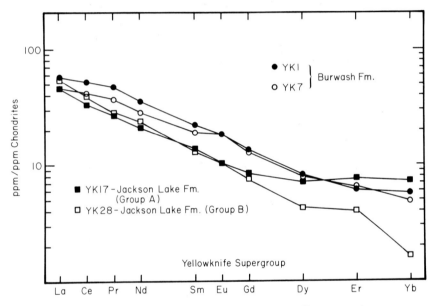

Fig. 7.13. Chondrite-normalized REE patterns from late Archean sedimentary rocks at Yellowknife, Canada. (Data from Table 7.5.) Differing La/Yb ratios indicate differing proportions of mafic volcanic rocks in the source. Note the absence of Eu-anomalies.

associated felsic volcanics such as the Banting Formation, to explain the very steep HREE-depleted patterns. However, the less fractionated REE patterns of Group A required a significant ($\approx 50\%$) mafic volcanic component mixed with a felsic volcanic or Na-granitic source.

The composition of greywacke–mudstone samples from the Burwash and Walsh Formation turbidites is similar and probably had the same provenance. REE patterns (Fig. 7.13) for 12 samples were very similar with: REE$=127\pm24$; $La_N/Yb_N=14.4\pm2.1$; Eu/Eu*$=1.03\pm0.05$ (uncertainties at 95% confidence level) [99, 110]. The most realistic source model [99], also consistent with the petrographical data, consisted of 80% felsic volcanic–Na-granitic rocks and 20% mafic volcanics. The high abundances of Ni and Cr in the sedimentary rocks, as compared to the model, could not easily be explained by a small ultramafic component, because such rocks do not occur in the Slave Province, and a secondary enrichment process may be required.

(c) *South Pass Greenstone Belt, Wind River Range, USA*. Greywackes of the South Pass greenstone belt were among the first Archean sedimentary rocks analysed for REE [95]. The South Pass greenstone belt has many of the features of Archean greenstone belts such as the thick sequences of mafic volcanics and turbidites, although it lacks an obvious vertical mafic to felsic trend in the volcanic rocks. The belt comprises a sequence of low-K tholeiitic mafic volcanics with minor siliceous clastic sediments (plus chert and iron formation) in fault contact with a thick sequence (>3 km) of greywackes with lesser amounts of intermediate-felsic volcanic rocks. The age of the greenstone belts is not well constrained but may have developed over as much as 200–400 m.y. during 3.0 to 2.8–2.6 Ae.

The petrography and major element geochemistry of these greywackes indicates a provenance of mixed granitic and recycled sedimentary and metamorphic debris [111]. There is a complete absence of volcanic rock fragments, although the presence of high Ni contents indicates a minor mafic volcanic component.

A typical analysis of greywacke from the South Pass greenstone belt is listed in Table 6.13 of the previous chapter. The REE characteristics are quite uniform: $\Sigma REE = 143 \pm 36$ ppm; $La_N/Yb_N = 13.4 \pm 3.3$, and $Eu/Eu^* = 0.9 \pm 0.06$ (uncertainties at 95% confidence) for six samples. The overall REE abundances and fractionation of the patterns (La_N/Yb_N) are fairly similar to post-Archean shales (e.g. for PAAS, $\Sigma REE = 183$ ppm, $La_N/Yb_N = 9.2$), probably reflecting the dominant granitic and metamorphic source. However, as with most other Archean sedimentary rocks, these greywackes lack the pronounced negative Eu-anomaly characteristics of post-Archean sedimentary rocks.

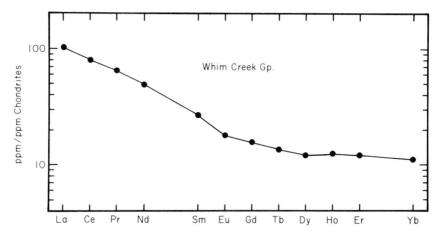

Fig. 7.14. Chondrite-normalized REE pattern from the Whim Creek Group, Pilbara Block, Western Australia. (Data from Table 7.5.) This is another typical example of a REE pattern from the late Archean greenstone belt sedimentary rocks.

(d) *Other areas.* Data are available from the Couchiching metasediments and Knife Lake Group of the Superior Province [112], Abitibi belt (Superior Province) [113] and Whim Creek Group (Pilbara Block) [92]. The sample from the Knife Lake Group was described in Section 6.4.3. The REE characteristics of most of these samples are typical of the other Archean sediments described above, without substantial Eu-anomalies. A good example for which complete data are available is from the Whim Creek Group [92]. An analysis is reproduced in Table 7.5 and the REE pattern is shown in Fig. 7.14. Rare earth element and other trace element data for other various late Archean sedimentary sequences are scattered throughout the literature. However, detailed interpretations are generally hampered by incomplete analyses, lack of geological and sedimentological detail and lack of precision in some older analyses.

7.8.3 High-grade terrains

(a) *Godthåb–Isukasia Region, West Greenland.* The early Archean high-grade terrains of the North Atlantic Craton (West Greenland–Labrador) are critical regions for the understanding of early crustal development. Table 7.2 lists the sequence of geological events in the Godthåb–Isukasia region [53, 114]. Ultramafic–mafic–felsic volcanics and volcaniclastic and chemical sedimentary rocks occur as enclaves within an extensive high-grade gneiss complex, dominated by tonalite–trondhjemite–granodiorite gneisses (Amîtsoq gneiss in West Greenland). The largest of these enclaves, the Isua supracrustal belt, is exposed at the head of Godthåbsfjord. Elsewhere in the Godthåb region, smaller enclaves of metamorphosed volcanic and sedimentary rocks, which are correlated with the Isua rocks, are grouped together as the Akilia association [63]. These are the oldest volcanic and sedimentary rocks known, approximately 3.75–3.80 Ae in age [32]. Metamorphic grades range from middle greenschist to amphibolite for the Isua area to amphibolite–granulite facies for the Akilia association. Sedimentological data are not readily available due to the metamorphic and structural complexity and the small extent of individual enclaves. No data are available on the Akilia association which could be used to constrain the sedimentary or tectonic environment of those ancient rocks. On the other hand, recent stratigraphic and geochemical evidence has been presented which indicates that the Isua sediments were probably deposited in a volcanic environment [75].

A younger generation of supracrustal rocks, termed the Malene supracrustals, also occur in the Godthåb region [114, 115] and make up 10–20% of the Archean gneiss complex. They comprise mafic amphibolites of probable volcanic origin with lesser amounts of pelitic, semi-pelitic and quartz-cordierite gneisses. The age of the Malene rocks is greater than 3.05 Ae [116]. Field evidence strongly suggests that at least some of the Malene units were laid down on a basement of Amîtsoq gneisses [115, 117]. However, the role of Amîtsoq gneiss as a source to Malene sediments is uncertain. Limited

Nd-isotopic data showed no evidence for earlier crust [32] and more isotopic work is needed to resolve the question.

There is a considerable body of geochemical data on metasedimentary rocks from the Isua and Akilia successions [63, 75, 100, 118]. Some representative analyses are presented in Table 7.6. The most significant feature of these rocks is the extremely wide range in REE patterns (Fig. 7.15) ranging from flat with LREE depletion, similar to modern mid-ocean ridge basalts, to very steep with HREE depletion, similar to many Archean felsic volcanics. Eu-anomalies are also variable. One suite of muscovite–biotite gneisses, some with severe depletion in Eu, probably were derived directly from a felsic volcanic source [118]. McLennan et al. [100] argued that the trace element characteristics of sedimentary rocks from the Isua and Akilia successions were best explained by variable mixtures of mafic and felsic volcanic rocks.

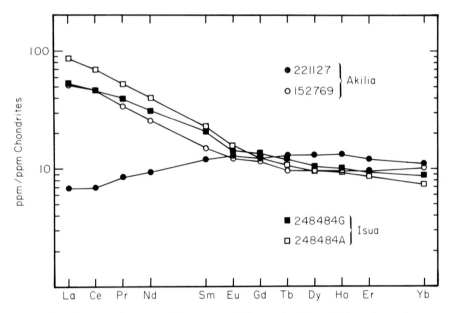

Fig. 7.15. REE patterns from metasedimentary rocks from early Archean high-grade terrains. Data from the Akilia and Isua sequences, West Greenland. (Data from Table 7.6.) Note the diverse patterns, interpreted as due to local provenance effects. The LREE depleted pattern is indicative of a basaltic source.

The composition of Malene metasedimentary rocks can be divided into high-SiO_2 (>75%) and low-SiO_2 (<60%) varieties [100] (Table 7.6). The same range of REE patterns, from flat with LREE-depletion to steep with HREE-depletion, is seen for the Malene samples as was seen for the Isua–Akilia samples (Fig. 7.16). In the case of the Malene sediments, negative Eu-anomalies are comparatively rare and typically Eu/Eu* >1.0. A mixing model of mafic and felsic rocks may explain the origin of the Malene sedimentary rocks [100]. It is possible that the more fractionated end-member was Na-rich granitic gneiss from the Amîtsoq basement [115, 117].

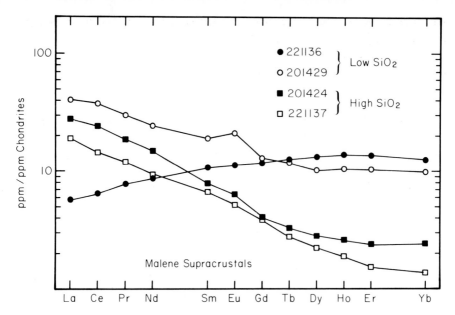

Fig. 7.16. REE patterns of metasedimentary rocks from the Archean high-grade Malene succession, West Greenland. (Data from Table 7.6.) Note some evidence of Eu enrichment and the diverse nature of the patterns, suggesting local provenance effects.

The Archean metasedimentary rocks of West Greenland are unique among Archean samples analysed, on account of the extreme range in REE patterns and the occurrence of negative Eu-anomalies in some samples. The variability strongly supports a lithologically complex provenance with inefficient mixing during sedimentation. In the case of the Isua samples, a volcanogenic source is most likely [75]. It is unlikely that much information about the average composition of the Archean upper continental crust can be gleaned from these rocks. However, in the following sections, we will see that some of these variations represent extreme cases of systematic differences in the composition of Archean sedimentary rocks which provides important evidence concerning their origin.

(b) *Other areas*. A number of samples with significant negative Eu-anomalies have also been reported from high grade schists and paragneisses from Ontario [112], India [119] and Wyoming [120]. It is an interesting observation that the rare Archean samples with negative Eu-anomalies appear to be restricted to high-grade terrains. There is not enough detailed work to make any generalizations, but if the depletion in Eu reflects intracrustal melting, it would have considerable implications for understanding the origin of Archean high grade terrains, the relationship between low- and high-grade terrains and the nature of the Archean crust.

7.9 Trace element systematics

In Chapter 2, the general uniformity of composition in post-Archean shales, for trace elements which reflect source rock composition (e.g. REE, Th, Sc) was documented. A notable feature of Archean clastic sedimentary data, summarized above, is the considerable variation in these elements, most notably for the REE. In the following section, we summarize the general features of trace element distribution in Archean sedimentary rocks. The most thorough studies have dealt with fine-grained rocks. We will demonstrate that the variability in Archean sedimentary composition is systematic and significantly constrains the geochemical nature of the source rocks.

7.9.1 Rare earth elements (REE)

Rare earth element abundances in Archean sedimentary rocks differ from those in post-Archean sedimentary rocks in four significant respects:
1 the patterns show considerably more variability, ranging from flat with occasional LREE-depletion to very steep with HREE-depletion;
2 Archean sedimentary rocks tend to have lower total rare earth abundances;
3 the REE patterns typically have lower La_N/Yb_N;
4 Archean sedimentary rocks are characterized by a lack of an Eu-anomaly ($Eu/Eu^* \approx 1.0$) although local positive Eu-anomalies occur, probably mainly derived from trondhjemitic intrusives. Significant negative Eu-anomalies are scarce and apparently restricted to high-grade terrains, possibly derived from volcanogenic sources (e.g. Isua).

By far, the most abundant patterns are those with intermediate La_N/Yb_N, between about 3 and 9 (La_N/Yb_N for PAAS is 9.2). Such patterns are common in the Yilgarn and Pilbara Blocks, Fig Tree and Moodies Groups, and in West Greenland. Less common, though still abundant, are steep REE patterns, with $La_N/Yb_N > 9$; these are found in almost all areas studied. The steepest of these patterns are similar to felsic volcanics and Na-rich plutonic rocks—one of the end members of the bimodal suite. Patterns with low values of ($La_N/Yb_N < 3$) are comparatively scarce. Those with LREE-depletion ($La_N/Sm_N < 1.0$) are restricted to West Greenland whereas those showing slight LREE enrichment ($La_N/Sm_N > 1.0$) are found in West Greenland, Kambalda (Western Australia), and in the Fig Tree Group (South Africa). These patterns are comparable to Archean mafic volcanics [121].

7.9.2 La–Th–Sc systematics

The coherence between Th and the LREE in sedimentary rocks [122] can be attributed to similar behaviour during sedimentary processes (see Chapter 2, Fig. 2.16). The abundances of Th and several other elements in the upper crust was successfully estimated using the La–Th systematics. Post-Archean shales have a La/Th=2.8 ± 0.2. For this analysis, only the Greenland values are excluded due to their uncertain origin and extreme variability. Inclusion of this data, however, would only reinforce the conclusions. The presently available data base for shales indicates an average Archean sedimentary ratio (shales) of La/Th=3.5 ± 0.3 (95% confidence). In igneous rocks, the La/Th ratio generally is compositionally dependent and the higher Archean ratio is consistent with an overall more mafic composition [122]. Ratios of elements with very different levels of incompatability are more useful in characterizing composition. The Archean shale data have average Th/Sc=0.43±0.07 and La/Sc=1.3±0.2, which contrasts with post-Archean shales which average Th/Sc=1.0±0.1 and La/Sc=2.7±0.3.

7.9.3 *Ferromagnesian trace elements*

Very high abundance of Cr and Ni in shales are observed for a number of early Archean sequences (Fig. 7.17), including the Fig Tree Group, Moodies Group and Gorge Creek Group. Many other late Archean sequences, such as Kalgoorlie [96] and various greenstone belts in the Canadian Shield [123] are either not or only slightly enriched in Cr and Ni, when compared to their

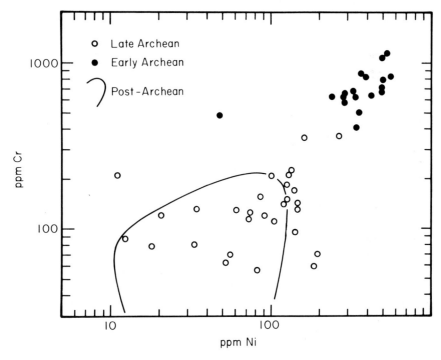

Fig. 7.17. Chromium and nickel abundances in Archean shales, showing their wide range (including high values) in comparison to the abundances in post-Archean shales. (Data from [92, 93, 96–98, 123].) Early Archean shales are considerably enriched in Cr and Ni compared with late Archean shales.

Table 7.7. Ratios of ferromagnesian trace elements in Archean and post-Archean shales [97].

	Early Archean	Late Archean	Post-Archean
Cr/V	5.3±1.0 (n=18)	1.5±0.3 (n=30)	0.9±0.1 (n=31)
Ni/Co	11.6±1.6 (n=17)	3.0±0.7 (n=27)	2.6±0.2 (n=31)
V/Ni	0.51±0.28 (n=18)	1.7±0.7 (n=29)	2.1±0.2 (n=32)

Uncertainties are 95% confidence intervals.

post-Archean counterparts. The enrichment does not extend to the other ferromagnesian trace elements, so that the data (Table 7.7) [97] show anomalously high Cr/V, Ni/Co and low V/Ni ratios for early Archean shales. In some sequences (e.g. Pilbara), the Cr and Ni abundances are too high to be explained simply by provenance and some secondary enrichment process is required (Section 7.8.1). The level of enrichment, compared to post-Archean sediments is so high that it must reflect differing source compositions to some extent. High Cr and Ni in Archean sedimentary rocks constitutes evidence against an island-arc origin since island-arc volcanics generally have very low Cr and Ni abundances due to olivine and spinel fractionation [124] (see Section 3.3.1).

7.9.4 A mixing model

In early Archean igneous terrains, and in the lower parts of many later Archean greenstone belt successsions, there is a scarcity of igneous rocks of intermediate composition. Such terrains contain a bimodal igneous suite of ultramafic–mafic volcanics and tonalite–trondhjemitic plutonic rocks or felsic volcanics (generally dacitic). These varieties of felsic plutonic and volcanic rocks have rather similar geochemistry. In Fig. 7.18, two typical REE patterns of the end members of the bimodal suite are shown.

Many of the Archean sedimentary REE patterns summarized above, with intermediate La_N/Yb_N, are indistinguishable from modern calc-alkaline andesites [125]. However, the petrographic and geological evidence is quite

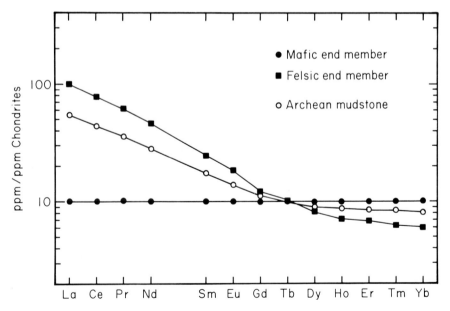

Fig. 7.18. Typical chondrite-normalized REE patterns for the basaltic and felsic end members of the Archean bimodal igneous rock suite and the average Archean mudstone. (Data from Table 7.8.) The average REE pattern of Archean mudstones is consistent with being derived from a 1:1 mixture of the mafic and felsic end members.

inconsistent with such rocks being a major source of Archean sediments, and variable mixtures of the Archean bimodal suite adequately explain the REE and petrographic data.

The general equation for two component mixing in geochemical calculations is a hyperbolic function [126] of the form:

$$Ax + Bxy + Cy + D = 0 \qquad (1)$$

where x and y are the general variables along the abscissa and ordinate, respectively. Ratio–ratio plots are a general case and if we let $x = P/b$ and $y = Q/a$ then:

$$A = a_2 b_1 y_2 - a_1 b_2 y_1 \qquad (2)$$

$$B = a_1 b_2 - a_2 b_1 \qquad (3)$$

$$C = a_2 b_1 x_1 - a_1 b_2 x_2 \qquad (4)$$

$$D = a_1 b_2 x_2 y_1 - a_2 b_1 x_1 y_2 \qquad (5)$$

where the subscripts refer to two points on the curve. Also:

$$r = a_1 b_2 / a_2 b_1 \qquad (6)$$

where the value r is a function of the extent of curvature of the mixing line such that if $r = 1$, a straight line is generated and if $r \gg 1$ or $r \ll 1$, the mixing line is highly curved. Ratio–element and element–element plots are simply special cases of the general equation (1). In ratio–element plots, $a = 1$ (resulting in $r = b_2/b_1$) or $b = 1$ (resulting in $r = a_1/a_2$). In an element–element plot, $a = b = 1$ (resulting in $r = 1$).

For igneous systems, all interelement relations must conform to the mixing in an internally consistent fashion and the fit to the mixing lines should be close [126]. Application of two-component mixing equations to sedimentary provenance studies is constrained by two major problems:
1 Secondary processes such as weathering, transport, and diagnesis can affect the distribution of many elements. Thus, one is restricted to elements which are the best provenance indicators (e.g. REE, Th).
2 End member compositions cannot always be specified with confidence. An alternative is to use extreme sedimentary compositions as end members and characterize the types of igneous source rocks they would represent.

A simple test of internal consistency is to demonstrate that intermediate points in a mixing curve have proportionately intermediate ratios of denominators. For example, assume that data on a ratio–ratio plot of P/a versus Q/b conform to a mixing curve. In this case, plots of P/a versus b/a and Q/b versus a/b should be straight lines [126].

This approach can be used to test a two-component mixing model for Archean sedimentary rocks, as suggested by the REE data. On Fig. 7.19, a ratio–ratio plot of Co/Th versus La/Sc is presented for the Archean sedimentary data. To maximize the dispersion of data, ratios were formed using incompatible and compatible elements. Superimposed on the plot is a

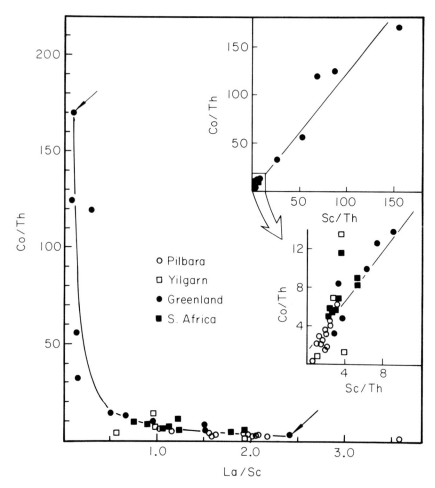

Fig. 7.19. A ratio–ratio plot of Co/Th versus La/Sc for Archean sedimentary rock data. The diagram is consistent with a two-component mixing model. (See text.)

mixing line with extreme sedimentary compositions as end members. The data are clearly consistent with two component mixing. Internal consistency is also demonstrated by the inset plot of Co/Th versus Sc/Th, which conform to a straight line.

Linear arrays on ternary diagrams also are consistent with two-component mixing, although one must ensure that such arrays are not simply slices through three-dimensional curved surfaces. In Fig. 7.20, two ternary diagrams of Th–Hf–Co and La–Th–Sc are presented. Post-Archean shales [e.g. 127] plot in restricted fields around or near the present-day upper continental crust (also see Chapter 2, Fig. 2.4). In contrast, the Archean sedimentary data plot on distinct linear arrays across large regions of the diagrams. The fact that linear trends occur on more than one ternary diagram is good supportive evidence for the two component mixing model. The data are consistent with the Archean bimodal suite as end members (AMV=Archean mafic volcanics; TT=tonalite–trondhjemite, felsic volcanics). The present-day upper continen-

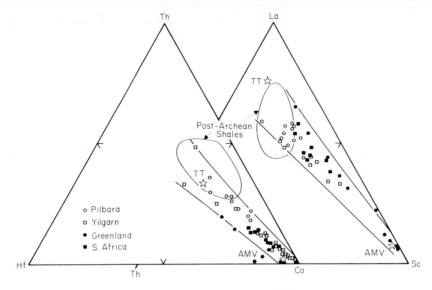

Fig. 7.20. Ternary diagrams for Th–Hf–Co and La–Th–Sc data for Archean and post-Archean shales. Note the restricted fields for the post-Archean data. In contrast, the Archean data are consistent with mixing of two components of Archean mafic volcanics (AMV) and average felsic igneous rock (TT) (see text for discussion).

tal crust could also act as one end member on these diagrams (see Fig. 2.4) but this possibility is excluded on account of the general lack of Eu depletion in Archean sedimentary rocks.

7.9.5 Comparisons between volcanic and sedimentary sequences

Enough data are available to make comparisons between the trace element geochemistry of sedimentary rocks deposited in the volcanic and sedimentary sequences of late Archean greenstone belts (see Table 7.5). At Kalgoorlie, sediments interbedded with volcanic rocks (Group I) have less fractionated REE patterns and lower Th/Sc and La/Sc ratios than those from the thick sedimentary sequences (Group II). This is consistent with a larger proportion of mafic volcanic debris in the provenance [96]. Similarly, at Kambalda [98], sediments are not laterally extensive and are interbedded with ultramafic–mafic rocks. The trace element geochemistry of the sediments cannot be derived from an ultramafic–mafic source alone and a significant felsic component is called for (see Section 7.8.2). The REE patterns are not as fractionated as many other late Archean sedimentary rocks, derived from the thick turbidite sequences (e.g. Group II from Kalgoorlie, Yellowknife, Wyoming) suggesting a larger mafic component in the source.

7.9.6 Comparisons between early and late Archean greenstone belts

There are insufficient data to examine differences in the trace element geochemistry of sedimentary rocks interbedded with the volcanic sequences of early and late Archean greenstone belts. For the thicker sedimentary sequences, no significant differences are apparent in the REE patterns (e.g. compare Fig Tree, Moodies, Pilbara with Yellowknife, Kalgoorlie (Group II) and Wyoming).

One difference which is clear is that early Archean sedimentary rocks are greatly enriched in Cr and Ni in comparison to their late Archean counterparts (Fig. 7.17). This enrichment does not extend to the other ferromagnesian trace elements V, Co and Sc (Table 7.7). As discussed above (Section 7.8.1) constraints on provenance, provided mainly by REE, Th and Sc data, indicate that the enrichment is not due simply to an ultramafic–mafic source and that the proportions of mafic and felsic components in the source are little different than for late Archean sequences. It is clear that some of the difference in Cr and Ni abundances is related to provenance. A likely explanation is that the mafic component of early Archean sedimentary rocks had a much higher proportion of high magnesium basalts and komatiitic rocks in the mafic component of the source. Such rocks contain much greater levels of Cr and Ni than Archean tholeiitic and calc-alkaline mafic volcanic rocks.

7.10 Composition of the Archean crust

7.10.1 *The upper crust*

We conclude from the geological, petrographical and geochemical data outlined above that many Archean sedimentary sequences have sampled a complex provenance of volcanic, plutonic, metamorphic and sedimentary rocks. Accordingly, we make some inferences from the composition of the sedimentary strata about the composition of the Archean upper crust, in the same way as for the post-Archean upper crust.

Trace element relationships in Archean sedimentary rocks provide persuasive evidence for an origin by mixing two components. End member compositions are consistent with those of the Archean bimodal suite. Based on a review of the literature, we have estimated average compositions for the two end member components. These are listed in Table 7.8. The variable individual compositions encountered together with the limited amount of data available for many elements limits the precision of such estimates. Based on the data discussed in this chapter and a review of the literature, an estimate of the average Archean pelitic rock is also given in Table 7.8.

Using the same reasoning by which we derived post-Archean upper crustal REE abundances (Chapter 2), we take the average Archean pelitic rock REE pattern to be representative of the Archean upper crust. This average pattern has $\Sigma REE = 105$ ppm, $La_N/Yb_N = 6.8$ and $Eu/Eu^* = 1.0$. It differs from PAAS in having lower total REE, La/Yb ratios and no Eu-anomaly. For the post-Archean upper crust REE abundances, the shale data were lowered by 20% to account for sedimentary rocks with lower REE abundances (e.g. carbonates, sandstones). We have refrained from this adjustment in the Archean because quartzites, carbonates and evaporites were less common in

Table 7.8. Chemical composition of Archean mafic and felsic end member compositions and average Archean mudstone.

	Mafic volcanic	Felsic	Mudstone
SiO_2	50.6	69.7	60.4
TiO_2	1.2	0.5	0.8
Al_2O_3	14.9	15.8	17.1
FeO	12.9	3.1	9.5
MnO	0.2	0.15	0.1
MgO	8.1	1.4	4.3
CaO	9.5	2.9	3.2
Na_2O	2.3	4.3	2.1
K_2O	0.3	2.2	2.3
Σ	100.0	100.1	99.8
Cs	—	—	6.0
Ba	130	400	575
Rb	10	65	60
Sr	175	300	180
Pb	—	—	20
La	3.7	37	20
Ce	9.6	75	42
Pr	1.4	8.5	4.9
Nd	7.1	33	20
Sm	2.3	5.7	4.0
Eu	0.87	1.6	1.2
Gd	3.0	3.7	3.4
Tb	0.58	0.6	0.57
Dy	3.8	3.1	3.4
Ho	0.85	0.6	0.74
Er	2.5	1.7	2.1
Tm	0.36	0.22	0.30
Yb	2.5	1.5	2.0
Lu	0.38	0.22	0.31
La_N/Yb_N	1.0	16.7	6.8
Y	21	14	18
Th	0.8	6.8	6.3
U	0.2	1.8	1.6
Zr	50	200	120
Hf	2	4	3.5
Sn	—	—	5.5
Nb	—	—	9.0
Mo	—	—	1.0
W	—	—	1.0
Cr	330	30	205 (675)*
V	345	45	135
Sc	40	5	20
Ni	185	20	100 (425)*
Co	40	10	40
Cu	—	—	150
Ga	—	—	15
Bi	—	—	0.1
B	—	—	65

*Values are for late Archean mudstones; values in parentheses for early Archean mudstones.

the Archean and there is no detectable difference in average REE contents for Archean shales and greywackes. The absolute REE values (and other incompatible element abundances derived from them) are probably upper limits and minimize the differences with the post-Archean crustal abundances. The sedimentary REE pattern indicates that a 1:1 mix of the bi-modal components best predicts the La_N/Yb_N ratio. REE abundances derived from these mixing proportions agree with the sedimentary REE pattern to within 20%. We take this as reasonable agreement and proceed to construct an Archean upper crustal abundance Table using this mixing model.

The upper crustal abundances for some elements can be determined solely from the sedimentary data, by equating upper crustal REE patterns to the average sedimentary pattern. This procedure makes no assumptions about provenance or source rock lithology. As discussed in Chapter 2, there is a correlation between Th and the LREE in post-Archean sedimentary rocks which can be used to estimate upper crustal Th abundances. Using the familiar interelement ratios (Th/U, K/U, K/Rb) for crustal rocks, it is possible to estimate other elemental abundances. In Section 7.9.2, the Archean sedimentary La/Th ratio was established at 3.5 ± 0.3 (also see Chapter 2, Fig. 2.16). Using an Archean upper crustal La value of 20 ppm, derived from the sediment data, the Archean upper crustal value for Th=5.7 ppm. Using Th/U=3.8, U is 1.5 ppm; for K/U=10^4, K_2O is 1.8%; and for K/Rb=300, Rb is 50 ppm. A value for Sc may be obtained from the following data: Th/Sc=0.43 ± 0.7, La/Sc=1.3 ± 0.2; a Sc average of 14 ppm results. These values and appropriate ratios are compared with those derived from the mixing model in Table 7.9. Considering the uncertainties, the agreement is remarkably good.

In Table 7.10, we present our estimate of the average composition of the Archean upper crust. REE are equated to the sedimentary data, Th, U, K, Sc and Rb are also calculated from the sedimentary data and the other elements are derived from a 1:1 mix of the Archean bimodal suite.

7.10.2 Total crustal compositions

The composition of the Archean upper crust has been estimated primarily by using the REE abundances in Archean sedimentary rocks as a guide to the

Table 7.9. Comparison of Archean upper crust estimates.

	Sediment data	Mixing model
La (ppm)	20.0	20.4
Yb (ppm)	2.0	2.0
La_N/Yb_N	6.8	6.9
K (%)	1.5	1.1
Th (ppm)	5.7	6.8
U (ppm)	1.5	1.8
Rb (ppm)	50	37.5
Sc (ppm)	14	22.5

Table 7.10. Chemical composition of the Archean crust.

	Upper crust (UC)	Total crust (TC)
SiO_2	60.1	57.0
TiO_2	0.8	1.0
Al_2O_3	15.3	15.2
FeO	8.0	9.6
MgO	4.7	5.9
CaO	6.2	7.3
Na_2O	3.3	3.0
K_2O	1.8	0.9
Σ	100.2	99.9

	UC	TC		UC	TC		UC	TC
Na (%)	2.45	2.23	Ni (ppm)	105	130	Gd (ppm)	3.4	3.2
Mg (%)	2.83	3.56	Cu (ppm)	—	80	Tb (ppm)	0.57	0.59
Al (%)	8.10	8.04	Rb (ppm)	50	28	Dy (ppm)	3.4	3.6
Si (%)	28.08	26.63	Sr (ppm)	240	215	Ho (ppm)	0.74	0.77
K (%)	1.50	0.75	Y (ppm)	18	19	Er (ppm)	2.1	2.2
Ca (%)	4.43	5.22	Zr (ppm)	125	100	Tm (ppm)	0.30	0.32
Sc (ppm)	14	30	Ba (ppm)	265	220	Yb (ppm)	2.0	2.2
Ti (ppm)	5000	6000	La (ppm)	20	15	Lu (ppm)	0.31	0.33
V (ppm)	195	245	Ce (ppm)	42	31	Hf (ppm)	3	3
Cr (ppm)	180	230	Pr (ppm)	4.9	3.7	Th (ppm)	5.7	2.9
Mn (ppm)	1400	1500	Nd (ppm)	20	16	U (ppm)	1.5	0.75
Fe (%)	6.22	7.46	Sm (ppm)	4.0	3.4			
Co (ppm)	25	30	Eu (ppm)	1.2	1.1			

average. This approach assumes that Archean sediments sampled a wide provenance and so could be used in a similar manner to post-Archean sedimentary rocks to provide a guide to upper crustal composition. The basis for this assumption is documented in this and the preceding chapter, from which we conclude that sedimentary processes in the Archean indeed provide an adequate sampling of the exposed Archean crust.

The next question to be addressed is the same as for the post-Archean crust. Is the surface composition representative of the whole crust? The Archean crustal thickness, as assessed in Section 7.7, is approximately comparable to that of the present-day crust, so this question can be addressed from heat flow considerations.

The upper crustal composition as derived in the preceding section will produce a heat flow of 40 mW m^{-2} for a 40 km thick crust. This exceeds the presently observed heat flow in Archean terrains by a factor greater than two (Section 5.5). Accordingly, it appears that the heat producing elements in the Archean crust were distributed in a manner analogous to that of the present crust. We conclude in subsequent chapters (see Chapter 9 especially) that fundamentally different mechanisms gave rise to the Archean and post-Archean crustal compositions. The Archean crust is composed dominantly of the bimodal basic–felsic suite. It is reasonable to assume that the lower density felsic components dominate the upper portions of the crust due to buoyancy effects, while the denser basic components lie deeper within the crust. Both components are derived from the mantle, in contrast to the post-Archean granodioritic upper crust, which is derived from intracrustal melting.

To obtain the composition of the total crust, we calculate the appropriate mixture of mafic and felsic rocks consistent with the observed heat flow. A two/one mixture of the mafic and felsic lithologies yields K values of 0.75%, with U=0.75 ppm and Th=2.9 ppm for K/U=10^4 and Th/U=3.8. This will produce a presently observed heat flow of 14 mW m^{-2} for a 30 km crust and 19 mW m^{-2} for a 40 km thick crust. The present crustal heat flow for Archean Provinces is 14±2 m Wm^{-2} (Table 5.5, Fig. 5.12). Considering the uncertainties involved, this is a reasonable agreement. The Archean heat flow data base is small (n=4). Additional factors include some erosional loss of upper felsic layers, and the fact that the compositional estimates for the Archean mafic and felsic rocks are themselves subject to more uncertainty than for the post-Archean.

The composition we derive by this means is close to that of average Archean volcanic rocks derived from large scale sampling of the Superior Province [128]. Accordingly, we use both this source, and the values from the bimodal averages, to arrive at the bulk composition for the Archean crust (Table 7.10).

7.10.3 *Differentiation of the Archean crust*

Despite the uncertainties involved in the Archean crustal compositions listed in Table 7.10, several inferences can be made with some confidence. The highly incompatible elements are enriched in the upper crust. For the most incompatible elements, K, U, Th and Rb, the values adopted indicate an enrichment by a factor of two over the bulk Archean crust. Conversely, most ferromagnesian elements, such as Fe, Cr, Ni, Co and Sc are somewhat depleted in the upper crust.

Of fundamental importance is the fact that the sedimentary data clearly indicate the mechanisms for differentiating the Archean crust were different to the post-Archean. The lack of a negative Eu-anomaly, so characteristic of the post-Archean upper crustal samples, shows that intracrustal partial melting, with plagioclase as a stable residual phase (<40 km), was relatively unimportant before the late Archean ($\approx$2.7 Ae). The presence of early Archean granitic rocks with negative Eu-anomalies and the comparatively rare occurrence of sedimentary rocks with negative Eu-anomalies in high-grade terrains, indicate this process did occur on a local scale. It was not, however, important enough to be reflected in the upper crustal averaging being carried out by sedimentary processes.

The felsic end members of the Archean bimodal suite (e.g. dacites, tonalites) have very steep REE patterns with HREE depletion indicative of garnet fractionation (e.g. see Fig. 7.6(a)). The lack of negative Eu-anomalies on average indicate plagioclase was not a significant residual phase. Thus the primary mechanism for differentiating the Archean crust probably was partial melting of high-grade mafic rocks (amphibolite, eclogite) at depths in excess of 40 km [e.g. 129]. The vertical distribution within the crust was controlled principally by density and buoyancy effects, with the less dense felsic rocks occurring nearer the surface.

7.11 Summary

1 Archean terrains can be divided into three major lithological associations: low-grade terrains, high-grade terrains, and cratonic terrains.

2 Archean volcanic rocks are commonly characterized by a bimodal association of mafic–ultramafic and felsic compositions. Andesitic rocks are scarce or absent in many greenstone belts. Where present, they are more abundant towards the top of greenstone sequences. Archean felsic volcanics typically have steep REE patterns with HREE-depletion indicating garnet was a fractionating phase. These felsic volcanics (mainly dacites) are chemically indistinguishable from the Archean tonalite–trondhjemite–granodiorite plutonic suite.

3 Sedimentary rocks in greenstone belts occur in two principal settings; as purely volcaniclastic and chemical sediments in close association with volcanic rocks, and as thick terrigenous clastic and volcaniclastic turbidite deposits lying above volcanic sequences.

4 No single tectonic model is applicable to Archean greenstone belts. Many thick sedimentary sequences indicate deposition at or near the edge of continents.

5 High-grade terrains commonly contain evidence of shallow water facies sediments and therefore probably are not metamorphosed facies equivalents of low-grade terrains.

6 The Archean continental crust attained thicknesses comparable to that of the present day crust at least by 3.0–2.7 Ae and probably much earlier.

7 Archean clastic sedimentary REE patterns differ from post-Archean patterns in being considerably more variable, having lower ΣREE and La/Yb and typically lacking Eu-anomalies. Archean sedimentary rocks also have lower Th/Sc, La/Sc ratios and higher La/Th ratios.

8 The trace element patterns of Archean sedimentary rocks indicate an origin dominated by mixing of the Archean igneous bimodal suite of mafic volcanics and felsic volcanic and plutonic rocks.

9 Sedimentary rocks from early Archean sedimentary sequences have much higher levels of Cr and Ni than their late Archean counterparts. This enrichment is due to a combination of provenance and secondary enrichment effects.

10 REE and other trace elements in the Archean upper crust can be modelled from the sedimentary data and indicate ΣREE=105 ppm, La_N/Yb_N=6.8, Eu/Eu*=1.0, K=1.5%, Th=5.7 ppm, U=1.5 ppm, Sc=14 ppm, Rb=50 ppm and Rb/Sr=0.21 ppm. These values are equivalent to a 1:1 mixture of the mafic and felsic end members of the Archean bimodal igneous

suits. Using these proportions, the upper crustal abundances of the other elements can be estimated.

11 The abundances of K, Th and U in the total Archean crust can be calculated from heat flow measurements. The abundances obtained (K=0.75%, Th=2.9 ppm, U=0.75 ppm) are equivalent to a 2:1 mixture of mafic and felsic end members of the bimodal suite and are also equivalent to the average volcanic rock from the Archean Superior Province. This allows the total crustal abundances of the other elements to be estimated. Key values include $SiO_2=57.0\%$, Rb=28 ppm, Rb/Sr=0.13, $La_N/Yb_N=4.6$, Cr=230 ppm, Ni=130 ppm.

12 The chemical differentiation of the Archean crust was related to melting at mantle depths. This contrasts sharply with the intracrustal processes responsible for differentiation of the post-Archean crust.

Notes and references

1 Condie, K.C. (1981) *Archean Greenstone Belts*. Elsevier.
2 Lowe, D.R. (1982) Comparative sedimentology of the principal volcanic sequences of Archean greenstone belts in South Africa, Western Australia and Canada: Implications for crustal evolution. *PCR*, **17**, 1.
3 Anhaeusser, C.R. (1982) Archean greenstone terrains: geological evolution and metallogenesis. *Rev. Brasil Geosc.*, **12**, 1.
4 Windley, B.F. (ed.) (1976) *The Early History of the Earth*. Wiley; (1977) *The Evolving Continents*. Wiley.
5 Tarling, D.H. (ed.) (1978) *Evolution of the Earth's Crust*. Academic Press.
6 Barker, F. (ed.) (1979) *Trondhjemites, Dacites and Related Rocks*. Elsevier.
7 Windley, B.F. & Naqvi, S.M. (eds) (1978) *Archean Geochemistry*. Elsevier.
8 Glover, J.E. & Groves, D.I. (eds) (1981) Archean geology. *Geol. Soc. Austr. Spec. Pub.* **7**.
9 Kroner, A. (ed.) (1981) *Precambrian Plate Tectonics*. Elsevier.
10 Tankard, A.J. et al. (1981) *Crustal Evolution of Southern Africa: 3.8 Billion Years of Earth History*. Springer-Verlag.
11 Gorman, B.E. et al. (1978) On the structure of Archean greenstone belts. *PCR*, **6**, 23.
12 De Wit, M.J. (1982) Gliding and overthrust nappe tectonics in the Barberton greenstone belt. *J. Struct. Geol.*, **4**, 117.
13 Lowe, D.R. (1980) Archean sedimentation. *Ann. Rev. Earth Planet. Sci.*, **18**, 145.
14 Bickle, M.J. & Eriksson, K.A. (1982) Evolution and subsidence of early Precambrian sedimentary basins. *Phil. Trans. Roy. Soc.*, **A305**, 225.
15 Goodwin, A.M. et al. (1972) The Superior Province. *Geol. Soc. Can. Spec. Pap.*, **11**, 527.
16 Henderson, J.B. (1981) Archean basin evolution in the Slave Province, Canada. In: [9], 213.
17 Anhaeusser, C.R. (1976) Archean metallogeny in Southern Africa. *Econ. Geol.*, **71**, 16.
18 Barager, W.R.A. (1966) Geochemistry of the Yellowknife volcanic rocks. *CJES*, **5**, 773.
19 Condie, K.C. (1980) Archean andesites. In: Thorpe, R.S. (ed.), *Orogenic Andesites and Related Rocks*. Wiley. 575.
20 Glikson, A.Y. & Hickman, A.H. (1981) Geochemical stratigraphy and petrogenesis of Archean basic–ultrabasic volcanic units, eastern Pilbara block, Western Australia. *Geol. Soc. Austr. Spec. Pub.*, **7**, 287.
21 Glikson, A.Y. & Hickman, A.H. (1981) Geochemistry of Archean volcanic successions, eastern Pilbara Block, Western Australia. *BMR Record 1981/36*.
22 Ayres, L.D. (1983) Bimodal volcanism in Archean greenstone belts exemplified by greywacke composition, Lake Superior Park, Ontario. *CJES*, **20**, 1168.
23 *Basaltic Volcanism on the Terrestrial Planets* (1981). Pergamon Press. 5–29.
24 Jahn, B-M. & Sun, S-S. (1979) Trace element distributions and isotopic composition of Archean greenstones. *PCE*, **11**, 597.
25 Arndt, N.T. & Nisbet, E.G. (eds) (1982) *Komatiites*. Allen & Unwin, London.

26 MacGeehan, P.J. & MacLean, W.H. (1980) An Archean sub-seafloor geothermal system, 'calc-alkali' trends, and massive sulphide genesis. *Nature*, **286**, 767.
27 Data used as representative analyses of Archean volcanic rocks in Figures 7.3–7.6 were taken from the following sources: Sun, S.-S. & Nesbitt, R.W. (1978). *CMP*, **65**, 301 (Fig. 7.3—87J, 331/347, 422/95; Fig. 7.4—331/475, 7360a; Fig. 7.5—RH-170); Jahn, B.-M. *et al.* (1980). *J. Pet.*, **21**, 201 (Fig. 7.3—575; Fig. 7.4—S145; Fig. 7.6—S29); Jahn, B.-M. *et al.* (1982). *CMP*, **80**, 25 (Fig. 7.3—5046); Hawkesworth, C.J. & O'Nions, R.K. (1977). *J. Pet.*, **18**, 487 (Fig. 7.4—ϕ9); Smith, I.E.M. (1980). *CJES*, **17**, 1292 (Fig. 7.4—595, 604); Jahn, B.-M. *et al.* (1981). *CGA*, **45**, 1633 (Fig. 7.5—3809; Fig. 7.6 —3815, 3822); Taylor, S.R. & Hallberg J.A. (1977). *CGA*, **41**, 1125 (Fig. 7.5—M-1, M-2; Fig. 7.6 —28, 31); Condie, K.C. & Baragar, W.R.A. (1974). *CMP*, **45**, 237 (Fig. 7.6—97); Arth, J.G. & Hanson, G.N. (1975). *CGA*, **39**, 325 (Fig. 7.6—MN-51, MN-53); Glikson, A.Y. (1976). *CGA*, **40**, 1261 (Fig. 7.6—78); Jenner, G.A. *et al.* (1981). *CGA*, **45**, 1111 (Fig. 7.6—YK27).
28 Gill, J.B. (1981) *Orogenic Andesites and Plate Tectonics*. Springer-Verlag.
29 Taylor, S.R. & Hallberg, J.A. (1977) Rare-earth elements in the Marda calc-alkaline suite: An Archean geochemical analogue of Andean-type volcanism. *GCA*, **41**, 1125.
30 McCulloch, M.T. & Compston, W. (1981) Sm–Nd age of Kambalda and Kanowna greenstones and heterogeneity in the Archean mantle. *Nature*, **294**, 322.
31 Hamilton, P.J. *et al.* (1978) Sm–Nd isotopic investigations of Isua supracrustals and implications for mantle evolution. *Nature*, **272**, 41.
32 Hamilton, P.J. *et al.* (1983) Sm–Nd studies of Archean metasediments and metavolcanics from West Greenland and their implications for the earth's early history. *EPSL*, **62**, 263.
33 Eriksson, K.A. (1980) Transitional sedimentation styles in the Moodies and Fig Tree Groups, Barberton Mountain Land, South Africa: evidence favouring an Archean continental margin. *PCR*, **12**, 141.
34 Walker, R.G. (1978) A critical appraisal of Archean basin–craton complexes. *CJES*, **15**, 1213.
35 Eriksson, K.A. (1977) Tidal deposits from the Archaean Moodies Group, Barberton Mountain Land, South Africa. *Sed. Geol.*, **18**, 255.
36 Eriksson, K.A. (1978) Alluvial and destructive beach facies from the Archaean Moodies Group, Barberton Mountain Land, South Africa and Swaziland. *In:* Miall. A.D. (ed.), *Fluvial Sedimentology*. Can. Soc. Petrol. Geol. Mem., **5**, 287.
37 Eriksson, K.A. (1979) Marginal marine depositional processes from the Archaean Moodies Group, Barberton Mountain Land, South Africa: evidence and significance. *PCR*, **8**, 153.
38 Eriksson, K.A. (1980) Hydrodynamic and paleogeographic interpretation of turbidite deposits from the Archean Fig Tree Group of the Barberton Mountain Land, South Africa. *GSA Bull.*, **91**, 21.
39 Eriksson, K.A. (1981) Archean platform-to-trough sedimentation, East Pilbara Block, Australia. *Geol. Soc. Austr. Spec. Pub.*, **7**, 235.
40 Eriksson, K.A. (1982) Sedimentation patterns in the Barberton Mountain Land, South Africa, and the Pilbara Block, Australia: evidence for Archean rifted continental margins. *Tectonophysics*, **81**, 179.
41 Eriksson, K.A. (1982) Archean and early Proterozoic sedimentation styles in the Kaapvaal Province, South Africa and Pilbara Block, Australia. *Rev. Brasil Geosc.*, **12**, 121.
42 Ojakangas, R.W. (1972) Archean volcanogenic greywackes of the Vermilion District, northeastern Minnesota. *GSA Bull.*, **83**, 429.
43 Walker, R.G. & Pettijohn, F.J. (1971) Archean sedimentation: analysis of the Minnitaki Basin, northwestern Ontario, Canada. *GSA Bull.*, **42**, 2099.
44 Hunter, D.R. (1974) Crustal development in the Kaapvaal Craton. I. The Archean. *PCR*, **1**, 259.
45 Hickman, A.H. (1981) Crustal evolution in the Pilbara Block, Western Australia. *Geol. Soc. Austr. Spec. Pub.*, **7**, 57.
46 Anhaeusser, C.R. (1973) The evolution of the early Precambrian crust of Southern Africa. *Phil. Trans. Roy. Soc.*, **A273**, 359.
47 Glikson, A.Y. (1978) On the basement of Canadian greenstone belts. *Geosci. Can.*, **5**, 3.
48 Hamilton, P.J. *et al.* (1979) Sm–Nd dating on Onverwacht Group Volcanics, Southern Africa. *Nature*, **279**, 298.
49 Carlson, R.W. *et al.* (1983) Sm–Nd age and isotopic systematics of the bimodal ancient gneiss complex, Swaziland. *Nature*, **305**, 701.

50 Gee, R.D. *et al.* (1981) Crustal development in the Archean Yilgarn Block, Western Australia. *Geol. Soc. Austr. Spec. Pub.*, **7**, 43.
51 Archibald, N.J. *et al.* (1981) Evolution of Archean crust in the Eastern Goldfields Province of the Yilgarn Block, Western Australia. *Ibid.*, **491**.
52 Baragar, W.R.A. & McGlynn, J.C. (1976) Early Archean basement in the Canadian Shield: A review of the evidence. *GSC Pap.*, **76-14**.
53 Bridgwater, D. *et al.* (1978) The development of the Archean gneiss complex of the North Atlantic region. *In:* [5], 19.
54 McGregor, V.R. (1979) Archean gray gneisses and the origin of the continental crust. *In:* [6], 169.
55 Sutton, J. & Watson, J.V. (1969) Scourian–Laxfordian relationships in the Lewisian of north-west Scotland. *Geol. Ass. Can. Spec. Pap.*, **5**, 119.
56 Tarney, J. *et al.* (1979) Geochemistry of Archean tonalitic and tronhjemitic gneisses from Scotland and East Greenland. *In:* [6], 275.
57 Nieuwland, D.A. & Compston, W. (1981) Crustal evolution in the Yilgarn Block near Perth, Western Australia. *Geol. Soc. Austr. Spec. Pap.*, **7**, 159.
58 Ermanovics, I.F. *et al.* (1979) Petrochemistry and tectonic setting of plutonic rocks of the Superior Province in Manitoba. *In:* [6], 323.
59 Park, R.G. (1981) Origin of horizontal structure in high-grade Archean terrains. *Geol. Soc. Austr. Spec. Pap.*, **7**, 481.
60 Collerson, K.D. *et al.* (1981) Geochronology and evolution of Late Archean gneisses in Northern Labrador: An example of reworked sialic crust. *Geol. Soc. Austr. Spec. Pap.*, **7**, 205.
61 Barker, F. *et al.* (1981) Tonalites in crustal evolution. *Phil. Trans. Roy. Soc.*, **A301**, 293.
62 Barker, F. & Arth, J.G. (1976) Generation of trondhjemitic–tonalitic liquids and Archean bimodal trondhjemite–basalt suites. *Geology*, **4**, 596; Barker, F. & Peterman, Z.E. (1974) Bimodal tholeiitic-dacitic magmatism and the early Precambrian crust. *PCR*, **1**, 1.
63 McGregor, V.R. & Mason, B. (1977) Petrogenesis and geochemistry of metabasaltic and metasedimentary enclaves in the Amîtsoq gneisses, West Greenland. *Am. Mineral.*, **62**, 887.
64 Gill, R.C.O. *et al.* (1981) The geochemistry of the earliest known basic metavolcanic rocks at Isua, West Greenland. *Geol. Soc. Austr. Spec. Pub.*, **7**, 313.
65 Windley, B.F. & Smith, J.V. (1976) Archean high grade complexes and modern continental margins. *Nature*, **260**, 671.
66 Kroner, A. (1984) Evolution, growth and stabilization of the Precambrian lithosphere. *PCE*, **15**, Chap. 4.
67 Green, D.H. (1972) Archean greenstone belts may include terrestrial equivalents of lunar maria. *EPSL*, **15**, 263.
68 The generation of basaltic lavas by impact melting, although frequently proposed, is not a viable hypothesis on several grounds. See pp.189–191. *In:* Taylor, S.R. (1975) *Lunar Science: A Post-Apollo View*. Pergamon Press.
69 Condie, K.C. (1981) *Plate Tectonics and Crustal Evolution* (2nd Ed.). Pergamon Press.
70 Taylor, S.R. & McLennan, S.M. (1981) The rare-earth element evidence in Precambrian sedimentary rocks: Implications for crustal evolution. *In:* [9], 527.
71 Crook, K.A.W. (1974) Lithogenesis and geotectonics. *SEPM Spec. Pub.*, **19**, 304.
72 Henderson, J.B. (1981) Archean basin evolution in the Slave Province, Canada. *In:* [9], 213.
73 Padgham, W.A. (1981) Archean crustal evolution—A glimpse from the Slave Province. *Geol. Soc. Austr. Spec. Pub.*, **7**, 99.
74 Dimroth, E. *et al.* (1982) Evolution of the south-central part of the Archean Abitibi Belt, Quebec. Part I, *CJES*, **19**, 1729; Part II, *CJES*, **20**, 1355; Part III, *CJES*, **20**, 1374.
75 Nutman, A.P. *et al.* (1984) Stratigraphic and geochemical evidence for the depositional environment of the early Archean Isua supracrustal belt, West Greenland. *PCR*, **25**, 365.
76 Young, G.M. (1978) Some aspects of the evolution of the Archean crust. *Geosci. Can.*, **5**, 140.
77 Fyfe, W.S. (1978) The evolution of the earth's crust: modern plate tectonics to ancient hot spot tectonics. *Chem. Geol.*, **23**, 89.
78 Hargraves, R.B. (1976) Precambrian geologic history. *Science*, **193**, 363. Although the views expressed in this paper have biblical sanction ('The waters stood above the mountains', Psalm 104, Verse 6, The Bible, King James Version), the common and widespread presence of sedimentary rocks in the earliest rocks appears to refute both authorities.

79 Tarney, J. & Windley, B.F. (1977) Chemistry, thermal gradients and evolution of the lower continental crust. *J. geol. Soc. Lond.*, **134**, 153.
80 Collerson, K.D. & Fryer, B.J. (1978) The role of fluids in the formation and subsequent development of early continental crust. *CMP*, **67**, 151.
81 Perkins, D. & Newton, R.C. (1981) Charnockite geobarometers based on coexisting garnet–pyroxene–plagioclase–quartz. *Nature*, **292**, 144.
82 Newton, R.C. (1978) Experimental and thermodynamic evidence for the operation of high pressures in Archean metamorphism. *In:* Windley, B.F. & Naqvi, S.M. (eds), *Archean Geochemistry*. Elsevier. 221.
83 Grambling, J.A. (1981) Pressures and temperatures in Precambrian metamorphic rocks. *EPSL*, **53**, 63.
84 Harris, N.B.W. *et al.* (1982) Geobarometry, geothermometry and late Archean geotherms from the granulite facies terrain of South India. *J. Geol.*, **90**, 509.
85 Saggerson, E.P. (1973) Metamorphic facies series in Africa: a contrast. *Geol. Soc. S. Afr. Spec. Pub.*, **2**, 227.
86 Dickinson, B.B. & Watson, J. (1976) Variations in crustal level and geothermal gradient during the evolution of the Lewisian complex of northwest Scotland. *PCR*, **3**, 363.
87 Bickle, M.J. (1978) Heat loss from the earth: a constraint on Archaean tectonics from the relation between geothermal gradients and the rate of plate production. *EPSL*, **40**, 301.
88 Nisbet, E.G. & Fowler, C.M.R. (1983) A model for Archean plate tectonics. *Geology*, **11**, 376.
89 Arndt, N.T. (1983) Role of a thin komatiite-rich oceanic crust in the Archean plate-tectonic process. *Geology*, **11**, 372.
90 Campbell, I.H. & Jarvis, G.T. (1984) Mantle convection and early crustal evolution. *PCR*, **26**, 15.
91 McLennan, S.M. & Taylor, S.R. (1984) Archean sedimentary rocks and their relation to the composition of the Archean crust. *In:* Kroner, A. (ed.), *Archean Geochemistry*. Springer-Verlag. 47.
92 McLennan, S.M. *et al.* (1983) Geochemistry of Archean shales from the Pilbara Supergroup, Western Australia. *GCA*, **47**, 1211.
93 McLennan, S.M. *et al.* (1983) Geochemical evolution of Archean shales from South Africa. I. The Swaziland and Pongola Supergroups. *PCR*, **22**, 93.
94 Condie, K.C. *et al.* (1970) Petrology and geochemistry of early Precambrian greywackes from the Fig Tree Group, South Africa. *GSA Bull.*, **81**, 2759.
95 Wildeman, T.R. & Condie, K.C. (1973) Rare earths in Archean greywackes from Wyoming and from the Fig Tree Group, South Africa. *GCA*, **37**, 439.
96 Nance, W.B. & Taylor, S.R. (1977) Rare earth element patterns and crustal evolution. II. Archean sedimentary rocks from Kalgoorlie, Australia. *GCA*, **41**, 225.
97 McLennan, S.M. (1981) *Trace Element Geochemistry of Sedimentary Rocks: Implications for the Composition and Evolution of the Continental Crust*. Ph.D. Thesis, Australian National University.
98 Bavinton, O.A. & Taylor, S.R. (1980) Rare earth element abundances in Archean metasediments from Kambalda, Western Australia. *GCA*, **44**, 639.
99 Jenner, G.A. *et al.* (1981) Geochemistry of the Yellowknife Supergroup. *GCA*, **45**, 1111.
100 McLennan, S.M. *et al.* (1984) Geochemistry of Archean metasedimentary rocks from West Greenland. *GCA*, **48**, 1.
101 Pidgeon, R.T. (1978) 3450 m.y.-old volcanics in the Archaean layered greenstone succession of the Pilbara Block, Western Australia. *EPSL*, **37**, 421; Hamilton, P.J. *et al.* (1981) Sm–Nd dating of the North Star Basalt, Warrawoona Group, Pilbara Block, Western Australia. *Geol. Soc. Austr. Spec. Pub.*, **7**, 187.
102 Oversby, V.M. (1976) Isotopic ages and geochemistry of Archean acid igneous rocks from the Pilbara, Western Australia. *GCA*, **40**, 817.
103 Trendall, A. (1983) *The Hamersley Basin*. (Preprint.)
104 Eriksson, K.A. (1984), pers. comm.
105 Danchin, R.V. (1967) Chromium and nickel in the Fig Tree Shale from South Africa. *Science*, **158**, 261.
106 Glikson, A.Y. (1971) Archean geosynclinal sedimentation near Kalgoorlie, Western Australia. *Geol. Soc. Austr. Spec. Pub.*, **3**, 443.

107 Gemuts, I. & Theron, A. (1975) The Archean between Coolgardie and Norseman—stratigraphy and mineralization. *In: Economic Geology of Australia and Papua New Guinea, 1, Metals.* Australas. Inst. Min. Metall., 66.

108 Bavinton, O.A. (1981) The nature of sulfidic metasediments at Kambalda. *Econ. Geol.*, **76**, 1606.

109 Henderson, J.B. (1975) Sedimentology of the Archean Yellowknife Supergroup at Yellowknife, District of Mackenzie. *GSC Bull.*, **246**.

110 Drury, S.A. (1979) Rare-earth and other trace element data bearing on the origin of Archean granitic rocks from Yellowknife, Northwest Territories. *CJES*, **16**, 809.

111 Condie, K.C. (1967) Geochemistry of early Precambrian graywackes from Wyoming. *GCA*, **31**, 2135.

112 Wildeman, T.R. & Haskin, L.A. (1973) Rare earths in Precambrian sediments. *GCA*, **37**, 419; Arth, J.G. & Hanson, G.N. (1975) Geochemistry and origin of the early Precambrian crust of northeastern Minnesota. *GCA*, **39**, 325.

113 Kerrich, R. & Fryer, B.J. (1979) Archean precious-metal hydrothermal systems, Dome Mine, Abitibi Belt. II. REE and oxygen isotope relations. *CJES*, **16**, 440.

114 McGregor, V.R. (1973) The early Precambrian gneisses of the Godthåb district, West Greenland. *Phil. Trans. Roy. Soc.*, **A273**, 343.

115 Nutman, A.P. & Bridgwater, D. (1983) Deposition of Malene supracrustal rocks on an Amîtsoq basement in outer Ameralik, southern West Greenland. *Rapp Grønlands geol. Unders.*, **112**, 43.

116 Taylor, P.N. *et al.* (1980) Crustal contamination as an indicator of the extent of early Archean continental crust. *GCA*, **44**, 1437.

117 Dymek, R.F. *et al.* (1983) The Malene metasedimentary rocks on Rypeø, and their relationship to Amîtsoq gneisses. *Rapp Grønlands geol. Unders.*, **112**, 53.

118 Boak, J.L. *et al.* (1982) Early crustal evolution: constraints from variable REE patterns in metasedimentary rocks from the 3800 Ma Isua supracrustal belt, West Greenland. *LPS XIII*, 51.

119 Naqvi, S.M. (1983) Geochemistry of some unusual Early Archean sediments from Dharwar Craton, India. *PCR*, **22**, 125.

120 Mueller, P.A. *et al.* (1982) Precambrian evolution of the Beartooth Mountains, Montana-Wyoming, USA. *Rev. Brasil. Geosc.*, **12**, 215.

121 Sun, S-S. & Nesbitt, R.W. (1978) Petrogenesis of Archean ultrabasic and basic volcanics: evidence from rare earth elements. *CMP*, **65**, 301.

122 McLennan, S.M. *et al.* (1980) Rare earth element–thorium correlations in sedimentary rocks, and the composition of the continental crust. *GCA*, **44**, 1833.

123 Cameron, E.M. & Garrels, R.M. (1980) Geochemical compositions of some Precambrian shales from the Canadian Shield. *Chem. Geol.*, **28**, 181.

124 Taylor, S.R. (1977) Island arc models and the composition of the continental crust. *AGU Ewing Series*, **I**, 325.

125 Jakěs, P. & Taylor, S.R. (1974) Excess Eu content in Precambrian rocks and continental evolution. *GCA*, **38**, 793.

126 Langmuir, C.H. *et al.* (1978) A general mixing equation with applications to Icelandic basalts. *EPSL*, **37**, 380.

127 Nance, W.B. & Taylor, S.R. (1976) Rare-earth element patterns and crustal evolution. I. Australian post-Archean sedimentary rocks. *GCA*, **40**, 1539.

128 Goodwin, A.M. (1977) Archean volcanism in Superior Province, Canadian Shield. *Geol. Ass. Can. Spec. Pap.* **16**, 205.

129 O'Nions, R.K. & Pankhurst, R.J. (1978) Early Archean rocks and geochemical evolution of the Earth's crust. *EPSL*, **38**, 211.

8

The Archean–Proterozoic Boundary

8.1 The Archean–Proterozoic boundary defined

The marked differences in composition, revealed by trace element abundances in clastic sedimentary rocks, between the Archean and post-Archean upper crust directs attention to the Archean–Proterozoic boundary. We interpret these changes as recording a marked change in the composition of the exposed crust, from more mafic to more felsic, during the late Archean. The contrast between the granite–greenstone belt terrains of the Archean and the relatively undeformed Proterozoic supracrustal sequences has been noted and discussed for over a century. In 1845, Sir William Logan, of the Geological Survey of Canada, first recognized the 'great unconformity' between what is now termed the Gowganda Formation and the underlying Archean granitic basement [1]. In Chapters 9 and 10, we discuss evolutionary models which have been proposed to account for compositional and tectonic changes associated with the Archean–Proterozoic boundary. In this chapter, we discuss the geology and geochemistry of some key early Proterozoic and late Archean sedimentary sequences that record the change in upper crustal composition.

The subdivision of Precambrian time has always met with considerable difficulty. Over the past 35 years, new (or reworked) schemes for classification of the Precambrian have appeared at the average rate of about two each year [2]. An acceptable definition of the Archean–Proterozoic boundary has proven particularly elusive. In this book we are not primarily concerned with stratigraphic subdivision or nomenclature, but some comments are appropriate because the fundamental change in upper crustal composition is associated with what conventionally has been termed the Archean–Proterozoic boundary.

The original definitions of Archean and Proterozoic were developed in the Canadian Shield. Traditionally, in the Canadian Shield, the Archean–Proterozoic boundary was taken to be the 'great unconformity' between the older generally more deformed granite–greenstone terrains and younger, generally less deformed epicontinental supracrustal successions, typified by the Huronian Supergroup and Animike Group around Lake Huron and Lake Superior. More precisely, Alcock [3] defined the boundary as the beginning of deposition of Huronian rocks, north of Lake Huron. Stockwell [4, 5] has taken a fundamentally different approach, using the cessation of intense volcanic–plutonic activity (Kenoran Orogeny) to mark the boundary, because this could be dated by isotopic methods. The most recent age determined for this event in the Canadian Shield is 2510–2500 m.y. [4].

Attempts have been made to peg the Archean–Proterozoic boundary exactly at 2500 m.y. [6] but this approach is inappropriate in the present context. Changes in trace element patterns in the upper crust are closely related to the major geological break which traditionally has been termed the Archean–Proterozoic boundary. These changes are not of the same age everywhere and in at least one case (Hamersley Basin) perhaps not even related to a major unconformity. In South Africa, the age of the major unconformity between granite–greenstone terrains and younger epicontinental supracrustals is several hundred million years older than the traditional date of 2500 m.y. (see discussion in [7], and below).

The sedimentary rocks from the Huronian Supergroup, Wopmay Orogen and Pine Creek Geosyncline are all younger than 2.5 Ae and accordingly are of early Proterozoic age by popular definition. In each case, they rest with a profound unconformity on granite–greenstone terrains >2.5 Ae (except the basement of the Pine Creek succession, which may be slightly younger). Thus the unconformities at the base of these sedimentary sequences mark the conventional Archean–Proterozoic boundary without ambiguity. The age and tectonic relations of the Mount Bruce Supergroup are less clear. The Fortescue Group, at the base of the Mount Bruce Supergroup, may be older than 2.7 Ae and it is not clear if a major tectono-thermal event immediately preceded deposition [8]. In this example, it is convenient to employ the traditional nomenclature and place the Archean–Proterozoic boundary for that locality at the base of the Mount Bruce Supergroup.

8.2   Geochemistry of early Proterozoic sedimentary rocks

8.2.1   Huronian Supergroup

The Huronian Supergroup is preserved as an arcuate belt on the north shore of Lake Huron, Canada. The sequence, containing up to 12 000 m of mainly terrigenous sedimentary rocks, rests unconformably on Archean granite–greenstone basement rocks (Fig. 8.1). The maximum thickness is found in the south and the entire sequence thins, with several units wedging out, to the north. The maximum limit for the age of the Huronian is probably about 2.5 Ae. The sequence is cut by the Nipissing Diabase dated at about 2.1 Ae [9]. The actual period of deposition was probably very much less than these limits. Frarey et al. [10] have determined an age of 2.33 Ae for the Creighton Granite which intrudes the lower portion of the Huronian Supergroup. These authors argue that this granite post-dates the Huronian Supergroup and thus 2.3 Ae is a minimum age. Alternatively, this granite may be an intrusive equivalent to the lower Huronian volcanics [11] which would make 2.3 Ae a maximum age for the Huronian.

The basal Elliot Lake Group has the most complex stratigraphy of the Huronian sequence. It is notable for the major deposits of uraniferous quartz-pebble conglomerates found, for example, in the Matinenda Formation near the town of Elliot Lake. This group is also unique in that it contains abundant, discontinuous volcanic and volcaniclastic rocks at and near the base (Thessalon, Frood-Stobie, Copper Cliff Formations). The Matinenda Forma-

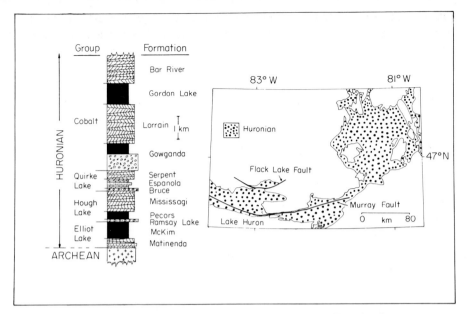

Fig. 8.1. Location and generalized stratigraphy of the early Proterozoic Huronian Supergroup exposed on the north shore of Lake Huron.

tion consists mainly of drab vari-coloured cross-bedded subarkoses, with lesser quartz-pebble conglomerates. A fluvial to fluvial–deltaic origin is likely [12]. The McKim Formation consists of argillites, siltstones and sandstones. Many features are indicative of turbidite deposition.

Higher in the succession, the Hough Lake, Quirke Lake and lower Cobalt Groups comprise three major sedimentary cycles [13]. Each cycle begins with a tillite unit (Ramsay Lake, Bruce, lower Gowganda Formations) of probable glacial origin. The thickest and most extensive of these, the Gowganda Formation, was probably derived from a continental glaciation covering most of the Canadian Shield [14]. A fine-grained unit is found in the Pecors, Espanola and upper Gowganda Formations above each tillite. The Espanola Formation is notable because it is the only Huronian formation which contains abundant carbonate material [15]. The Mississagi and Serpent Formations are of fluvial origin and the Lorrain is probably of mixed fluvial–marine origin.

The upper two Huronian formations have not been studied extensively. The Gordon Lake Formation consists of mudstones with minor evaporite minerals, cherty mudstones, siltstones and minor sandstones. A shallow marine depositional environment, possibly on tidal flats, has been proposed for this unit. The Bar River Formation comprises orthoquartzite with lesser siltstone, haematitic sandstones and siltstones and arkose. This formation may have been deposited by shallow marine, beach and aeolian processes.

Much petrographical and palaeocurrent data are available which constrain the source of Huronian sedimentary rocks. Palaeocurrent directions indicate that source rocks were in the Archean Superior Province, to the north, west and east [12, 14, 16]. Petrographic features support this interpretation. Young

193

[16] has proposed an aulocogen model for the tectonic and depositional history of early Proterozoic sedimentary successions around Lake Huron and Lake Superior. In this model, the Huronian represents the northern edge of an easterly-trending fault-bounded trough that may have opened into an ocean where the Grenville Province is now located. Others have suggested deposition on the southern margin of an east-trending continental margin [e.g. 17] or an intracratonic rift [18].

Table 8.1. Average chemical composition of fine-grained sedimentary rocks from the Huronian Supergroup [19, 20].

	McKim (1)	Pecors (2)	Espanola (3)	Serpent (4)	Gordon Lake (5)
SiO_2	62.66	63.92	60.31	74.31	71.69
TiO_2	0.81	0.80	0.51	0.47	0.41
Al_2O_3	21.47	19.45	14.15	13.01	15.96
FeO	6.46	6.72	4.70	3.59	3.36
MnO	0.06	0.06	0.15	0.02	0.02
MgO	2.31	2.86	6.22	1.45	1.67
CaO	0.76	0.71	8.61	0.98	0.26
Na_2O	1.40	1.66	1.64	1.80	1.35
K_2O	3.98	3.72	3.61	4.28	5.15
P_2O_5	0.08	0.10	0.10	0.09	0.12
Σ	99.99	100.00	100.00	100.00	99.99
LOI	4.31	4.01	11.24	2.13	3.25
Ba	830	838	1254	—	1037
Rb	147	139	142	—	201
Sr	160	146	83	—	25
La	29	36	28	33	49
Ce	67	82	60	77	115
Nd	29	34	24	30	51
Sm	5.7	6.5	4.3	5.4	9.3
Eu	1.5	1.6	0.93	1.2	1.8
Gd	5.1	5.6	3.7	4.5	7.6
Dy	4.6	4.6	2.8	3.0	5.8
Er	2.7	2.8	1.5	1.9	3.9
Yb†	2.4	2.6	1.3	1.8	3.8
ΣREE	157	187	133	167	263
La_N/Yb_N	8.2	9.4	14.6	12.4	8.7
Eu/Eu*	0.85	0.81	0.71	0.74	0.65
Y	27	27	16	—	52
Zr	152	186	129	—	284
Cr	161	163	103	—	56
Ni	75	81	43	—	14
Cu	65	71	24	—	46
Zn	91	90	42	—	23
Li	37	35	52	—	18
Ga	26	24	18	—	19

†Estimated from chondrite-normalized diagrams.
1. Average of 8 mudstones from McKim Fm.
2. Average of 13 mudstones (8 for REE) from Pecors Fm.
3. Average of 4 calcareous siltstones from Espanola Fm.
4. Average of 3 mudstones (2 for major elements) from Serpent Fm.
5. Average of 7 mudstones (6 for REE) from Gordon Lake Fm.

Recognition of the difference between Archean and post-Archean upper crustal REE patterns led to a detailed trace element study of this classical early Proterozoic sedimentary succession [19, 20]. In Table 8.1 the average chemical compositions of selected fine-grained units from the Huronian Supergroup are

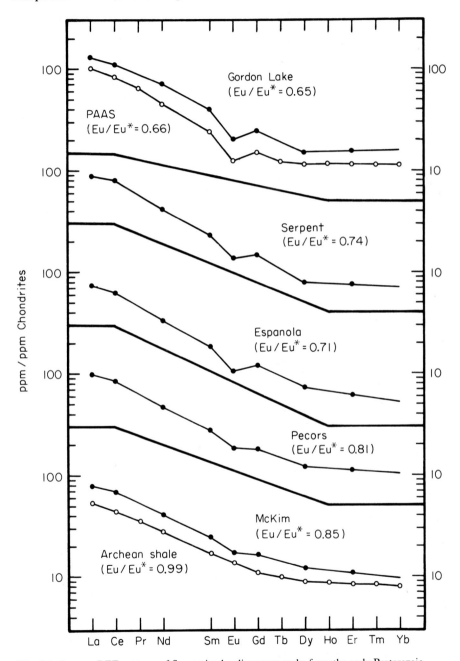

Fig. 8.2. Average REE patterns of fine-grained sedimentary rocks from the early Proterozoic Huronian Supergroup, presented in stratigraphic order. (Data from Table 8.1.) Lower Huronian formations are similar to average Archean shales whereas REE patterns from the top of the Huronian are indistinguishable from PAAS.

listed. The Gowganda Formation is not considered because of its distinct origin from continental glaciation (see [20]). Only the middle siltstone member of the carbonate-rich Espanola Formation was analysed because it contains the least amount of carbonate (<15%) and is therefore most comparable to other Huronian mudstone units [19].

Average REE patterns of the various units are shown, in stratigraphic order, in Fig. 8.2. It is clear that there is a gradual change in REE patterns from the bottom to top of the sequence. Those in the stratigraphically lower McKim and Pecors Formations have Archean-style patterns. At the top, in the Gordon Lake Formation, REE patterns are typical of post-Archean shales (PAAS).

8.2.2 Wopmay Orogen

The Wopmay Orogen, part of the Bear structural province, is an early Proterozoic orogenic belt which flanks the western edge of the Archean Slave Province, in the Canadian Shield [21]. At the bottom of this supracrustal sequence is the Akaitcho Group, comprising 8–10 km of metasedimentary and metavolcanic rocks [22], inferred to overlie Archean crust [22]. The exact age of these rocks is uncertain but they were deposited between 2.1 and 1.8 Ae [21]. The lower part of the Akaitcho Group is dominated by a suite of volcanic rocks of bimodal composition. This is overlain by 1–2 km of pelitic rocks of mixed continental and volcanogenic origin. Within the volcanic sequence is a 1–2 km thick arkosic turbidite unit (Zephyr Formation) of continental origin. Geochemical and stratigraphic evidence indicate this sequence was deposited in a continental rift [22]. The Epworth Group conformably overlies the Akaitcho Group and is comprised of sedimentary rocks of mainly continental origin. This group has been interpreted as a continental platform, terrace and rise deposit affected in the later stages of development by collision with a small continental mass [21].

The volcaniclastic sediments possess trace element characteristics indicative of their volcanic precursors [23]. However, all sedimentary rocks with continental sources have REE patterns characteristic of PAAS with high ΣREE, La/Yb (13–15) and Eu/Eu*=0.6. These samples also have high levels of Th and U, greater than 10 ppm and 3 ppm, respectively. Thus, the early Proterozoic sedimentary rocks deposited in this area appear to have typical post-Archean trace element characteristics.

8.2.3 Pine Creek Geosyncline

The Pine Creek Geosyncline, Northern Territory, Australia, contains up to 14 km of early Proterozoic sedimentary rocks. In contrast to the Huronian Supergroup, stratigraphic subdivision and correlations are controversial due to very poor exposure, structural and metamorphic complexity and several apparent facies changes. The sequence contains mainly pelitic and arenaceous sedimentary rocks with minor carbonates, conglomerates and volcanics; it rests

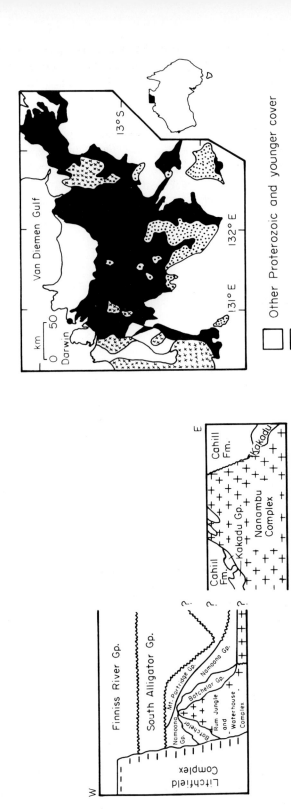

Fig. 8.3. Generalized geological map and stratigraphic relationships of the Pine Creek Geosyncline, Northern Territory, Australia [25].

unconformably on Archean granite–gneiss complexes (Rum Jungle, Waterhouse and Nanambu Complexes). Geochronological work on these [24] indicates ages in excess of 2.4 Ae. The Nanambu Complex is best dated, by U–Pb zircon and Rb–Sr whole rock methods, at 2470 m.y. The entire sequence is older than the Nimbuwah Complex which is about 1870 m.y. The most likely time of deposition of the sedimentary rocks is c. 2.2–2.0 Ae [24]. Stratigraphic relationships are shown in Fig. 8.3 [25]. Sedimentation is thought to have occurred in an intracratonic basin in three major stages [26]:

1 rifting of Archean basement to form a north- to north-west-trending trough with shallow marine sandstones and carbonates (Batchelor and Kakadu Groups) deposited at the margins; pelitic and lesser chemical and volcanic rocks of the Namoona Group deposited in the centre of the basin;

2 uplift of Archean basement, with partial erosion of early Proterozoic sedimentary rocks, and influx of fluvial–alluvial sandstones of the lower Mount Partridge Group grading into littoral and subtidal finer grained clastics of the upper Mount Partridge Group (Wildman Siltstone; Nourlangie Schist); subtidal deposits later transgressed onto alluvial fan deposits;

3 major uplift and peneplanation followed by a shallow marine transgression and deposition of algal dolomite and carbonaceous shale of the south Alligator Group during an extended period of subsidence; continued subsidence and uplift in the west resulted in an influx of flysch (Finnis River Group).

McLennan & Taylor [27] examined the REE distribution in a number of early Proterozoic fine-grained sedimentary rocks from the Pine Creek Geosyncline (Fig. 8.4). They divided the sequence into two groups, on the

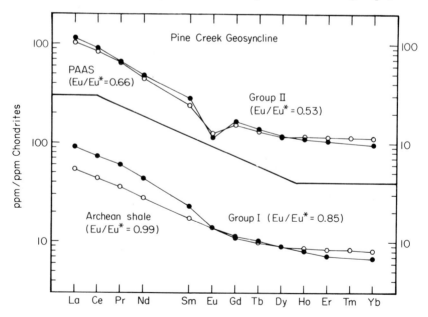

Fig. 8.4. Average REE patterns of fine-grained sedimentary rocks from the early Proterozoic succession of the Pine Creek Geosyncline [27]. Group I represents the lower part of the succession and Group II represents the upper part. These data are consistent with the pattern of evolution seen for sedimentary rocks from the Huronian Supergroup.

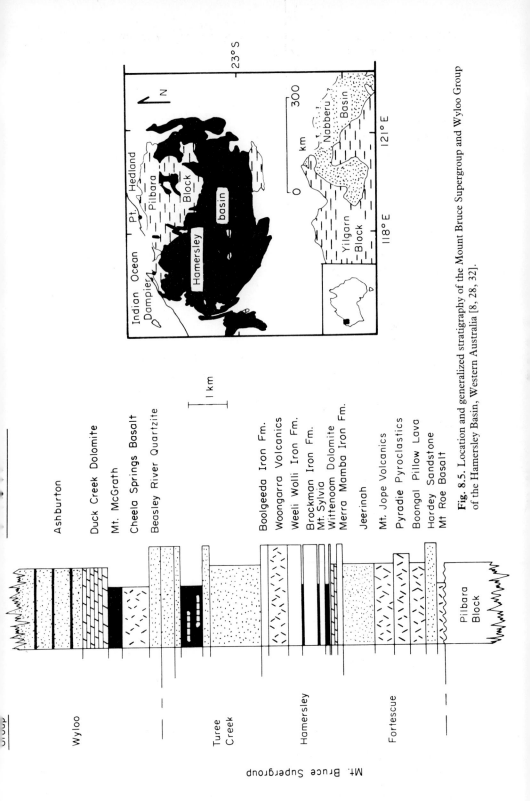

Fig. 8.5. Location and generalized stratigraphy of the Mount Bruce Supergroup and Wyloo Group of the Hamersley Basin, Western Australia [8, 28, 32].

basis of trace element composition, with the boundary between the lower and upper Mount Partridge Groups. Group I, from the lower part of the succession, has Archean-like REE patterns, similar to those seen in the lower part of the Huronian Supergroup (McKim and Pecors Formations). Group II, from the upper part, has REE patterns indistinguishable from typical post-Archean sedimentary rocks and similar to those in the top of the Huronian Supergroup (Gordon Lake Formation).

8.2.4 Hamersley Basin

Approximately 10 km of mainly clastic sedimentary rocks, with lesser iron formations, carbonates and volcanics, are preserved as the Mount Bruce Supergroup in the Hamersley Basin of Western Australia (Fig. 8.5). Recent work has resulted in a major revision of the stratigraphic framework of the Mount Bruce Supergroup [8, 28]. An age of 2490 m.y. has been published for a tuff horizon in the Hamersley Group and Pidgeon reports an age of 2768 ± 16 m.y. for a porphyry which intrudes near the base of the Fortescue Group [29]. An upper age limit is less well constrained but 2.3 Ae is generally cited [8].

Trendall [8] does not recognize any major unconformity between the Mount Bruce Supergroup and the underlying Archean supracrustals of the Pilbara Supergroup. Instead, it appears that granite magmatism and cratonization of the Pilbara Block continued well after the initiation of deposition of the Mount Bruce Supergroup. This situation is dramatically different from the geology of the north shore of Lake Huron and the Pine Creek Geosyncline, where a profound unconformity is found. Indeed, immediately to the south, in the Yilgarn Block, a major granite 'event' occurred at about 2.7–2.6 Ae leading to cratonization. However, a major period of stabilization of the granite–greenstone terrains immediately preceding deposition of the Fortescue Group, has been postulated from sedimentological evidence [30].

With the exception of the iron formations, there has been little work on the sedimentary rocks in the Hamersley Basin. An alternating sequence of mainly mafic volcanics and clastic sedimentary rocks is found at the base of the Fortescue Group. The sediments are undoubtedly of shallow water origin. The lower Hardey Sandstone is unique in that it has clear evidence of a major granitic (with minor volcaniclastic) component in the source ([8]; also see [30]). The intervening sedimentary units, the Kuruna Siltstone and Tumbiance Formation, are predominantly volcaniclastic (tuffs, lapilli tuffs, etc.) but minor stromatolitic carbonate intercalations are developed. The uppermost, mainly pelitic, Jeerinah Formation is closely associated with volcanic rocks but there is no clear evidence of a major volcaniclastic component.

The well studied Hamersley Group is a sequence of alternating shale or dolomite and banded iron formation. The various shale units have many common characteristics including minor chert, carbonate and iron formation. The upper part of the Mount McRae Shale contains thin units of volcanic ash. A volcaniclastic (ash) origin for much of the shale in the Hamersley Group is likely [8, 31]. The deposition of the Turee Creek Group heralded the return to epiclastic sedimentation involving deposition of shale, siltstone, greywacke and quartzite with minor carbonate near the top. A diamictite unit (Meteorite Bore Member) may be of glacial origin [32]. The Wyloo Group unconformably overlies the Mount Bruce Supergroup. Originally considered as part of the Hamersley Basin succession, it is now thought to have been deposited in the younger and tectonically separate Ashburton Trough.

Trace element geochemistry of fine-grained sedimentary rocks of the Hamersley Supergroup and Wyloo Group was investigated by McLennan [33] and average analyses are given in Table 8.2. The lower Hardey Sandstone from the Fortescue Group is a critical unit in this sequence because it is near the base and has a continental origin (in contrast to the Hamersley Group). Samples from this formation show a range of REE patterns, but most have Eu/Eu*$\leq$0.79. The average REE pattern is quite similar to typical post-Archean shales with an average Eu/Eu*=0.72 (Fig. 8.6). The epiclastic sediments of the Turee Creek Group and Wyloo Group, also have sedimentary REE patterns with post-Archean characteristics (Fig. 8.6).

Table 8.2. Average chemical composition of fine-grained sedimentary rocks from the Hamersley Basin and Wyloo Group [33].

	Fortescue (1)	Turee Creek (2)	Wyloo (3)
SiO_2	58.58	62.98	71.29
TiO_2	0.98	0.68	0.51
Al_2O_3	24.38	23.49	15.95
FeO	6.59	4.52	3.53
MnO	0.06	0.13	0.06
MgO	2.97	1.75	1.56
CaO	0.22	0.51	1.05
Na_2O	0.58	0.79	2.33
K_2O	5.65	5.12	3.71
Σ	100.01	99.97	99.99
LOI	4.95	3.77	2.27
Cs	8.51	8.08	6.70
Ba	1130	1125	1200
Pb	28.4	13.8	28.7
La	54.2	77.2	52.2
Ce	113.6	166.9	108.6
Pr	12.6	20.6	13.1
Nd	47.5	82.0	47.8
Sm	8.87	14.8	7.17
Eu	1.71	2.62	1.15
Gd	5.94	13.2	4.78
Tb	0.99	2.06	0.71
Dy	5.96	10.8	4.08
Ho	1.33	1.75	0.84
Er	3.75	4.39	2.54
Yb	3.44	3.50	2.13
ΣREE	260.9	400.9	245.7
La_N/Yb_N	10.6	14.9	16.6
Eu/Eu*	0.72	0.57	0.60
Y	39.5	60.1	26.8

Table 8.2 (*continued*)

	Fortescue (1)	Turee Creek (2)	Wyloo (3)
Th	16.2	20.4	25.6
U	3.90	5.72	3.98
Zr	211	202	196
Hf	5.24	4.91	5.57
Sn	10.8	6.91	5.48
Nb	20.4	20.0	17.1
Mo	1.16	0.38	0.22
W	1.19	1.46	0.81
Th/U	4.2	3.6	6.4
La/Th	3.3	3.8	2.0
Cr	303	81	61
V	192	110	44
Sc	21	15	8.7
Ni	198	36	21
Co	32	10	9
Cu	83	19	11
Ga	18	19	8.8
La/Sc	2.6	5.1	6.0
Th/Sc	0.77	1.4	2.9
Bi	0.42	*	0.34
B	21	64	20

1. Average of 5 shales (4 for ferromagnesian trace elements and B) from the Hardey Sandstone, lower Fortescue Group, Hamersley Basin.
2. 54944, siltstone from Turee Creek Group, Hamersley Basin.
3. Average of greywacke (R2100) and siltstone (R2098) from the Ashburton Fm., Wyloo Group, Ashburton Trough.

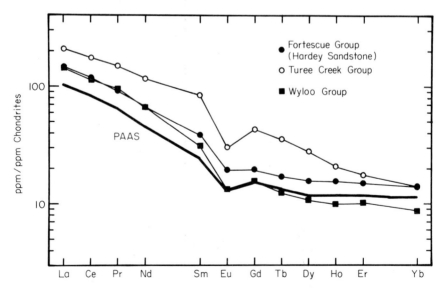

Fig. 8.6. Average REE patterns of fine-grained sedimentary rocks from the Fortescue (Hardey Sandstone), Turee Creek and Wyloo Groups. (Data from Table 8.2.) PAAS plotted as heavy solid line for comparison.

8.3 Late Archean igneous activity and crustal evolution

The REE patterns in early Proterozoic sedimentary rocks clearly show that the change in upper crustal composition was very rapid and closely related to the Archean–Proterozoic boundary. In the Huronian and the Pine Creek Geosyncline, a change in REE patterns was recorded at the base of the successions. No such gradual change is noted for the Epworth Group and Fortescue Group, and post-Archean patterns occur at the base. Early Proterozoic sedimentary rocks are recording, to a large extent, the erosion of the late Archean upper crust, so that we must seek explanations for these effects by considering igneous activity in the late Archean.

Most Archean terrains underwent a particularly active period of volcanism during the late Archean, usually at about 2.8–2.6 Ga. In most places this volcanism was followed closely by an extraordinary period of large-scale intrusions of granodiorites and granites (see Chapter 7). The chemical composition of these rocks is discussed in Chapter 9. This activity is commonly associated with cratonization, marking the close of the Archean Era. Of the cases examined here, only the Pilbara Block may have had a somewhat different history. Here, granitic plutonism occurred continuously throughout, even following the initiation of sedimentation in the Hamersley Basin [8]. Notwithstanding this evidence, Blake [30] has suggested a major period of stabilization of the Pilbara Block immediately preceding deposition of the Mount Bruce Supergroup.

In contrast to early Archean granitic activity (see Section 9.5.1), which was mainly of a Na-rich variety (tonalite–trondhjemite–granodiorite), late Archean granitic activity (see Section 9.5.2), was predominantly K-rich (granodiorite–quartz–monzonite–granite). In contrast to all other Archean igneous rock suites, these are characterized by Eu depletion of varying magnitude and are derived via intracrustal melting where plagioclase is a residual phase. Many of these granitic rocks also display low $^{87}Sr/^{86}Sr$ initial ratios suggestive of short crustal residence times. Nd-isotopic evidence further supports this. The most likely origin for much of this granitic material is through remelting of greenstone belt material derived from the mantle at 2.8–2.6 Ae. In this model, the change in upper crustal composition recorded in early Proterozoic sedimentary rocks is thus related to an intense period of crustal growth and differentiation at the end of the Archean.

8.3.1 Recycling and the sedimentary mass

The rapid change in the chemical composition of sediments at the Archean–Proterozoic boundary also places important constraints on the growth rate of the continental sedimentary mass [34]. Such changes are at least three to four times more rapid than allowed by present day sedimentary recycling models (see Chapter 5). They can be explained, in a steady-state system, by much faster sedimentary recycling rates or by a much less efficient pattern of recycling (70% open rather than the present 35% [34]). The most likely explanation, however, is that the sedimentary mass is not in steady-state but grows at a rate related to the recycling rate, so that a major increase in the sedimentary mass occurred at the Archean–Proterozoic boundary. The sedimentary mass is related to the extent and freeboard of the exposed continental crust and so the increase in sedimentary mass is consistent with an increase in continental mass [34, 35] (see Chapter 10).

8.4 Late Archean of Southern Africa

If this view of the Archean–Proterozoic boundary is correct then the late Archean of Southern Africa represents a special case. In this region, widespread potassic granitic magmatism began much earlier at about 3.2 Ae and continued to about 2.6–2.5 Ae [36]. Typical Archean greenstone belts are preserved in the 3.5–3.3 Ae Swaziland Supergroup in the eastern Transvaal Province. Proterozoic-style sedimentation probably began at about 3.1 Ae with the Pongola Supergroup and continued with the deposition of the Godwin Formation and Witwatersrand Supergroup, the latter being deposited at about 2.7 Ae [37]. Thus, the late Archean of Southern Africa provides an interesting test for our model of crustal evolution. Since the K-rich granitic magmatism and cratonization occurred much earlier in Southern Africa (3.2–2.9 Ae), the change in sedimentary REE patterns should also occur earlier.

The trace element geochemistry of Archean terrigenous sedimentary rocks from the Fig Tree and Moodies Groups of the Swaziland Supergroup was discussed in the previous chapter and REE patterns were found to be typical of the Archean.

The Pongola Supergroup rests unconformably on crystalline basement south of the Barberton Mountain Land. The lower Nsuze Group comprises up

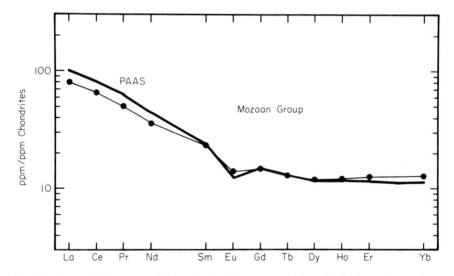

Fig. 8.7. Average REE pattern of fine-grained sedimentary rocks from the Archean Mozaan Group (Pongola Supergroup), South Africa. (Data from [40].) PAAS plotted as heavy solid line for comparison. The Mozaan REE pattern is well evolved towards PAAS.

to 4 km of interbedded, volcanics, volcaniclastic and quartzose sandstone with minor shale, conglomerate and dolomite. Geochemical data on volcanics indicate a graben-like setting [38] or intraplate volcanism [39]. The overlying Mozaan Group is 3 km of alternating sandstone and argillite. The average REE patterns [40] for shales from the Mozaan Group is shown in Fig. 8.7. Much of the Nsuze Group contains volcaniclastic material, which may give equivocal information regarding upper crustal composition. The Mozaan patterns are not so depleted in Eu (Eu/Eu*=0.74–0.79) as typical post-Archean shales (e.g. PAAS) but are clearly well evolved towards such patterns and probably represent an intermediate stage.

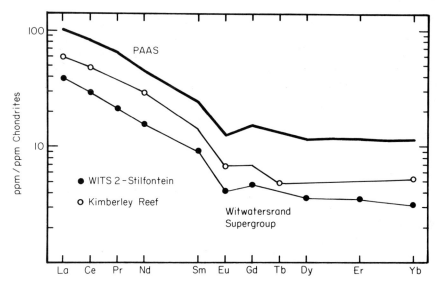

Fig. 8.8. Chondrite-normalized REE diagram of a quartzite (WITS2) from the Stilfontein Quartzite [40] and the average of nine quartzites from the Kimberley Reef [41], Witwatersrand Supergroup. PAAS plotted as heavy solid line for comparison. The quartzites have much lower ΣREE abundances than PAAS but the shape of the patterns (including the negative Eu-anomalies) is very similar.

REE data for the Witwatersrand are confined to quartzite samples [40, 41], but the patterns closely resemble that of PAAS (Fig. 8.8). The absolute abundances are much lower, reflecting the quartzose nature of these sedimentary rocks. Thus by the time of deposition of the Witwatersrand sequence, the upper crust in this region appears to have had typical post-Archean characteristics.

8.5 Early Proterozoic greenstone belts

Local 'greenstone belt' development also occurred after 2.5 Ae. Examples may exist in North America, South America and Africa but they do not appear to be common or extensive and have not been well documented. Sedimentary REE data are available for the early Proterozoic of northern Guyana [42].

Here, first-cycle volcaniclastic sedimentary rocks have REE patterns similar to typical Archean shales and of post-Archean sediments deposited in fore-arc basins, and are consistent with derivation from underlying intermediate-felsic volcanics. Such findings are consistent with our understanding the composition of the post-Archean upper crust. REE patterns of first-cycle volcanogenic sediments simply reflect those of the volcanic precursor and do not bear any necessary relationship to that of the average upper crust. This is true for Phanerozoic, Proterozoic and Archean Eons. Volcanogenic sediments are also common in the Archean. However, many Archean sequences studied show abundant petrographical and geochemical evidence for derivation from a wide variety of source rocks, including mafic and felsic volcanic, granitic and recycled sedimentary and metamorphic rocks. Moreover, first-cycle volcaniclastic sediments in the Archean have REE patterns indistinguishable from sediments derived from terrigenous sources, indicating Archean volcanics and plutonics of felsic composition have a similar origin. This contrasts with post-Archean volcanogenic sediments which are typically andesitic (Chapter 6) and bear little relationship to the post-Archean upper crust [43]. It would not be particularly surprising to find some Archean-like REE patterns in localized early Proterozoic 'greenstone belts' if widespread K-rich granitic intrusions had not occurred in that region, or if such rocks had not been eroded.

8.6 Summary

1 REE data from early Proterozoic terrigenous sedimentary rocks provide persuasive evidence that the upper continental crust changed in composition during the late Archean.

2 This change in composition was related to a major period of growth and differentiation of the continental crust. In particular, the widespread appearance of K-rich granitic rocks of intracrustal origin was the main cause in creating a more felsic upper crust.

3 Evidence from late Archean sedimentary successions from South Africa indicates that while the change in upper crustal composition was very rapid in any given place, it was not isochronous on a worldwide scale. Although it occurred over a rather protracted period from about 3.2–2.5 Ae it was probably concentrated during the period 2.8–2.5 Ae, when most late Archean greenstone belts were formed.

Notes and references

1 Logan, W.E. (1849) Report on the north shore of Lake Huron. *GSC*; the name 'Huronian' was introduced in 1855 on a geological map of Canada at the Paris Exhibition (Logan, W.E. & Sterry Hunt, T. (1855) *Esquisse Geologique du Canada*. Bossange et fils, Paris). As pointed out by M. J. Frarey & S. M. Roscoe ((1970) *GSC Pap.*, **70-40**, 143), 'the Huronian has great historical significance as the first valid stratigraphic subdivision within the Canadian Shield, or probably anywhere in Precambrian terrains...' (p. 144).

2 Unquestionably, the most original scheme was proposed by A. F. Trendall (*The Australian Geologist*, 25 Mar. 1983, p. 23), who suggested, 'a sweepstake and associated lottery to establish a subdivision for the Australian Precambrian. Twenty-seven numbers, from 800 to

3400 m.y. at intervals of 100 m.y. . . . with each winner having the right to name one of four consequent sections of Precambrian time . . . such events could be repeated (every) four years'. Trendall also observed that such a lottery was in keeping with Australian tradition 'and would impede research on the Precambrian no more than the present arid debate on its formal stratigraphic subdivision'.

3. Alcock, F.J. (1934) Report of the National Committee on Stratigraphic Nomenclature. *Trans. R. Soc. Can.* (Ser. 3), Sect. IV, **28**, 113.
4. The most recent being: Stockwell, C.H. (1982) Proposals for time classification and correlation of Precambrian rocks and events in Canada and adjacent areas of the Canadian Shield. Part 1: A time classification of Precambrian rocks and events. *GSC Pap.*, **80-19**.
5. Stockwell's approach was recently reviewed by: Harland, W.B. (1983) Precambrian geochronology in Canada. *Geol. Mag.*, **120**, 195.
6. James, H.L. (1978) Subdivision of the Precambrian—a brief review and a report on recent decisions by the Subcommission on Precambrian stratigraphy. *PCR*, **7**, 193.
7. Cloud, P. (1976) Major features of crustal evolution. *Trans. Geol. Soc. S. Afr.*, **79**, Annex.
8. Trendall, A.F. (1983) *The Hamersley Basin*. Preprint.
9. Van Schmus, R. (1965) The geochronology of the Blind River–Bruce Mines area, Ontario, Canada. *J. Geol.*, **73**, 755; Fairbairn, H.W. *et al.* (1969) Correlation of radiometric ages of Nipissing diabase and Huronian metasediments with Proterozoic orogenic events in Ontario. *CJES*, **6**, 489.
10. Frarey, M.J. *et al.* (1982) A U–Pb zircon age for the Creighton Granite, Ontario. *GSC Pap.*, **81-1C**, 129.
11. Card, K.D. *et al.* (1972) General geology of the Sudbury–Elliot Lake region. *24th IGC Field Excursion C28*.
12. Card, K.D. *et al.* (1972) The Southern Province. *Geol. Ass. Can. Spec. Pap.*, **11**, 335.
13. Roscoe, S.M. (1973) The Huronian Supergroup, a Paleoaphebian succession showing evidence of atmospheric evolution. *Geol. Assoc. Can. Spec. Pap.*, **12**, 31.
14. Young, G.M. (1969) Geochemistry of Early Proterozoic tillites and argillites of the Gowganda Formation, Ontario, Canada. *GCA*, **33**, 483; (1970) An extensive early Proterozoic glaciation in North America? *Palaeogeog. Palaeoclim. Palaeoecol.*, **7**, 85; (1973) Tillites and aluminous quartzites as possible time markers for middle Precambrian (Aphebian) rocks of North America. *Geol. Ass. Can. Spec. Pap.*, **12**, 97.
15. Young, G.M. (1973) Origin of carbonate-rich early Proterozoic Espanola Formation, Ontario, Canada. *GSA Bull.*, **84**, 135.
16. Young, G.M. (1983) Tectono-sedimentary history of early Proterozoic rocks of the northern Great Lakes region. *GSA Mem.*, **160**, 15.
17. Van Schmus, W.R. (1976) Early and middle Proterozoic history of the Great Lakes area, North America. *Phil. Trans. Roy. Soc.*, **A280**, 605.
18. Sims, P.K. *et al.* (1980) The Great Lakes tectonic zone—a major crustal structure in central North America. *GSA Bull.*, **91**, 690; (1981) Evolution of early Proterozoic basins of the Great Lakes Region. *GSC Pap.*, **81-10**, 379.
19. McLennan, S.M. *et al.* (1979) The geochemistry of the carbonate-rich Espanola Formation (Huronian) with emphasis on the rare earth elements. *CJES*, **16**, 230.
20. McLennan, S.M. *et al.* (1979) Rare earth elements in Huronian (Lower Proterozoic) sedimentary rocks: composition and evolution of the post-Kenoran upper crust. *GCA*, **43**, 375.
21. Hoffman, P.F. (1980) Wopmay Orogen: a Wilson Cycle of early Proterozoic age in the northwest of the Canadian Shield. *Geol. Ass. Can. Spec. Pap.*, **20**, 523.
22. Easton, R.M. (1980) Stratigraphy and geochemistry of the Akaitcho Group, Hepburn Lake map area, District of Mackenzie: an initial rift succession in Wopmay Orogen (early Proterozoic). *GSC Pap.*, **80-1B**, 47; (1981) Stratigraphy of the Akitcho Group and the development of an early Proterozoic continental margin, Wopmay Orogen, Northwest Territories. *GSC Pap.*, **81-10**, 79; (1983) Crustal structure of rifted continental margins: geological constraints from the Proterozoic rocks of the Canadian Shield. *Tectonophysics*, **94**, 371.
23. Easton, R.M. (1981) REE, U and Th contents of Proterozoic and Archean sedimentary rocks from the Bear and Slave structural provinces, NWT, Canada (abst.). *Geol. Ass. Can. Prog. Abst.*, **6**, A-16.

24 Page, R.W. et al. (1980) Geochronology and evolution of the late-Archean basement and Proterozoic rocks in the Alligator Rivers Uranium Field, Northern Territory, Australia. In: Ferguson, J. & Goleby, A.B. (eds), Uranium in the Pine Creek Geosyncline. IAEA, 39.
25 Needham, R.S. et al. (1980) Regional geology of the Pine Creek Geosyncline. Ibid., 1.
26 Stuart-Smith, P.G. (1980) Evolution of the Pine Creek Geosyncline. Ibid., 23.
27 McLennan, S.M. & Taylor, S.R. (1980) Rare earth elements in sedimentary rocks, granites and uranium deposits of the Pine Creek Geosyncline. Ibid., 175.
28 Trendall, A.F. (1979) A revision of the Mount Bruce Supergroup. W. Aust. Geol. Surv. Ann. Rept. 1978; (1980) A progress review of the Hamersley Basin of Western Australia. Geol. Surv. Finland, Bull., **307**, 113.
29 Compston, W. et al. (1981) A revised age for the Hamersley Group (abst.). Geol. Soc. Austr. Ann. Conv. Abst., **3**, 40; Pidgeon, R.T. (1984) Geochronological constraints on early crustal volcanic evolution of the Pilbara Block, Western Australia. Aust. J. Earth Sci., **31**, 237.
30 Blake, T.S. (1984) Evidence for stabilization of the Pilbara Block, Australia. Nature, **307**, 721.
31 La Berge, G.L. (1966) Altered pyroclastic rocks in iron-formations in the Hamersley Range, Western Australia. Econ. Geol., **61**, 147; MacDonald, J.A. & Grubb, P.L.C. (1971) Genetic implications of shales in the Brockman iron formation from Mount Tom Price and Wittenoom Gorge, Western Australia. J. Geol. Soc. Austr., **18**, 81; Button, A. (1975) The Gondwanaland Precambrian project. Econ. Geol. Res. Unit, Univ. Witwatersrand, 16th Ann. Rept., 37.
32 Trendall, A.F. (1981) The Lower Proterozoic Meteorite Bore Member, Western Australia. In: Hambrey, M.J. & Harland, W.B. (eds), Earth's Pre-Pleistocene Glacial Record. Cambridge University Press. 555.
33 McLennan, S.M. (1981) *Trace Element Geochemistry of Sedimentary Rocks: Implications for the Composition and Evolution of the Continental Crust*. Ph.D. Thesis, Australian National University.
34 Veizer, J. (1984) Recycling on the evolving earth: geochemical record in sediments. Proc. IGC, **11**, 325.
35 Veizer, J. & Jansen, S.L. (1979) Basement and sedimentary recycling and continental evolution. J. Geol., **87**, 341.
36 Hunter, D.R. (1974) Crustal development in the Kaapvaal craton. 1. The Archean. PCR, **1**, 259; Anhaeusser, C.R. & Robb, L.J. (1981) Magmatic cycles and the evolution of the Archean granitic crust in the eastern Transvaal and Swaziland. Geol. Soc. Austr. Spec. Pub., **7**, 467.
37 A review of the geochronological data can be found in ref. [40].
38 Hegner, E. et al. (1981) Geochemie und petrogenese archaischer vulkanite der Pongola-Gruppe in Natal, Südafrika. Chem. Erde, **40**, 23.
39 Armstrong, N.V. et al. (1982) Stratigraphy and petrology of the Archean Nsuze Group, northern Natal and southeastern Transvaal, South Africa. PCR, **19**, 75.
40 McLennan, S.M. et al. (1983) Geochemical evolution of Archean shales from South Africa. I. The Swaziland and Pongola Supergroups. PCR, **22**, 93.
41 Rasmussen, S.E. (1977) Activation analysis and classification to source of samples from the Kimberley Reef conglomerate. Nat. Inst. Metall., Randburg, S. Afr., Rep. **1874**.
42 O'Day, P.A. & Gibbs, A.K. (1982) Volcanic provenance of metapelites from Proterozoic greenstone belts, Guyana: rare-earth element evidence (abst.). GSA Abst. Progr., **14**, 578.
43 We would anticipate that Phanerozoic volcanogenic sediments derived from ignimbritic sources at volcanic arcs would be similar to PAAS and carry the negative Eu-anomaly characteristic of felsic volcanics of intracrustal origin (e.g. Mahood, G. & Hildreth, W. (1983) Large partition coefficients for trace elements in high-silica rhyolites. GCA, **47**, 11).

9

Models for the Origin of the Continental Crust

9.1 Introduction

In previous chapters, we assessed the information available on the composition of the Archean and post-Archean upper crusts, and speculated about the nature and composition of the lower crust. It is clear that there is a major change in upper crustal composition marked by the Archean–Proterozoic boundary. In this chapter, we explore models for the growth of the crust and examine the causes of the episodic change in upper crustal composition. We begin by discussing mechanisms for the growth of the Archean crust, followed by an examination of the causes responsible for the change in upper crustal compositions in the late-Archean. We conclude with a discussion of the present-day processes responsible for additions to the continental crust. *En route*, we discuss the formation of granites, on account of their importance in the upper crust. We also devote some space to massif anorthosites which form large but dimly understood components of the crust. The rate of growth of the crust is addressed in the next chapter and the overall crust–mantle relationships are discussed in Chapter 11. The question of early crusts is examined in Chapter 12.

9.2 The Archean crust

The REE evidence assessed in Chapter 7 could, at first glance, be consistent with derivation of the Archean crust from calc-alkaline island-arc volcanic sources. Does this provide evidence for uniformitarianism, so that present-day igneous processes at subducting plate margins were also operating in the Archean? The answer is no. Closer examination of the REE patterns in Archean sedimentary rocks show that, in contrast to post-Archean patterns, there is much diversity. Although the average pattern resembles that of present-day calc-alkaline volcanic rocks, an extreme range from LREE enriched to flat chondritic normalized patterns exists, consistent with the derivation of these clastic sediments from two principal igneous sources, basic and felsic.

The evidence for large-scale 'calc-alkaline' volcanism of conventional type is not persuasive. A number of examples occur [1] but these do not appear to comprise a large fraction of the Archean igneous rocks exposed at present. Present-day andesitic stratovolcanoes are rapidly eroded to form volcanogenic sandstones, and so andesites might be under-represented in the geological record. The evidence both from the greenstone belts and the Archean 'granitic' terrains, however, is that of a bimodal suite of two principal rock components [2, 3, 4]. These are tholeiitic basalts and felsic volcanics, tonalitic or trondhjemitic granitic rocks. The former have mainly flat REE patterns (see Fig. 7.3, 7.4) whereas the latter are typified by steep LREE enriched-HREE depleted patterns generally with no europium anomalies (Fig. 9.1; also see Fig. 7.6). REE patterns from sedimentary rocks of localized provenance show these distinctive patterns.

We assume that the dominant rock types produced in the Archean were the bimodal suite [2–4] with minor production of calc-alkaline rocks indistinguishable from their modern analogues (Fig. 9.2; also see Fig. 7.5). Quantitative estimates of the various lithologies are inherently difficult to

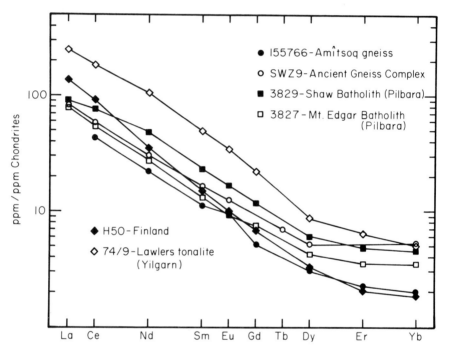

Fig. 9.1. Chondrite-normalized REE patterns for typical Archean tonalites and trondhjemites and Na-granodiorites. Note the steep patterns and the general absence of Eu-anomalies. (Data from [5, 6]).

make but some constraints are available. In Chapter 7, the sedimentary trace element data were used to constrain the proportion of mafic–felsic end members at 1:1. For the bulk crust, the heat flow data were used to obtain abundances of the heat producing elements (K, Th, U) which in turn constrained the proportions of mafic to felsic end members at 2:1. These results indicated an upward concentration of felsic material in the Archean crust (see Section 7.10).

These calculations enable some tests to be made on current models of crustal evolution and growth. Hargraves [7] has proposed that by about 3.5 Ae the Earth had a globe-encircling sialic crust which was completely submerged beneath the oceans. During the period 3.5–2.5 Ae greenstone piles dominated

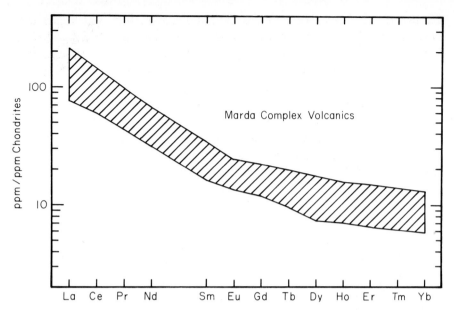

Fig. 9.2. Chondrite-normalized REE patterns for Archean andesites and dacites from the Marda complex, Western Australia [1]. These patterns are not distinguishable from those of present-day andesites (Fig. 3.2) and are much less steep than the patterns shown in Fig. 9.1.

the land masses above sea-level. This model has been supported by Fyfe [8]. Such a scenario is in conflict with the sedimentary REE data. It is clear that significant amounts of greenstone and granitic debris were incorporated into sedimentary rocks throughout the Archean, so that the upper exposed crust was chemically and lithologically complex for as far back as the record is preserved in sedimentary rocks.

What were the geological conditions which gave rise to the Archean crust? It appears that extrusion of basalts took place under conditions little different from today, except for higher heat flow and more rapid mantle convection. The only viable method of dissipating this additional heat is through the oceanic crust. At the present time, 50% of the total heat loss occurs at the spreading centres [9, 10]. A realistic scenario thus involves many microplates in the Archean. Formation of basaltic oceanic crust and subduction back into the mantle is assumed to be a viable process in the early Archean. The high heat flow would encourage large degrees of partial melting, and hence the production of magnesium-rich basalts or komatiites. Such conditions imply that sea-floor spreading occurred, a view reinforced by the lateral extent of greenstone terrains. Possibly large volcanic domes were also constructed. Sinking of dense basaltic material could lead to the formation of mafic amphibolites, garnet granulites or eclogites depending on P-T conditions. Partial melting of such materials at depth produce the tonalitic and trondhjemitic intrusives. Their steep REE patterns with HREE depletion are a consequence of equilibration with garnet as a residual phase. Shallower patterns with less HREE depletion indicate lesser amounts of garnet.

Weathering and erosion of both these terrains (which must have been above sea-level [cf. 7]), produced the typical Archean sedimentary REE patterns. Such a model involves a distinctive tectonic regime commencing with extrusion of tholeiitic basalt, possibly followed by sea-floor spreading, piling up of tholeiitic masses and sinking of these which, in turn, produce the more acidic intrusives and volcanics by partial melting. The intrusive masses coalesce, or are pushed together, to form the Archean cratonic nuclei.

This model indicates that all material in the Archean crust was derived from mantle sources by one episode of partial melting (for the basic suite) or two (for the felsic suite), the latter being derived by partial melting of basaltic precursors at depths where garnet is stable, and below depths where plagioclase is stable.

The tectonic emplacement of this material will be affected by the higher heat-flow in the Archean. As was noted earlier, thermal gradients in Archean crustal rocks were probably not very different from those of later geological periods, from the metamorphic evidence. Therefore, the major loss of heat must have taken place through the ocean floor [11]. Such heat loss occurs principally at spreading centres, and consequently plates would have been much smaller [10, 11]. These speculations on the conditions during the early Archean thus provide a scenario of an oceanic basaltic crust, comprised of many small plates, with a life of perhaps 10–20 m.y. [10]. Sinking of this material into the mantle is aided by the higher heat flux.

Apparently the production of felsic material was localized. Why were the Archean continental nuclei apparently scattered in numerous discrete centres? Then, as now, the secondary melting processes required to generate the more felsic or silicic rocks of the continental crust are less efficient by orders of magnitude than those which produce primary basaltic magmas; otherwise, extensive early sialic crusts would have evolved very quickly. The distribution of the continental nuclei was apparently sporadic. Were there variations in the distribution of the heat-producing elements in the mantle, which localized the secondary melting of the sinking eclogite masses? If rapid recycling of basalt was occurring (the life of an average plate was calculated by Pollack [10] at 10–20 m.y.), then most basalt must have been recycled into the mantle and sunk without undergoing secondary melting to produce tonalites, trondhjemites and their volcanic equivalents.

These conditions persisted from 3.8 Ae throughout the Archean, and the area of crust slowly grew. Parts of the crust were above sea-level from the beginning, for clastic sediments exist in the oldest terrains. Localized segments of thick crust may have accreted as small occurrences of buoyant felsic material were swept together by the vigorous plate motions. The lighter, more felsic components are likely to dominate toward the top of the crust, producing the variations in sedimentary REE patterns noted earlier. In some regions, true granitic rocks were produced by partial melting within the crust. These rocks, with their distinctive depletion in europium, are recorded in a few isolated sediments (e.g. West Greenland) but never comprised more than about 5–10% of the upper crust, for otherwise the signature of Eu depletion would be widespread in the sedimentary record. Instead, the known occurrences of Archean sedimentary or metasedimentary rocks displaying depletion in Eu is limited to minor occurrences in Greenland (Section 7.8.3).

It is thus clear that tectonic and igneous activity in the Archean did not resemble that of the present-day plate tectonic regime. Evidence for conventional plate tectonics does not extend back beyond about 1000 m.y., although the occurrence of calc-alkaline rocks in the mid-Proterozoic [12] suggests that the process was operating somewhat earlier. There is a general absence of characteristic features of convergent plate boundaries such as paired metamorphic belts, ophiolites, and blueschists, in the Archean. The model adopted here, of many small plates, with sinking of basaltic and komatiitic lavas and production of felsic igneous rocks, accords with the evidence from the sedimentary rocks. Towards the end of the Archean, a change occurred in this scenario. In the next section we explore the causes of this change and speculate on some of the reasons for it.

9.3 The Archean–Proterozoic transition

An underlying theme in this discussion has been the distinction between Archean and post-Archean sedimentary REE patterns and, by inference, the difference in upper crustal compositions. Some understanding of the reasons for this is now available. Detailed studies of lower Proterozoic sequences in Canada and Australia reveal a rapid evolution of sedimentary REE patterns associated with the Archean–Proterozoic boundary. Details of these changes are given in Chapter 8 and reveal the rather sudden change from Archean to post-Archean sedimentary REE patterns. A key to the cause of this change is provided by the significant depletion in Eu in post-Archean sedimentary rocks. This Eu depletion must be intracrustal in origin. No common volcanic rocks derived from the mantle exhibit Eu depletion of the magnitude (Eu/Eu*=0.65) typical of the average upper crust. Mid-ocean ridge basalts, intraplate volcanics and island-arc volcanics alike are characterized by the scarcity of positive or negative europium anomalies [13]. The occasional occurrences are associated with the presence or absence of cumulate plagioclase.

The basic cause of the difference in behaviour between Eu and the other rare earth elements is due to the change in valency and ionic radius between Eu^{3+} and Eu^{2+} [14]. The trivalent radius for Eu forms part of the monotonic decrease in radius exhibited by the other trivalent REE. Europium, halfway through the REE sequence, is more readily reduced than the neighbouring REE. The increase in radius and change in valency of the divalent ion causes it to enter different crystal sites from those available to the smaller trivalent ions. The Ca sites in feldspars readily accept Eu^{2+}, which closely mimics Sr^{2+} in radius. The most likely mechanism to produce Eu depletion is partial melting where feldspar is a residual phase. Since plagioclase is not a stable phase below about 40 km (10 kbar), Eu anomalies due to this cause must be produced by shallow intracrustal processes. This explanation involves crystal–liquid equilibria rather than other processes (e.g. aqueous transfer). The Eu depletion in post-Archean sedimentary rocks is consequently the signature of intracrustal melting events.

The change in the composition of the post-Archean upper crust, as documented by the REE patterns, is inferred to be a consequence of massive intracrustal partial melting, to form magmas predominantly of granodioritic composition. These rise into and intrude the upper crust, principally as batholiths. During this intracrustal melting the melt becomes enriched in the light REE and depleted in Eu. Eu^{2+}, as noted above, is preferentially incorporated in feldspar, particularly plagioclase. Charge-balance difficulties and smaller ionic radii tend to exclude the trivalent REE. Thus granodiorite melts forming from material of island-arc composition within the stability field of plagioclase (<10 kbar) will develop a significant depletion in europium.

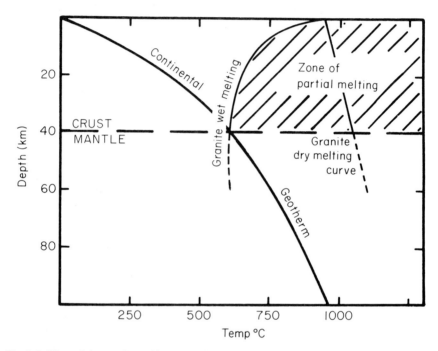

Fig. 9.3. Wet and dry granite melting curves [17] and the zone of partial melting. A typical continental geotherm is shown.

Experimental evidence indicates that the production of granitic rocks occurs within the P-T range typical of crustal rather than mantle regions (Fig. 9.3) (Section 4.2). Models for the production of the upper crustal granodiorites are not discussed in detail, since their production by partial melting processes seems established. It is significant in this context that the formation of granodiorites, rather than granite, is favoured by increasing pressure. However, the experimental evidence shows that 'liquids of tonalite composition cannot be generated by crustal anatexis under conditions of normal regional metamorphism' [16, p. 66] so that intracrustal melting is not a viable mechanism to produce the Archean tonalite suites. These come from mantle depths. The relative importance of removal of elements from the lower crust by metamorphic processes or by fluid phases is inherently difficult to evaluate (see Section 4.2). The processes responsible for the generation of the upper crust must involve crystal–liquid equilibrium to produce the observed LREE enrichment and Eu depletion as well as the large volumes of granodiorites. The sedimentary sampling of the upper granodioritic crust records this Eu depletion, even though the europium may be oxidized to Eu during weathering. The crucial point is that the development of the upper-crustal Eu depletion is *intracrustal* in origin.

The REE evidence for a major episodic change at about the Archean–Proterozoic boundary is supported by the Sm–Nd isotopic systematics of crustal rocks, which indicate a massive increase in crustal growth at that time. Also significant is the observation that the $^{87}Sr/^{86}Sr$ ratio in carbonates (reflecting sea-water composition) is rather low in the Archean but shows a major increase (0.0025) over a short time interval at about 2.5 Ae, with a more gradual increase up to the present. These trends are consistent with a large addition of Rb to the upper-crustal weathering regime during the late Archean (see Chapter 10). The change in upper-crustal composition is not expected to be isochronous on a world-wide scale, but to occur at somewhat different times (within 0.5 Ae) in different regions. The change in the REE patterns in sedimentary rocks follows the unroofing of the granodioritic batholiths, which depends on local rates of uplift and erosion.

K-rich granites and granodiorites appear sporadically throughout the Archean. How much of this type of granite was present in the early Archean crust? Apart from LREE enrichment their most diagnostic feature is the signature of Eu depletion which the granitic debris contributes to the Archean sedimentary pattern. The presence of more than about 10% K-granite will produce a discernible depletion in Eu in the sedimentary record. The lack of such a signature, except very locally, in the Archean sedimentary record, enables us to place an upper limit of about 10% on the surface exposure of K-rich granodiorites and granites in the Archean.

In the model adopted here, massive intracrustal melting produces a granodioritic upper crust and the continental crust toward the end of the Archean assumes its present character. This event is conventionally dated at 2500 Ma, although the actual age varies for each Archean nucleus (see Chapter 8). The buoyant continental regions now form massive barriers to sea-floor spreading largely as a consequence of increasing size. Initiation of modern linear tectonic subduction regimes occurs when the oceanic lithosphere spreading from ridges encounters the buoyant continental masses.

If the Archean crust prior to about 2.8–2.7 Ae comprised only about 15% of the present crustal volume (see Chapter 10), a major question is the source of the material which undergoes intracrustal melting in the late Archean. The change in upper crustal composition is accompanied by a major increase in the volume of continental crust (Chapter 10). Isotopic studies indicate that the crustal material was derived from the mantle shortly (<100–200 m.y.) before the intracrustal melting took place. What was the nature of this mantle addition? Various processes are possible. Two which appear reasonable are: (1) initiation of orogenic calc-alkaline volcanism; (2) an increase in the rate of the basic–felsic igneous activity, typical of the earlier Archean.

We suggest that the latter was the most probable (see Chapter 3). This has the consequence that the present bulk crustal composition contains a large component with a composition equivalent to that of the Archean crust. As noted earlier, this is similar to that of modern day average calc-alkaline rocks, except that Cr and Ni are more abundant. (See Section 3.2.1.)

9.4   The post-Archean crust

Present-day addition of new material to the continental crust is dominated by island-arc and orogenic zone igneous activity. Processes such as continent–continent collisions (e.g. India–Asia collision to produce the thickened crust of Tibet and the Himalayas or the lateral accretion of 'suspect terrains') do not generate new crust, nor does underthrusting of slices of crust (Section 1.2). Although some oceanic crust may be incorporated in such tectonic events, the continental crust involved is mainly pre-existing material. Underplating by tonalitic intrusives is a possible method of growth in the Archean but occurrence of metasediments in many deep crustal sections makes underplating a difficult concept. Intrusion of tonalites, which are close in composition to andesites, will result in a crust of similar composition. Thus models which call for crustal growth by underplating of tonalitic intrusives are not fundamentally different from the 'andesite' model (see Chapter 3), which always allowed for intrusive as well as volcanic additions to the crust. In general, our models of crustal growth call for relatively minor additions in post-Archean times. Perhaps 25% of the crust is so added, and additions at subducting plate boundaries via calc-alkaline volcanism seem adequate to account for this growth. Essentially lateral accretion of island-arcs to existing continental masses, during the Phanerozoic especially, is in agreement with much geological evidence. Such considerations indeed formed the basis for the 'andesite' model. Likewise, production of granites and granodiorites, by intracrustal melting remains a common event throughout post-Archean time, so that additions to the upper crust have remained the same throughout this period, as is shown by the REE record in the clastic sedimentary rocks (see Chapter 2). The Sm–Nd data from sedimentary rocks is quite unequivocal in this respect (see Sections 2.5 and 10.4.3).

If this model is correct, then the latter-day crust will eventually become a little less basic. Ni and Cr are both notably low in present-day island-arc volcanic rocks and their abundance is not sufficient to account for the observed abundances in the bulk crust. Both elements should be concentrated in the lower crust, in residual phases during partial melting (see Section 3.3.1). Eventually a secular decrease in Ni and Cr abundances in the crust will occur if additions to the continents continue to be made via lateral accretion of island-arcs.

9.5 The formation of granites

The question of the origin of granites has been one of the great geological debates of the twentieth century [e.g. 18]. The problem of *in situ* versus intrusive origin has been generally settled in favour of the latter. Models deriving granite from fractional crystallization of basalt have mostly been replaced by partial melting models involving crustal anatexis [e.g. 16]. However, many of the arguments for mantle origin in preference to the generally accepted derivation by intracrustal melting have been revived [e.g. 19]. The whole question is of fundamental importance since K-rich granodioritic and granitic rocks, with negative Eu anomalies, make up the bulk of the earth's upper crust (Chapter 2). The overall composition of the upper continental crust approximates to granodiorite. Such rocks are the major source of heat production within the crust since lower crustal granulites generally have very low abundances of the heat producing elements (K, Th, U) due either to melt extraction and/or loss in a fluid phase. During the Phanerozoic, granodiorite formation has been concentrated in continental crust near plate margins. This is probably merely a consequence of crustal growth through additions from island-arc sources.

We consider five observations about the role of granites in crustal evolution to be particularly pertinent:

1 There is a temporal trend in the bulk composition of granitic rocks. The sodium-rich varieties (tonalite–trondhjemites–granodiorite) are abundant, and probably dominate during the early Archean. During the late Archean and subsequently, the potassium-rich varieties (granodiorite–monzonite–adamellite–granite) become dominant.

2 There may be a temporal trend in the areal pattern of granitic rocks [20]. Archean granite batholiths tend to be roughly circular; post-Archean batholiths tend toward elongate outcrop patterns. This reflects a change in the site of granite generation from intracontinental to continental margin [20] and is interpreted here as due to a fundamental change in crust-building processes between the Archean and later times.

3 Many granitic rocks display mantle-like or only slightly evolved isotopic characteristics [21–25] indicating short crustal residence times prior to formation. Granitic rocks which are unequivocally derived from reworked older continental crust are common only from the late Archean onward [e.g. 26, 27].

4 Major episodes of production of K-rich granites appear to occur during several relatively short events during earth history, at about 2.8–2.5, 2.0–1.6, 1.2–0.9 and <0.5 Ae [22].

5 Experimental and geochemical evidence indicates that the majority of the potassium-rich varieties of granitic rocks formed at crustal depths (<40 km) where Ca-plagioclase was a stable residual phase [e.g. 16, 28–30]. Less evidence exists concerning the origin of the early Archean Na-granite suites but higher P-T conditions with garnet as a residual phase indicate source regions at mantle depths.

9.5.1 Early Archean granites

The recognition that sodium-rich granitic rocks are major components of early Archean (pre-3 Ae) high-grade and low-grade terrains has provided a major insight into Archean geology [e.g. 2–4, 31–34]. These rocks, along with the extrusive dacites, form one end member of an Archean bimodal suite with mafic volcanics as the other end member. This bimodal suite, with a notable paucity of intermediate compositions, is dominant in early Archean terrains.

Glikson [34] has reviewed much of the geological and geochemical data available although many of his interpretations are controversial [e.g. 35]. The REE have been found to be the most useful index. In Fig. 9.1, some typical REE patterns are shown. The patterns are characteristically very steep with variable Eu-anomalies. Geochemical modelling of such patterns [e.g. 2–4, 33, 37] indicate an origin through 10–20% partial melting of a basaltic amphibolite–granulite–eclogite grade rock with garnet and clinopyroxene as residual phases indicating depths possibly in excess of 60 km. In some cases fractional crystallization of basaltic liquids, with minerals such as hornblende acting as the fractionating phase can be invoked. These models are probably inappropriate for the larger granitic masses because of the extreme amounts of fractional crystallization required [e.g. 34, 37]. These models are generally supported by experimental evidence and O-isotope evidence [38]. Recently, Campbell & Jarvis [39] have suggested that komatiitic compositions are a more likely source.

The isotopic characteristics [25, 34] show primitive (mantle-like) initial isotopic ratios and indicate that early Archean tonalite–trondhjemite–granodiorite gneisses and plutons have either been derived directly from the mantle or more likely from sources which have had short crustal histories (see recent review [34]). These rocks thus represent fresh additions of material to the continents. (See [40] for an alternative view).

Nd and Hf isotopic data indicate essentially the same conclusions [e.g. 25]. A striking feature is that many of the gneiss complexes have $\varepsilon_{Nd} > 0$ indicating derivation from a mantle source which had already been depleted in Nd relative to Sm (i.e. prior melt extraction). Also, in some of the younger gneiss complexes, evidence for considerable crustal reworking is apparent. Thus, Collerson & McCulloch [41] found that the older gneisses of Labrador (Uivak Gneiss) was derived from a slightly depleted mantle source with little crustal prehistory at about 3.6–3.7 Ae but the younger K-rich gneisses, 2.8–3.0 Ae in age, from the same region (Kiyuktok Gneiss) formed by mixing of juvenile mantle derived material with reworked 3.6–3.7 Ae crust.

9.5.2. Late Archean potassium granites

The characteristic of late Archean igneous activity is the development of K-granite–greenstone terrains. This was particularly common during the period 2.8–2.5 Ae, but extends back to about 3.4 Ae at some locations in the southern hemisphere. During this period, Na-granitic rocks (tonalites–

trondhjemites) were less abundant, comprising typically 10% (up to 15–25%) of these terrains. The major occurrence of granitic rocks is of the potassium-granitic varieties (granodiorite–monzonite–adamellite–granite). Volumetrically, these rocks are the most common in many shield areas. For example, in the Barberton Mountain land, the episode of granitic activity '...is considered to represent the main event contributing to cratonization of the early continental masses and the one during, and subsequent to which, tectonic stability prevailed' [42, p. 464].

Potassium granites of this type were intruded during the period 3.4–2.5 Ae the majority forming during the period 2.75–2.5 Ae [43, 44]. In general, these rocks followed the development of Na-granitic rocks and greenstone belts. Such rocks were intruded into the upper crust either very late in or following the development of greenstone belts.

Some typical REE patterns are shown in Fig. 9.4. They are characterized by high but variable La/Yb, flat to moderately steep HREE and significant Eu depletion. Such patterns are best explained by moderate degrees of partial melting (20%) of crustal material, either pre-existing Na-granitic rocks or greenstone belt material (e.g. sedimentary rocks) where plagioclase is a dominant residual phase [36, 43]. Oxygen-isotope data are also consistent with a crustal origin [43, 45].

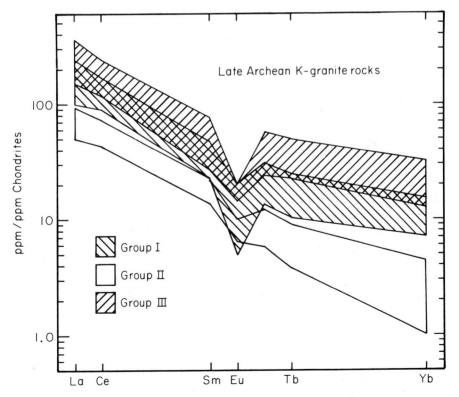

Fig. 9.4. Chondrite-normalized REE data for late Archean potassium granites. Note the well-developed depletion in europium. Group II, showing depletion in the HREE are uncommon [43].

The Sr-isotopic data for Archean potassium-rich granites and quartz monzonites plot on or near the mantle growth curve suggesting derivation from source rocks with short crustal pre-histories [43]. This interval of granite generation is the first in which many initial $^{87}Sr/^{86}Sr$ ratios fall well above the mantle growth curve, indicating significant crustal histories, or in the case of some extremely high values, selective contamination with radiogenic Sr. The ease with which Rb moves makes the Rb/Sr method less reliable than the Sm–Nd method. Nd-isotopic data, however, are less abundant. McCulloch & Wasserburg [46] examined composite samples representing large areas of the Canadian Shield. Granitic material comprised the major fraction of these composites (see Chapter 10, Table 10.6). The Nd model age results indicated that the sources of much of the surface rock of the Superior, Slave and Churchill Provinces were derived from the mantle at 2.7–2.5 Ae.

Late Archean granitic magmatism is the single most important period of granite generation recorded [e.g. 25, 46, 47].

9.5.3 Proterozoic granites

A common feature of the Proterozoic Eon is the widespread occurrence of high-grade metamorphic (so-called 'mobile') belts which show considerable evidence of crustal reworking. Satisfactory tectonic models to account for these

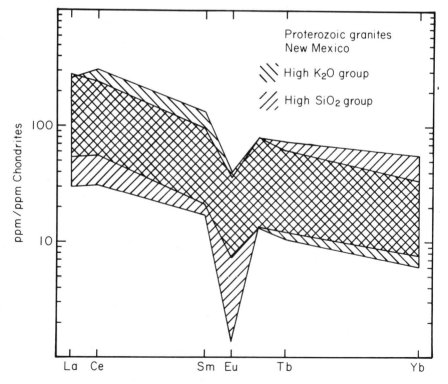

Fig. 9.5. Chondrite-normalized patterns for mid-Proterozoic potassium-rich granites from New Mexico [48]. Note the development of a significant depletion in europium.

terrains have proven particularly elusive. Potassium granitic rocks are common in the Proterozoic, although there is debate about their tectonic setting. The widespread occurrence of rapakivi granites during the mid-Proterozoic is notable. They are commonly associated with anorthosites.

Trace element studies of Proterozoic granitic rocks are scarce. Figure 9.5 shows REE patterns from a sequence of Proterozoic plutons from New Mexico, USA which were intruded over a period of about 1.8–1.3 Ae [48]. The composition of the rocks range from granodiorite to granite with minor trondhjemite. Although the history of this granite province is complex, the geochemical and isotopic data are consistent with a crustal origin from a mafic source with a short crustal prehistory ($^{86}Sr/^{87}=0.702$–0.705).

Isotopic data are similarly scarce, but two recent studies are revealing [26, 49]. Studies of the Colorado Front Range, by DePaolo [49] show that a succession of granodiorite–granite plutons were intruded into a sequence of metavolcanics at 1670, 1400 and 1000 Ga. An excellent case can be made for deriving these granitic rocks from a single source (the mafic volcanic country rocks) extracted from a depleted mantle source about 1800 m.y. ago.

9.5.4 Phanerozoic granites

The major sites of granitic batholiths during the Phanerozoic are in linear belts where the crust is thick and where geothermal gradients are high. The greatest volumes of modern granitic rocks are found at Cordilleran tectonic settings [20]. Typical lithological proportions in Cordillera-type batholiths are as follows: gabbro–diorite, 7–16%; tonalite, 50–58%; granodiorite–adamellite, 25–35%; granite, <1–4% [50]. In most other orogenic belts, granodiorites and granites are dominant.

The Palaeozoic granitic batholiths of south-eastern Australia have been divided into two major types on the basis of chemistry, mineralogy and field relations (Table 9.1) [51]. Average chemical compositions for I- and S-type granites from south-eastern Australia are given in Table 9.2. I-type granitic rocks are thought to have been derived from lower crustal material of igneous origin whereas S-type granitic rock are derived from a source which has been through at least one cycle of weathering (and thus bear a sedimentary signature). The applicability of the I-S classification, outside of south-eastern Australia, remains to be fully documented. Rare earth element data for I- and S-type granites from the late Palaeozoic New England Batholith of eastern Australia are shown in Fig. 9.6 [52]. The I-types are notable for their more variable REE patterns with respect to La/Yb ratios and for the magnitude of negative Eu-anomalies.

Some recent isotopic studies of Phanerozoic granitic rocks [26, 27, 49, 53] indicate a wide range in both Sr and Nd isotopic characteristics (Fig. 9.7). The range of values are consistent with mixing an old continental crust component and a depleted and/or undepleted mantle component. The amounts of each component are highly variable both among batholiths [27] and within many batholiths [53]. The presence of a component with mantle-like isotopic ratios

Table 9.1. Mineralogical and chemical differences between I- and S-type granites [51].

	Mineralogical criteria	
	I-type	S-type
	1. Hornblende present 2. Muscovite rare 3. Cordierite, garnet, andalusite and sillimanite absent	1. Hornblende absent 2. Muscovite common 3. Cordierite, garnet, andalusite and sillimanite may be present

	Geochemical criteria			
		I-type		S-type	
		Characteristic value	Explanation	Characteristic value	Explanation
1.	SiO_2	Wide range 53–76%	Relatively mafic source rocks	Within range 64–74%	Derived from SiO_2-rich source
2.	K_2O/Na_2O	Low	Na has not been removed by weathering	High	K adsorbed by clays on weathering, whereas Na is removed
3.	$\dfrac{Al_2O_3}{Na_2O+K_2O+CaO}$	Normally low	Only minimum temperature melts or fractionated I-type rocks may be peraluminous	High (>1.05) and increases as the rocks become more mafic	Weathering increases Al relative to Na+K+Ca
4.	$\delta^{18}O$	Low	Primary igneous source rocks	High	Oxygen isotopes fractionate during production of clays during low temperature weathering
5.	$^{87}Sr/^{86}Sr$	Generally low	Mantle-derived igneous source rocks. Some high values for granitoids derived from old source rocks with high Rb/Sr	High (normally >0.708)	Rb concentrated relative to Sr during weathering and sedimentation
6.	Cr and Ni	Low	Source rocks relatively low in Cr and Ni, indicating prior fractionation	High relative to I-types	Cr and Ni incorporated into clays during weathering

Table 9.2. Average chemical composition of I- and S-type granites [51].

	I-type	S-type
n	532	316
SiO_2	69.1	70.5
TiO_2	0.46	0.56
Al_2O_3	14.8	14.6
FeO	3.78	3.97
MgO	1.78	1.86
CaO	3.85	2.54
Na_2O	3.00	2.24
K_2O	3.11	3.70
Σ	99.9	100.0
	ppm	ppm
Ba	520	480
Rb	132	180
Sr	253	139
Pb	16	27
La	29	31
Ce	63	69
Nd	23	25
Y	27	32
Zr	143	170
Nb	9	11
Cr	27	46
V	74	72
Sc	15	14
Ni	9	17
Co	12	13
Cu	11	12
Zn	52	64
Ga	16	17

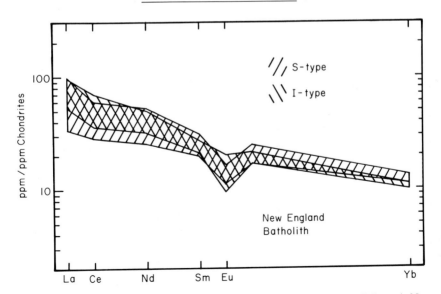

Fig. 9.6. Chondrite-normalized REE patterns for I- and S-type granites from the Palaeozoic New England Batholith of eastern Australia (from compilation in [52]).

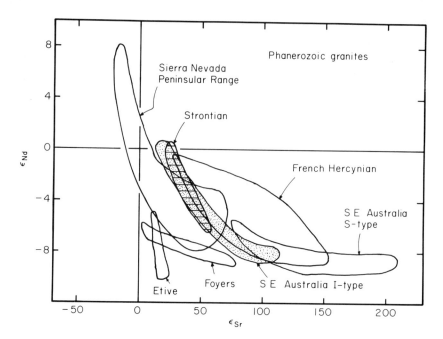

Fig. 9.7. ε_{Nd}–ε_{Sr} diagram for Phanerozoic granites [53]. Most of the granites plot in the lower right-hand quadrant, indicating derivation from sources composed of significant amounts of older crustal materials. The only 'granitic' rocks which plot in the upper left-hand quadrant are from the Sierra Nevada and Peninsular ranges of California, representing derivation from island-arc sources.

does not imply a direct mantle origin, but indicates a short crustal residence time for the mantle-derived igneous material.

S-type granitic rocks are characterized by low $^{143}Nd/^{144}Nd$ and high and variable $^{87}Sr/^{86}Sr$ [53]. I-types showed the reverse (Fig. 9.7). These data are consistent with S-types being derived predominantly ($\approx$70–100%) from an older (c. 1.0 Ae) crustal source. The I-type data also suggest a significant crustal component (25–80%) but this is apparently in conflict with the geochemical data [53]. Resolution of this contradiction could lie in I-types being derived from igneous sources of various ages.

A summary diagram based on Sm–Nd isotopic data (Fig. 9.8) shows significant variations with geological age [26]. Older granites have initial Nd ratios rather close to the chondritic evolution line. Younger granites have ratios showing that increasing amounts of older continental crust were involved. These data support models specifying continental growth through time, with intracrustal partial melting in more recent times involving previously formed continental crust. No evidence is apparent for a direct mantle derivation of granites.

9.6 Anorthosites and crustal growth

Anorthosites form a substantial, but dimly understood component of the continental crust. Massif anorthosites occur as batholithic masses, commonly

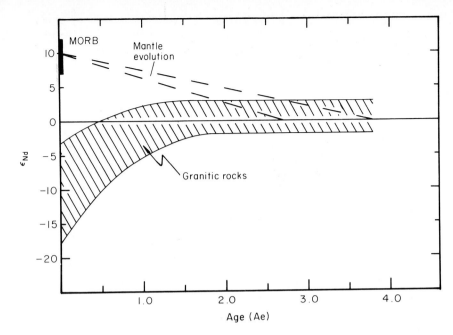

Fig. 9.8. Plot of ε_{Nd} versus geological age, showing the chondrite ($\varepsilon_{Nd}=0$) and mantle evolution lines and ε_{Nd} values for present-day MORB. The data for granitic rocks are consistent with an increasing component of older crustal material in the generation of the younger granites [26, 27].

exceeding 10 000 km² in outcrop areas. Their source and origin is relevant to any discussion of the evolution of the continental crust. There is a clear distinction between Archean and Proterozoic anorthosites, reinforcing the evidence from many other sources that the Archean–Proterozoic boundary marks a major dichotomy in Earth history.

9.6.1 Archean anorthosites

These generally occur in association with other components of layered igneous complexes of overall basaltic composition. A typical example is the Fiskenaesset complex in Greenland, which covers an area of about 500 km². Perhaps the most distinctive signature of their origin is the frequent association of chromite seams (up to 20 m thick), but many other units, typical of basaltic layered intrusions, commonly are present. The plagioclase crystals tend to be large (up to 30 cm), and equant typically in the range An_{80-95}, so that they may be strictly regarded as anorthosites, unlike their Proterozoic counterparts. This composition is similar to that of feldspars crystallizing in layered igneous intrusions, following the appearance of olivine and clinopyroxene. The bulk composition of the complexes, given in Table 9.3, typically contains about 20% Al_2O_3. Possible parental liquids are estimated to contain less than 20% alumina [54].

The origin of Archean anorthositic complexes is thus reasonably well understood. They are derived by fractional crystallization of basaltic melts, albeit somewhat aluminous. They do not form a major part of the crust, but

Table 9.3. Typical anorthosite compositions

	Archean			Proterozoic	
	1	2	3	4	5
SiO$_2$	50.2	50.1	47.2	55.1	54.0
TiO$_2$	0.11	0.56	0.41	0.15	1.70
Al$_2$O$_3$	30.8	19.7	23.5	27.1	21.5
FeO	0.99	8.1	6.4	1.97	6.66
MgO	1.12	6.9	6.4	1.14	2.55
CaO	12.5	11.5	14.3	10.3	9.49
Na$_2$O	3.49	2.45	1.6	4.50	4.14
K$_2$O	0.85	0.18	0.2	0.43	0.71
Σ	100.1	99.5	100.0	100.6	100.8
	ppm	ppm		ppm	ppm
Cs	3.1	0.6	—	—	—
Ba	—	—	—	265	374
Rb	—	—	—	1.72	8.90
Sr	300	—	—	695	515
La	0.87	—	—	—	—
Ce	1.78	7.7	—	7.81	19.8
Nd	0.72	—	—	3.42	10.6
Sm	0.19	1.41	—	0.53	2.14
Eu	0.37	0.54	—	0.754	1.53
Gd	—	—	—	0.429	2.07
Tb	0.033	0.26	—	—	—
Dy	—	—	—	0.305	1.88
Er	—	—	—	0.15	1.07
Yb	0.09	—	—	0.138	0.985
Lu	0.014	0.16	—	0.022	0.15
Th	0.05	—	—	—	—
U	0.1	—	—	—	—
Hf	0.1	0.9	—	—	—
Ta	0.6	1.1	—	—	—
Cr	1.0	220	—	—	—
Sc	1.4	27	—	—	—
Ni	14	—	—	—	—
Co	12.2	50	—	—	—

1. Archean anorthosite 148B, Bad Vermilion Lake, Ontario [54].
2. Calculated bulk composition of Bad Vermilion Lake Complex [54].
3. Bulk composition, Archean Fiskenaesset Complex, Labrador [54].
4. Massif-type Proterozoic anorthosite, Nain Complex, Labrador, CS 53A [55].
5. Gabbroic anorthosite, CS 54B [55].

are considered here for two reasons: firstly to distinguish them from the volumetrically important Proterozoic anorthosites, and secondly to emphasize that they are not analogous to the lunar highlands crust, and are not remnants of an early terrestrial anorthositic crust (see Chapter 12) [56].

9.6.2 Proterozoic anorthosites

In contrast to their Archean counterparts, the massif-type anorthosites of

Proterozoic age represent a major component of the continental crust [57–61]. The term 'anorthosite', although firmly entrenched, is a misnomer, for the plagioclase compositions lie typically in the range An_{40}–An_{65} [62]. These feldspars are thus much less primitive than those of Archean anorthosites. The Mg/Mg+Fe numbers of the bulk complexes are likewise evolved, in the range 40–50. Bulk compositions are highly feldspathic and aluminous, Al_2O_3 contents being generally above 20% (Table 9.3). There is clear evidence for the intrusion of these anorthosites as melts [59]. The absence of water is notable and the temperatures of these magmas are estimated to be about 1300°C, from the associated metamorphic effects [60].

The question of the origin of these dry feldspathic melts has intrigued petrologists and no agreed solution has been reached. Their field occurrence is in batholithic masses, of very large surface outcrop, up to 30 000 km² in extent [63]. Whether they occur in flat sheets up to 5 km thick or have more basic roots depends partly on the interpretation of the gravity data. Their intrusion is not connected with orogenic episodes. The depth of emplacement as shown by oxygen isotope studies [64], is shallow, typically in the range 5–13 km, although some are deeper. The common association with 'granitic' rocks (mangerites, charnockites, rapakivi granites) appears to be non-genetic and the granitic melts are generated independently, as shown by field, chemical and isotopic evidence [55, 65]. Such evidence disposes of the hypotheses that suggest a direct complementary relationship between anorthosites and granites.

A unique feature of the massif anorthosites is that most are found in the mid-Proterozoic. Although they are difficult to date on account of very low Rb/Sr ratios, their ages range from about 2300 m.y. down to the lower Palaeozoic occurrence at Sept Iles, Quebec. Mostly they were intruded between about 1.7 and 0.9 Ae, and so represent a major episode in the evolution of the continental crust [57, 60]. They are notably depleted in Rb, with low Rb/Sr ratios, but have $^{87}Sr/^{86}Sr$ ratios in the range of basaltic rocks, of 0.704–0.705 [65]. The Nd isotopic data are scattered with ε_{Nd} values from −6 to +4 (see review in [66]). Typically they display Eu enrichment. This appears to be a primary feature of the magma, rather than due to local accumulation of feldspar (Fig. 9.9).

This interesting set of properties elevates the question of the origin of massif anorthosites to a status close to that previously occupied by the 'granite problem'. In common with similar geological puzzles, many theories have been advanced. They are clearly magmatic so that the location and source of the melts becomes a major problem. The lack of volcanic equivalents, their restriction to the continental crust, and to a particular, although long, period of geological time all pose considerable constraints. Other significant constraints include Eu enrichment, Rb depletion, low Mg/Mg+Fe values and the lack of water. Fractional crystallization of basaltic magma is unlikely to produce monomineralic masses of anorthosite, and the apparent absence of the large masses of basic and ultramafic cumulates beneath anorthosite plutons makes the existence of a primary anorthosite magma more likely. Derivation of such a magma from the mantle is unlikely on account of the absence of such

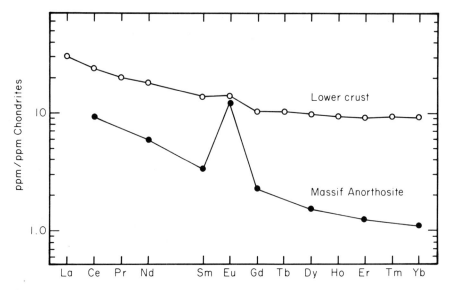

Fig. 9.9. Chondrite-normalized REE patterns for the lower crust (Chapter 4) and a typical Proterozoic massif anorthosite (Table 9.3, column 4).

magmas in oceanic environments and the low Mg numbers. Continental tectonic settings, involving rifting or 'failed rifting' are popular to explain the somewhat localized scale of the production of massif anorthosites, on a global context.

Partial melting of high-Al basaltic parents is also popular, although this too begs the question of the source of such large volumes of parent material. Crustal contamination has usually been invoked to explain the unsupported radiogenic ^{87}Sr relative to the present very low Rb/Sr ratios, but the extremely low amounts of Rb and high Sr values surely rule out much addition of typical upper crustal material. It thus seems necessary to invoke the production of a primary anorthositic magma to solve the question.

'Such a unique magma association implies that unique tectonic and thermal settings are responsible for the generation of Proterozoic anorthosite massifs' [59, p. 567]. One possible region from which such melts can be derived is the lower continental crust. Subsequent to the massive extraction of granite around 2.5 Ae, regions will exist with residual compositions low in Si, Rb (but with higher ^{87}Sr/^{86}Sr ratios), high in Al, Eu and Sr and with variable ε_{Nd}. The frequent occurrence of xenoliths with these compositional characteristics (see Chapter 4) encourages this speculation. The production of high-Al magmas from this source later in the Proterozoic, during periods of rifting, seems capable of explaining many of the temporal, spatial, chemical and isotopic features of the massif anorthosites. In this scenario, their origin had to await the prior development of a source region with the appropriate chemical and isotopic signatures. However, such a region will have been depleted in the heat-producing elements K, U and Th, by prior extraction. Accordingly, mantle sources of heat are required to generate anorthositic magmas. Possibly an event such as the migration beneath a continent of a mid-ocean ridge might supply the necessary heat and would also be consistent with the concepts of rifting or 'failed rifting' noted above [66].

9.7 The development of the present-day plate-tectonic regime

The dramatic increase in continental volume at the end of the Archean changed the geography of the world from one essentially of oceanic aspect, with a few scattered large continental islands, to a ratio of continent to ocean very similar to that of the present. There was a large increase in the areal extent of buoyant unsinkable or unsubductible continental masses. These would produce several effects. Firstly there is an insulating effect on the mantle due to their low thermal conductivity. At the same time, there was a massive transfer of radioactive heat-producing elements from the mantle into the continental crust. Thirdly, the multi-plate scenario of the Archean was transformed by the formation of large continental blocks. It is tempting to place the beginning of modern style plate tectonics at this time. Spreading basaltic sea floor, encountering this massive barrier or barriers to subduction, perforce was directed into the mantle along linear trends, the precursors of the present day island-arcs.

It is possible that all the continental material was driven into one large unit during the early Proterozoic. Evidence for a Proterozoic Pangaea appears from the palaeomagnetic data [67] although the interpretation is not unique [68]. In this reconstruction, a lenticular shaped crustal area results and apparently is stable from about 2600 m.y. down to the Cambrian, at about 570 m.y. This view is not consistent with multi-continent plate tectonics in the Proterozoic. In this model [67], the continents formed a stable unit until late in the Proterozoic, when brittle fracture began, probably due to accumulating thermal stresses, and the present-day style of continental drift began. Conventional plate tectonics in this reconstruction is thus essentially a late Proterozoic and Phanerozoic phenomenon. Whether this scenario is valid, or whether present-day plate tectonic regimes extended further back into the Proterozoic Eon, remains to be tested.

9.8 Summary

1 The dominant mechanism for producing the Archean crust was the derivation of the bimodal basic–felsic igneous suite from mantle sources. Calc-alkaline (island-arc) igneous activity was minimal, and the similarity of bulk crustal Archean REE pattern to that of present-day andesites is coincidental.
2 The ratio of basic–felsic igneous rocks varies from 1:1 for the Archean upper crust to 2:1 for the bulk crust.
3 Granites and granodiorites with significant Eu depletion comprised less than 10% of the exposed early Archean crust.
4 The change in composition between the Archean and post-Archean upper

crusts is caused by massive intracrustal melting resulting in the intrusion of granites and granodiorites with significant depletion in Eu (average Eu/Eu* ≈ 0.65).

5 Addition of material to the continental crust from the mantle since the Archean has been mainly through orogenic zone igneous activity.

6 The production of granitic rocks plays a key role in crustal evolution. Mantle derived Na-granites dominate the early Archean crust. In the late Archean, massive intracrustal melting produces an upper crust dominated by K-rich granitic rocks. Younger granites show increasing evidence of crustal recycling.

7 Massif-type anorthosites appear as important crustal components in the Proterozoic. Archean examples are generally parts of smaller layered intrusions. It is proposed that the massif-type anorthosites represent intracrustal melts derived from an Al-rich lower crust, developed following granitic melt extraction.

Notes and references

1 Taylor, S.R. & Hallberg, J.A. (1977) Rare-earth elements in the Marda calc-alkaline suite: An Archean geochemical analogue of Andean-type volcanism. *GCA*, **41**, 1125.
2 Barker, F. & Arth, J. (1976) Generation of trondhjemitic–tonalitic liquids and Archean bimodal trondhjemite–basalt suites. *Geology*, **4**, 596.
3 Arth, J.G. & Hanson, G.N. (1975) Geochemistry and origin of the early Precambrian crust of northeastern Minnesota. *GCA*, **39**, 325.
4 Barker, F. & Peterman, Z.E. (1974) Bimodal tholeiitic–dacitic magmatism and the early Precambrian crust. *PCR*, **1**, 1.
5 O'Nions, R.K. & Pankhurst, R.J. (1974) Rare-earth element distribution in Archean gneisses and anorthosites, Godthaab area, West Greenland. *EPSL*, **22**, 328; Martin, H. *et al.* (1983) Major and trace element geochemistry and crustal evolution of Archean granodioritic rocks from eastern Finland. *PCR*, **21**, 159; Hunter, D.R. *et al.* (1978) The geochemical nature of the Archean Ancient Gneiss Complex and granodiorite suite, Swaziland. *PCR*, **7**, 105.
6 Jahn, B-M. *et al.* (1981) REE geochemistry and isotopic data of Archean silicic volcanics and granitoids from the Pilbara Block, Western Australia: implications for early crustal evolution. *GCA*, **45**, 1633; Foden, J.D. *et al.* (1984) The geochemistry and crustal origin of the Archean acid intrusive rocks of the Agnew Dome, Lawlers, Western Australia. *PCR*, **23**, 247.
7 Hargraves, R.B. (1976) Precambrian geologic history. *Science*, **19**, 363. (See Chapter 7, note 78.)
8 Fyfe, W.S. (1978) The evolution of the earth's crust: modern plate tectonics to ancient hot spot tectonics? *Chem. Geol.*, **23**, 89.
9 Sclater, J.G. *et al.* (1980) The heat flow through oceanic and continental crust and the heat loss from the Earth. *RGSP*, **18**, 269.
10 Pollack, H.N. (1980) The heat flow from the Earth: a review. *In:* Davies, P.A. & Runcorn, S.K. (eds), *Mechanisms of Continental Drift and Plate Tectonics*. Academic Press. 183.
11 Bickle, M.J. (1978) Heat loss from the Earth: a constraint on Archean tectonics from the relation between geothermal gradients and the rate of plate production. *EPSL*, **40**, 301.
12 Wilson, I.H. (1982) *Petrology and Geochemistry of Selected Proterozoic Volcanics from the Mt. Isa Inlier, Queensland*. Ph.D. Thesis, University of Queensland.
13 *Basaltic Volcanism on the Terrestrial Planets* (1981) Pergamon.
14 See Appendix 3.
15 Fyfe, W.S. (1973) The granulite facies, partial melting and the Archean crust. *Phil. Trans. Roy. Soc.*, **A273**, 457.
16 Wyllie, P.J. (1977) Crustal anatexis: An experimental review. *Tectonophysics*, **43**, 66.

17 Brown, G.C. & Mussett, A.E. (1981) *The Inaccessible Earth.* Allen & Unwin.
18 Gilluly, J. (1948) Origin of granite. *GSA Mem.,* **28**; Read, H.H. (1957) *The Granite Controversy.* Thomas Murby, London.
19 Brown, G.C. (1977) Mantle origin of Cordilleran granites. *Nature,* **265**, 21.
20 Brown, G.C. & Hennessy, J. (1978) The initiation and thermal diversity of granite magmatism. *Phil. Trans. Roy. Soc.,* **A288**, 631.
21 Moorbath, S. (1975) Evolution of Precambrian crust from strontium isotopic evidence. *Nature,* **254**, 395.
22 Moorbath, S. (1976) Age and isotope constraints for the evolution of Archean crust. *In:* Windley, B.F. (ed.), *The Early History of the Earth.* Wiley. 351.
23 Moorbath, S. (1977) Ages, isotopes and evolution of Precambrian continental crust. *Chem. Geol.,* **20**, 151.
24 Moorbath, S. (1978) Age and isotope evidence for the evolution of the continental crust. *Phil. Trans. Roy. Soc.,* **A288**, 401.
25 Moorbath, S. & Taylor, P.N. (1981) Isotopic evidence for continental growth in the Precambrian. *In:* Kroner, A. (ed.), *Precambrian Plate Tectonics.* Elsevier. 491.
26 Allegre, C.J. & Ben Othman, D. (1980) Nd–Sr isotopic relationship in granitoid rocks and continental crust development: A chemical approach to orogenesis. *Nature,* **286**, 335.
27 Hamilton, P.J. *et al.* (1980) Isotopic evidence for the provenance of some Caledonian granites. *Nature,* **287**, 279.
28 Brown, G.C. & Fyfe, W.S. (1970) The production of granitic melts during ultrametamorphism. *CMP,* **28**, 310.
29 Hanson, G.N. (1980) Rare earth elements in petrogenetic studies of igneous systems. *Ann. Rev. Earth Planet. Sci.,* **8**, 371.
30 See also Section 4.2.
31 McGregor, V.R. (1973) The early Precambrian gneisses of the Godthaab district, West Greenland. *Phil. Trans. Roy. Soc.,* **A273**, 343.
32 Viljoen, M.J. & Viljoen, M.P. (1969) *Geol. Soc. South Africa Spec. Pub.,* **2** (this publication contains numerous papers by this team).
33 Barker, F. *et al.* (1981) Tonalites in crustal evolution. *Phil. Trans. Roy. Soc.,* **A301**, 293.
34 Glikson, A.Y. (1979) Early Precambrian tonalite–trondhjemite sialic nuclei. *Earth Sci. Rev.,* **15**, 1.
35 Bettenay, L.F. *et al.* (1981) Evolution of the Shaw Batholith—an Archean granitoid gneiss dome in the eastern Pilbara. *Geol. Soc. Austr. Spec. Pub.,* **7**, 361.
36 Hanson, G.N. (1978) The application of trace elements to the petrogenesis of igneous rocks of granitic composition. *EPSL,* **38**, 26.
37 O'Nions, R.K. & Pankhurst, R.J. (1978) Early Archean rocks and geochemical evolution of the Earth's crust. *EPSL,* **38**, 211.
38 Barker, F. *et al.* (1976) Oxygen isotopes of some trondhjemites, siliceous gneisses and associated mafic rocks. *PCR,* **3**, 547.
39 Campbell, I.H. & Jarvis, G.T. (1984) Mantle convection and early crustal evolution. *PCR* **26**, 15.
40 Collerson, K.D. & Fryer, B.J. (1978) The role of fluids in formation and subsequent development of early continental crust. *CMP,* **67**, 151.
41 Collerson, K.D. & McCulloch, M.T. (1982) The origin and evolution of Archaean crust as inferred from Nd, Sr and Pb isotopic studies in Labrador (abst.). *5th Int. Conf. Geochron. Cosmochron. Isotope Geol. Abst.,* **61**.
42 Anhaeusser, C.R. & Robb, L.J. (1981) Magmatic cycles and the evolution of the Archean granitic crust in the eastern Transvaal and Swaziland. *Geol. Soc. Austr.,* **7**, 457.
43 Condie, K.C. (1981) Geochemical and isotopic constraints on the origin and source of Archean granites. *Geol. Soc. Austr. Spec. Pub.,* **7**, 469; (1981) *Archean Greenstone Belts.* Elsevier.
44 Ayres, L.D. & Cerny, P. (1982) Metallogeny of granitoid rocks in the Canadian shield. *Can. Mineral.,* **20**, 439.
45 Taylor, H.P. (1977) Water/rock interaction and the origin of H_2O in granitic batholiths. *J. geol. Soc. Lond.,* **133**, 509.
46 McCulloch, M.T. & Wasserburg, G.J. (1978) Sm–Nd and Rb–Sr chronology of continental crust formation. *Science,* **200**, 1003.
47 O'Nions, R.K. *et al.* (1979) Geochemical modelling of mantle differentiation and crustal growth. *JGR,* **84**, 6091.

48 Condie, K.C. (1978) Geochemistry of Proterozoic granitic plutons from New Mexico, USA. *Chem. Geol.*, **21**, 131.
49 DePaolo, D.J. (1980) Sources of continental crust: neodymium isotope evidence from the Sierra Nevada and Peninsular Ranges. *Science*, **209**, 684; (1981) Neodymium isotopes in the Colorado Front Range and crust-mantle evolution in the Proterozoic. *Nature*, **291**, 193.
50 Cobbing, E.J. & Pitcher, W. (1972) The Coastal Batholith of Peru. *J. geol. Soc. London*, **128**, 421.
51 Chappell, B.W. & White, A.J.R. (1983) Granitoid types and their distribution in the Lachlan Fold Belt, southeastern Australia. *GSA Mem.*, **159**, 21.
52 Shaw, S.E. & Flood, R.H. (1981) The New England Batholith, eastern Australia: geochemical variations in time and space. *JGR*, **86**, 10530.
53 McCulloch, M.T. & Chappell, B.W. (1982) Nd isotope characteristics of S and I-type granites. *EPSL*, **58**, 51.
54 Ashwal, L.D. *et al.* (1983) Origin of Archean anorthosites: Evidence from the Bad Vermilion Lake anorthosite complex, Ontario. *CMP*, **82**, 259.
55 Simmons, E.C. & Hanson, G.N. (1978) Geochemistry and origin of massif-type anorthosites. *CMP*, **66**, 119.
56 Phinney, W.C. (1982) Petrogenesis of Archean anorthosites. *LPI Tech. Rpt. 82-01*, Lunar Planet. Inst., Houston, Texas, 121.
57 Herz, N. (1969) Anorthosite belts, continental drift and the anorthosite event. *Science*, **164**, 944.
58 Emslie, R.F. (1978) Anorthosite massifs, Rapakivi granites and Late Proterozoic rifting of North America. *PCR*, **7**, 61.
59 Wiebe, R.A. (1980) Anorthositic magmas and the origin of Proterozoic anorthosite massifs. *Nature*, **286**, 564.
60 Morse, S.A. (1982) A partisan review of Proterozoic anorthosites. *Am. Mineral.*, **67**, 1087.
61 Ashwal, L.D. (1982) Proterozoic anorthosite massifs: a review. *LPI Tech. Rpt. 82-01*, Lunar Planet. Inst., Houston, Texas, 40.
62 The mean plagioclase composition (An_{50}) of massif anorthosites inconsiderately falls on the boundary between andesine and labradorite. Names such as andesinite or labradoritite or some combination thereof are apparently shunned even by petrologists. The fact that Proterozoic massif-type anorthosites are not named from some obscure hamlet is possibly due to the barren and infertile nature of anorthosite terrains, which has discouraged close settlement. There is apparently a correlation between regions of anorthosite outcrop in Norway and the occurrence of strongly fundamentalist sects. (Henrich Neumann, pers. comm.)
63 Emslie, R.F. (1980) Geology and petrology of the Harp Lake complex, Central Labrador. *GSC Bull.*, **293**.
64 Valley, J.W. & O'Neill, J.R. (1982) Oxygen isotope evidence for shallow emplacement of Adirondack anorthosite. *Nature*, **300**, 497.
65 Duchesne, J.C. & Demaiffe, D. (1978) Trace elements and anorthosite genesis. *EPSL*, **38**, 249.
66 Taylor, S.R. *et al.* (1984) The origin of massif-type anorthosites. *Nature*, **311**, 372.
67 Piper, J.D.A. (1982) The Precambrian palaeomagnetic record: the case for the Proterozoic supercontinent. *EPSL*, **59**, 61.
68 McElhinny, M.W. & McWilliams, M.O. (1977) Precambrian geodynamics—a palaeomagnetic view. *Tectonophysics*, **40**, 137.

10
The Growth Rate of the Crust

10.1 The nature of the problem

The task of estimating the rate of growth of the continental crust in quantitative terms has fallen historically to the isotopic geochemists. More recently, attempts to measure Phanerozoic crustal growth rates directly have been attempted [1, 2] but this approach is fraught with difficulty because of uncertainties in recycling rates and in extrapolation both of the growth and recycling rates back through time. Crustal rocks with mantle-like initial isotopic ratios (Sr, Nd, Hf, Pb) are conventionally taken to represent new crustal additions [3]. Even when an intracrustal origin for such rocks is clear (e.g. K-rich granites with large negative Eu-anomalies), it generally is argued that only a short period in a crustal environment (high Rb/Sr, U/Pb, Th/Pb; low Sm/Nd, Lu/Hf) followed extraction from the mantle [3]. In this model of crustal growth, continental crust, once formed, is irreversibly differentiated and because of its low density cannot be returned and remixed with the mantle on a large scale.

There is an alternative interpretation of the isotopic data [4–6]. If it is accepted that crustal material can be returned and efficiently remixed with the mantle in sufficient quantities, then the isotopic results are consistent with a model in which the entire crust was formed very early in earth history (>4.0 Ae) and continually recycled through the mantle, with no overall change in crustal mass. This may be termed the steady-state or crustal recycling model.

The isotopic data provide fundamentally different information for these two extreme models. In the crustal growth model the isotopic ratios allow estimates of growth rates, whereas in the steady-state model they allow estimates of crustal recycling rates. Obviously, some combination of these models, where growth and recycling are taking place, is also possible. Interpretation of the isotopic evidence is difficult for such scenarios [7–11] and an independent estimate of either growth rate or recycling rate is necessary. Because of this non-uniqueness in the interpretation of the isotopic record, we require independent evidence to constrain crustal growth rates or to eliminate one of the extreme models of crustal evolution (i.e. growth versus steady-state). In this chapter we examine some of the evidence relevant to this question.

10.2 Models of crustal growth

A number of extreme models of early crustal formation have been proposed. One is that the continents represent rim debris of huge meteorite impact sites, now ocean basins, which formed in an analogous fashion to the lunar multi-ringed basins [12]. An equally spectacular model [13] suggests the continents were remnants of meteoritic material (of appropriate sialic composition) accreted to the Earth during the latest stages of its formation. Such catastrophic models have received virtually no support. More popular models propose that most of the continents differentiated from the mantle early in Earth history and have been continuously recycled through the mantle ever since [4, 5]. The rate of recycling is linked to the Earth's heat production and accordingly diminishes through time as the Earth cools [6, 14–16]. An extreme variant of this model is that the mass of the continents have diminished through time [15].

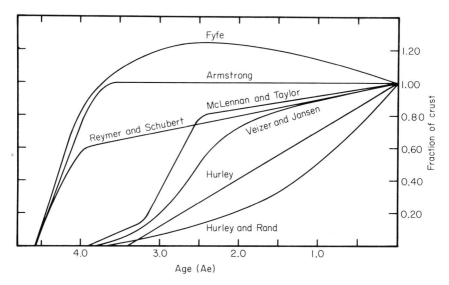

Fig. 10.1. A selection of crustal growth models. Models shown include those of Reymer & Schubert [2], Armstrong [5], Fyfe [15], Hurley [17], Hurley & Rand [18], Veizer & Jansen [20, 23] and McLennan & Taylor [24].

The concept of continuous or quasi-continuous growth of the continental crust has received much support from the isotopic geochemists. On the basis of early modelling of Sr isotopes, a uniform growth rate over the past 3–4 billion years was proposed [17]. With increased knowledge of crustal Rb–Sr and K–Ar age patterns, this model quickly gave way to a model of accelerated growth through geological time [18]. As the potential resetting of K–Ar and Rb–Sr isotopic systems and the intrinsically cannibalistic nature of sedimentation were appreciated, such models were re-examined [19–22]. Currently, many adherents of crustal growth models agree that the majority of crust (up to 90%) was formed by the end of the Archean, at about 2.5 Ae. Figure 10.1 summarizes some of the various models of crustal growth.

10.3 Geological constraints

10.3.1 The freeboard argument

Continental freeboard is the elevation of the continents relative to mean sea-level. Egyed [25] was the first to produce a freeboard curve for the Phanerozoic, based on global palaeogeographic maps [26, 27]. The curve indicated continual emergence of the continents throughout the Phanerozoic, and was later used as evidence for the Earth expansion hypothesis [25, 28] (see

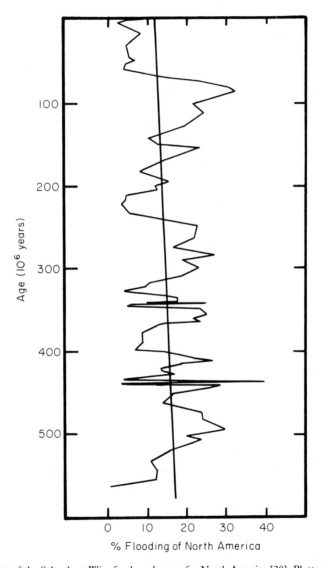

Fig. 10.2. Plot of the Schuchert-Wise freeboard curve for North America [30]. Plotted is the percentage flooding of North America against time (determined on palaeogeographical maps). The slight slope towards greater flooding with increasing age is considered to be an artefact inherent in the data and this curve has been used to hypothesize constant freeboard of the continents during the Phanerozoic [30].

Section 12.8). It has since been recognized that the time intervals used for these palaeogeographic reconstructions increased as a function of age and so could result in a trend towards greater continental flooding with the larger time intervals [29]. Wise [30] suggested that such biases invalidated Egyed's treatment. As an alternative, he used the palaeogeographic data for one continent (North America) [30], where the length of time intervals were more uniform and the palaeogeographic data were better constrained. In this analysis, most of the evidence for changes in freeboard vanished (Fig. 10.2), the data showing that the freeboard of the continents was within 60 m of the base level (20 m above present mean sea-level) for about 80% of Phanerozoic time. In Fig. 10.3, graphical solutions to the constant freeboard equations [30] are given, for generally accepted models of crustal structure [31].

A re-examination [32, 33] of the record of continental freeboard, using the most recent palaeogeographic data from the USSR [34] and North America [35] suggests that there has been a gradual emergence of the continents during the Phanerozoic (Fig. 10.4). Such evidence does not exclude the arguments for long-term constant freeboard because extensive continental emergence is also documented (but not in a rigorous quantitative fashion) for the late Precam-

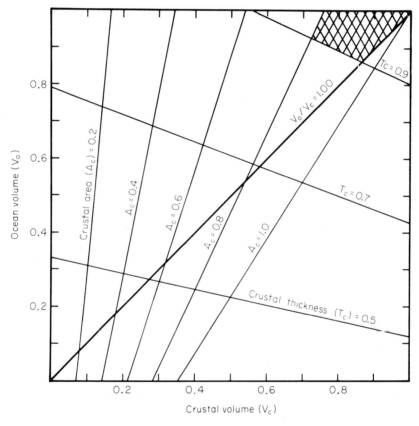

Fig. 10.3. Graphical solutions of the constant freeboard equations [30, 31]. The available constraints indicate that the system has remained within the hatched area for the past 2500 m.y.

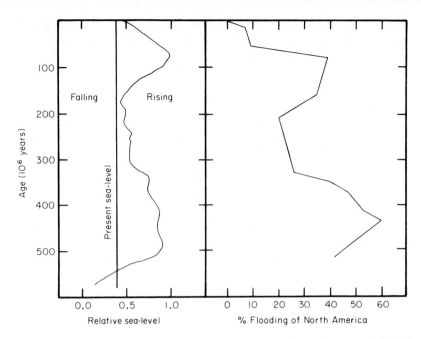

Fig. 10.4. *Left.* First-order cycles of relative change in sea-level during the Phanerozoic [36]. Note that the late Precambrian was a period of low sea-level and continental emergence. *Right.* Plot of the Hallam freeboard curve for North America [32]. This curve differs from that of Wise [30] and suggests a trend towards continental emergence through the Phanerozoic. Because of the convincing evidence for periods of emergence during the Precambrian, the Phanerozoic trend is thought to be a second-order feature controlled by changes in the volume of mid-ocean ridges or thickening of the continental crust during orogenic activity [32].

brian [30, 32, 36] (see Fig. 10.4) and early Proterozoic [5, 37, 38]. The secular change in freeboard during the Phanerozoic may be a second order feature controlled by the volume of mid-ocean ridges [e.g. 39] or thickening of the continental crust during orogenic activity [32, 33].

Most of the models of crustal growth depicted in Fig. 10.1 would be compatible with constant freeboard, but other data, related to crustal thickness, area and ocean volume evolution, are available which can be used to provide additional constraints. It has been argued by many authors that the continental crust has remained at approximately its present thickness for at least 2500 m.y. [1, 40, 41]. Seismic studies in Archean shield areas which have been stabilized for the past 2.5 Ae indicate thicknesses in the range 30–50 km, averaging close to 40 km. A compilation of much of the seismic data (Table 10.1) indicates the continents have maintained their thickness to within ±10% over the past 2500 m.y. [41]. Later widespread underplating of sialic material which thickens the crust uniformly is considered unlikely because of the large areas involved and the lack of surface expression [42]. An illuminating example, which argues against underplating as a mechanism for thickening Archean crust at a later date is found in Western Australia. Seismic profiles [43] across the Pilbara and Yilgarn Blocks indicate crustal thicknesses increasing from about 30 km in the Pilbara to about 50 km in the Yilgarn (see

Table 10.1. Estimates of mean crustal thickness as a function of age [41].

Age (m.y.)	Number of seismic sections	Crustal thickness (km)*
<225	37	38±4
225–600	21	40±4
600–1200	12	39±5
1200–2500	46	39±2
>2500	13	38±4

* Uncertainties are 95% confidence invervals.

Fig. 4.2). The additional thickness in the Yilgarn is due to a thick lower crustal high velocity layer (P-wave velocity of about 7.0 km s^{-1}). This layer has been attributed to underplating of mafic material during the breakup of Gondwanaland in the Phanerozoic [44]. However, palaeomagnetic data indicate that the Yilgarn and Pilbara Blocks have maintained their relative positions for 2400 m.y. [45] and so underplating models must explain why the Yilgarn was selectively affected by Phanerozoic plate tectonic processes. The high velocity layer can equally well be explained by the more realistic alternative of simple metamorphic layering [43].

The only other reliable estimate of Archean crustal thickness is from geobarometric studies of metamorphic assemblages. Ideally, pressures obtained for a metamorphic assemblage should provide a minimum crustal thickness. The data for the Precambrian suggested there was a slight secular decrease of mean metamorphic pressure with geologic age [46]. The interpretation of this trend is not clear but an important constraint on crustal thickness is that high metamorphic pressures (>6 Kbar) are not restricted to young rocks and burial depths of up to 40 km are recorded in the Archean. The P-T relationships of Archean granulite facies rocks indicate that pressures of 6–10 Kbar (and possibly up to 13 Kbar) are required to produce the granulite mineralogy [47].

Another constraint on continental volume comes from estimates of crustal area through time. This has been modelled on the basis of the REE distribution in sedimentary rocks. The upper crust underwent a major change in composition during the late Archean, as shown by changing sedimentary rock compositions. Such dramatic changes in composition cannot easily be explained by recycling processes and an increase in sedimentary mass (related to an increase in continental crust) is required [22]. Since the composition of the upper crust has remained constant during the post-Archean (Chapter 5), additions to the upper crust during that time must have been indistinguishable from the upper crust itself [24]. This view is supported by Phanerozoic sediment data which represent crustal additions from 1.7–1.0 Ae but whose REE patterns are uniform and identical to Early Proterozoic and Recent sedimentary rocks of upper crustal origin. The rapid change in upper crustal composition, if related to crustal growth, can be modelled as an increase in exposed crustal area [24]. The results indicate an increase in crustal area by a factor of at least 5 to >10 during the late Archean. For reasonable estimates of early Archean crustal areas, this indicates that at least 50% and more realistically 80–100% of the area of the continents were present by 2.5 Ae.

Cook & Turcotte [48] examined the differentiation of radioactive isotopes into the continents and proposed that the fraction of K, Th, U in the continents is proportional to the surface area. This seems reasonable, on account of the upward concentration of heat producing elements. Preliminary results indicate that in excess of 70% of the continental area was present by 2.5 Ae.

A final constraint comes from the degassing history of the earth [24]. The two extreme models of catastrophic early degassing [e.g. 49] and continuous degassing, at a rate proportional to crustal growth [e.g. 50] provide boundary conditions on the evolution of the ocean volume. For catastrophic early degassing, the fractional volume of the ocean (V_O) would be 1.00 throughout earth history. For oceans degassing at a rate proportional to crustal growth, $V_O/V_C=1.00$ throughout earth history, where V_C is the fractional volume of the continental crust (see Fig. 10.2).

The overall weight of evidence indicates that neither the thickness nor the area of continental crust has changed appreciably since the late Archean (2.7–2.5 Ae). The average thickness of the continental crust has remained within 10% of its present value (assuming that underplating is not a viable process on a large-scale) and the average area has probably remained within about 20–25% of its present value. These constraints in conjunction with those from the constant freeboard model, place severe restrictions on crustal growth models (Fig. 10.2). Those requiring linear [17] or accelerated [18] growth are unlikely because they predict ocean volumes at 2.5 Ae about 10% greater than present. Although a model of hydrosphere shrinking has been proposed [51] there is no independent evidence to support it. Linear or accelerated growth also cannot be reconciled with the evidence for both substantial thickness and area of continental crust. The available constraints indicate that crustal growth was nearly complete ($\geq 70\%$) by about 2.5 Ae.

10.3.2 *Sedimentation rates and crustal recycling*

The mechanism of recycling following very early crust formation involves the large-scale subduction of sediments of continental crustal origin. In the following chapter, geochemical and isotopic evidence relevant to crust–mantle interaction will be discussed. Evidence from Nd and Be isotopes show that in some subduction zones, sediment is subducted and is involved in arc volcanism. In this section we will examine the volumetric constraints on sediment subduction to see if sufficient amounts of continental crust are delivered to subduction zones to sustain a steady-state recycling model [52].

The isotopic record can be consistent with a steady state model for the continental crust [4, 5]. About $1-3\,\mathrm{km^3\,a^{-1}}$ of continental crust must be recycled and remixed with the mantle at present, with the rate increasing exponentially with geologic age. DePaolo [6] calculated recycling rates averaging about $2.3 \pm 1.0\,\mathrm{km^3}$ over the past 2.0 Ae to reproduce the Hf and Nd isotopic characteristics of the mantle–crust system [53]. Two questions are obvious. Are present-day sedimentation rates sufficient to sustain these recycling rates? Were sedimentation rates different in the past?

The greatest proportion of material delivered to the oceans from the continents is by rivers, either as the particulate load or in solution. The most exhaustive analysis estimates total particulate load (suspended, bedload, flood discharge) at about 16×10^{15} g a^{-1} [54]. This value is less than many previous calculations [55]. The dissolved load in rivers (including ground waters) is less than 4.5×10^{15} g a^{-1} [56, 57]. Other inputs include glacial erosion, marine erosion and aeolian material, which in total account for about 2.4×10^{15} g a^{-1} [56]. Recycling of sea salts amounts to about 0.3×10^{15} g a^{-1} [56]. Thus the total flux of crustal material to the oceans is 22.6×10^{15} g a^{-1} (see Table 2.4). Assuming an average crustal density of 2.8 g cm^{-3} this results in a volume flux of about 8.1 km^3 a^{-1} of crustal material.

Human activity has had a massive effect on erosion and sedimentation rates [58–63]. Estimates of man's influence on the sediment flux from the continents range from factors of two to five. If we adopt the lowest value and apply it to riverine and aeolian flux rates only (i.e. not to glacial or marine erosion fluxes) this results in a pre-man sedimentation rate of 12.3×10^{15} g a^{-1} or 4.4 km^3 a^{-1} of crustal material. Pre-human global sediment yields to the oceans using various constraints are listed in Table 10.2. A consensus on the magnitude of

Table 10.2. Estimates of pre-man annual global sediment yield to oceans.

Mass (10^{15} g)	Crustal volume (km^3)*	Method of estimate	Reference
9.3	3.3	Global extrapolation of erosion rates of areas relatively unaffected by man	[59]
12.9†	4.6	Sodium mass balance in rivers	[61]
10.0	3.6	Method not explicitly stated	[60]
7.5‡	2.7	Model assuming constant sedimentary mass and sedimentary recycling rate for Phanerozoic time	[62]
10.0§	3.6	Similar to above	[64]
12.3	4.4	Uses measured sedimentation rates and estimates of the influence of man	[52]

* Assumes crustal density of 2.8 g cm^{-3}.
† Includes 2.4×10^{15} g to account for glacial, aeolian and marine erosion.
‡ Represents a Phanerozoic average.
§ Represents average since Devonian.

sediment yield is apparent, with values ranging from 2.7 to 4.6 km^3 a^{-1}. Much of the sediment delivered to the oceans remains on the continental shelf. For reasonable sedimentation rates, of >0.003 cm a^{-1} for continental terraces [65], the amount of continental crust removed to the oceanic crust is unlikely to exceed about 3 km^3 a^{-1}.

A further difficulty is that major repositories of oceanic sediment are not closely related to subduction zones. The main suppliers of sediment to the oceans are rivers and big rivers form in specific tectonic settings that

Table 10.3. Sediment discharge from major rivers [54] and their tectonic setting.

River	Average sediment discharge (10^{15} g)	Tectonic setting of ocean basin
1. Ganges/Brahmaputra	1.670	Continental collision
2. Yellow (Huangho)	1.080	Back-arc
3. Amazon	0.900	Trailing-edge
4. Yangtze	0.478	Back-arc
5. Irrawaddy	0.285	Continental collision
6. Magdalena	0.220	Back-arc
7. Mississippi	0.210	Trailing-edge
8. Orinoco	0.210	Trailing-edge
9. Hungho (Red)	0.160	Back-arc
10. Mekong	0.160	Back-arc
11. Indus	0.100	Continental collision
12. MacKenzie	0.100	Trailing-edge

commonly are unrelated to subduction in space and time [66]. In Table 10.3, we list the sediment discharge for the world's 12 major rivers (in terms of sediment supply) and the tectonic environment of the ocean basins into which they run. Many major river systems form at passive continental margins (e.g. Amazon, Niger), behind island-arcs (e.g. Yellow, Yangtze) or adjacent to continental collision basins (e.g. Ganges, Indus). This results in immense accumulations of sediment on oceanic crust remote from regions where subduction is currently taking place. It seems unlikely that such thicknesses of sediment (up to 12–15 km) could be removed efficiently by some later episode of subduction [67], given the graben-fill mechanism of sediment subduction generally appealed to [5, 68]. The magnitude of this process can be appreciated from the observation that the Bengal Fan has accreted at an average rate of about 0.4 km^3 a^{-1} during the past 55 m.y. [69].

It is commonly suggested that the bulk of subducted crustal material is pelagic sediment. Recent estimates of pelagic sedimentation rates are less than 3.0×10^{15} g a^{-1} [70] which is equivalent to about 1.1 km^3 a^{-1} of continental crust (Table 10.4). The terrigenous fraction of pelagic sediments is variable but averages less than 60%, with the remainder comprising biogenic siliceous and carbonate material (Table 10.4). The concentrations of many elements, of isotopic importance, such as Th, U and REE are significantly lower in biogenic than in terrigenous sediment; for example, the REE content in siliceous and carbonate biogenic sediment is typically 5–10 times lower than

Table 10.4. Estimates of global rates of pelagic sedimentation [70].

Sediment type	Atlantic		Pacific		Indian		Total	
	10^{15} g	%	10^{15} g	%	10^{15} g	%	10^{15} g	%
Terrigenous	0.642	52.2	0.784	67.9	0.304	50.9	1.730	58.0
Biogenic carbonate	0.543	44.1	0.305	26.5	0.231	38.7	1.079	36.2
Biogenic siliceous	0.045	3.7	0.065	5.6	0.062	10.4	0.172	5.8
Total	1.230	100.0	1.154	100.0	0.597	100.0	2.981	100.0

crustal abundances. Thus, the sedimentation rate of crustal 'equivalent' material for the REE is less than 0.8 km^3 a^{-1}.

Crustal recycling can also be modelled using the present-day distribution of areas of continental radiometric age provinces, sedimentary rocks and ore deposits [e.g. 20, 56, 71]. The theory and mathematical formulation of the models treated here can be found in Veizer & Jansen [20]. The best estimate of the recycling constant b for the area of continental crust is $b=4\pm10^{-10}$ a^{-1} [22, 72]. If this is adopted for bulk crustal recycling, it amounts to about 3 km^3 a^{-1}. Such a value includes both intracrustal recycling and crust–mantle recycling and so the amount of crust recycled through the mantle must be less than 3 km^3 a^{-1} [73]. A more stringent limit on the magnitude of sediment subduction comes from estimates of recycling rates of ocean sediments. Veizer [22] estimates $b=131\times10^{-10}$ a^{-1}. For an ocean sediment mass of 215×10^{21} g, this suggests a recycling rate of 2.8×10^{15} g a^{-1} or 1.0 km^3 of crustal material. This value also includes intracrustal sedimentary recycling and sediment–mantle recycling and so the latter value must be less than 1.0 km^3 a^{-1}. The low concentration of many elements of isotopic importance in pelagic sediments (see above) must also be taken into account.

The actual fate of sediment delivered to the subduction zone remains speculative. It may be preserved in the accretionary wedge, subcreted (underplated) to the base of the crust or subducted to mantle depth. Based on the sediment distribution in accretionary wedges, it has been proposed that about 85% of the sediment delivered to the subduction zone must be subducted [1]. Measurements along part of the Middle America Trench [74] suggest one-third of the sediment is accreted and the other two-thirds subducted (with this being equally divided between subcreted material and deeply subducted material). Karig & Kay [75] argue that most sediment delivered to subduction zones is either accreted or underplated to the upper plate and less than 10% is subducted to mantle depths. Of this material, a small fraction may return rapidly to the surface via arc-magmatism (see following chapter) [5]. We consider estimates of sediment subduction, based on sediment preservation in accretionary wedges, to be upper limits because they fail to take into account the well-documented process of intracrustal sedimentary recycling, which is especially rapid at continental margins [20, 22].

In summary, the amount of continental crust delivered to subduction zones (≤ 1 km^3 a^{-1}) is only barely enough to sustain a no-growth crustal recycling model at the present time. Of this, some material is accreted [75] and recycled as sediment [22]. There is growing evidence that some material which is subducted, is added (subcreted) to the base of the plate [74, 75]. Thus, we consider, on the basis of present evidence, the amount of crustal material which is subducted to mantle depth to be very much less than this amount, perhaps ≤ 0.3 km^3 a^{-1}.

The volumetric arguments for steady-state crustal recycling are thus not convincing. The problem is exacerbated by the requirement, in such models, of exponentially increasing recycling rates with geologic age [4–6]. There is no convincing evidence for higher sedimentation rates in the past [50], and if anything, they were lower [62, 64]. An additional source of uncertainty is that apparent sediment accumulation rates are biased towards large values unless very long time intervals are considered [29].

10.3.3 Measured and inferred growth rates

There have been a number of attempts to measure the growth rates of continents directly by quantifying the net additions of igneous material (Table 10.5). For the most part, this approach is only valid for the Cenozoic and in some cases the Mesozoic, although extrapolation to the entire Phanerozoic is not uncommon. The simplest approach is to examine the igneous additions at island-arcs, which indicate growth from about 0.5 km^3 a^{-1} [77, 78] to 1.1 km^3 a^{-1} [2]. Several workers have attempted more sophisticated estimates of

Table 10.5. Estimates of Phanerozoic crustal growth rates (in km^3 a^{-1})*

Reference	Addition	Subtraction	Growth
Hurley [17]	—	—	≈2
Hurley & Rand [18]	—	—	≈3
Moorbath [76]	≈1.05	Negligible	≈1.05
Fyfe [15]	0.5	1–2	<0
Brown [77]	<0.5	Negligible	<0.5
Veizer & Jansen [20]	—	—	0.76
Armstrong [5]	1–3	1–3	0
Dewey & Windley [1]	1.40	0.81	0.59
Karig & Kay [75]	0.73	Small	≈0.7
McLennan & Taylor [24]	—	—	0.3–0.7
Reymer & Schubert [2]†	1.65	0.59	1.02
Reymer & Schubert [2]‡	—	—	0.9

* Based on compilation in [2].
† Based on measured and inferred rates.
‡ Based on freeboard constraints.

Phanerozoic crustal growth rates based on balancing measured or inferred crustal additions (e.g. arc volcanism, ophiolites, continental volcanism) and crustal losses (e.g. sediment subduction, tectonic erosion, spilitization). The studies indicate that additions outweigh losses by a factor of about two to three. Dewey & Windley [1] estimated a net growth rate of 0.6 km^3 a^{-1} and Reymer & Schubert [2] estimated a net growth rate of 1.0 km^3 a^{-1}. There are at least three major sources of uncertainty in these calculations.
1 the losses to the mantle via sediment subduction (and other processes) are uncertain;
2 the additions by arc volcanism are uncertain with estimates varying from about 0.5–1.1 km^3 a^{-1};
3 measured rates of any episodic geological process (such as volcanism) are biased towards large values unless time intervals considerably in excess of the frequency of the event are considered [22, 29].

Changes in continental area during the Phanerozoic have been modelled using the inferred secular decrease in ocean ridge volume [2]. It is generally accepted that the secular increase in heat production and terrestrial heat flow in the past resulted in hotter, shallower ocean basins [2, 39, 79]. Changes in continental area can be modelled as a function of the decay in terrestrial heat production if one assumes constant freeboard [2]. The results indicate a Phanerozoic growth rate of 0.9 km^3 a^{-1}. However, this value represents a maximum if the evidence for a secular change in freeboard during the Phanerozoic is accepted [32, 33] (see Fig. 10.4).

10.4 Isotopic constraints

In this section, we make the conventional interpretation of the isotope data. Accordingly, the role of crustal recycling through the mantle during the past 3.8 Ae will be considered to be small in comparison to crustal growth.

10.4.1 Early crustal growth (>3.2 Ae)

The oldest rocks are found in the early Archean high-grade terrains (3.8–3.6

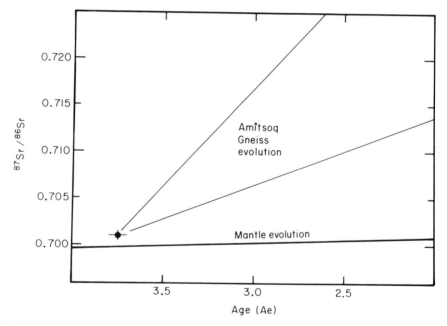

Fig. 10.5. Sr isotope evolution diagram for Amîtsoq gneisses from West Greenland (adapted from [3]). Plotted is the initial ^{87}Sr/^{86}Sr ratio against time along with the possible ^{87}Sr/^{86}Sr evolution (growth curves) inferred from the Rb/Sr ratios. The low ^{87}Sr/^{86}Sr ratios and high Rb/Sr ratios (resulting in rapid evolution of ^{87}Sr/^{86}Sr) suggest these gneisses were derived from the mantle shortly before (<100–200 m.y.) their last isotopic rehomogenization (this is calculated by extrapolation of the growth curves back to the mantle evolution growth curve). This treatment assumes that Rb and Sr are in a closed system during the transformation of the granitic protoliths into the presently observed gneisses. Such an assumption has been questioned by some workers for the Rb–Sr and U–Th–Pb systems [80].

Ae). The best studied examples are found in the North Atlantic area (Labrador, West Greenland). Moorbath [3, 76] has proposed a two-stage model to explain the origin of this differentiated material. Amîtsoq Gneiss (West Greenland), and the probably equivalent Uivak Gneiss (Labrador) are characterized by low initial $^{87}Sr/^{86}Sr$ ratios (0.700–0.702) and Pb–Pb isotopic characteristics, consistent with derivation from a homogeneous U/Pb source predicted from a single stage mantle Pb growth curve. Such data indicate that these rocks were derived from a mantle source (low Rb/Sr, U/Pb) with only a short crustal residence in a high Rb/Sr, U/Pb environment (Fig. 10.5). Nd and Hf isotopic data also support the derivation of such material at about 3.8–3.6 Ae [81].

An important period of crustal growth was associated with the period of greenstone–gneiss formation represented in the Barberton Mountain Land [e.g. 82] and Pilbara Block [83, 84]. The greenstone sequences (Swaziland Supergroup; Pilbara Supergroup) have been dated at about 3.6–3.4 Ae [85–87]. The associated Na-rich granitic gneissic terrains are of similar age or slightly younger [82, 88, 89].

Is there any evidence, direct or indirect, for any earlier crustal formation? The Nd-isotopic composition of 3.78 Ae gneissic rocks from eastern India indicates the ultimate mantle source had an initial $^{143}Nd/^{144}Nd$ ratio corresponding to $\varepsilon_{Nd}=+3$ [90]. Similarly, mantle sources for Isua supracrustals had

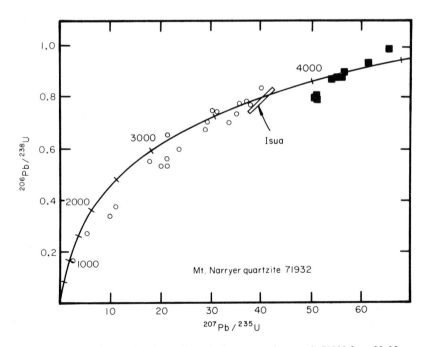

Fig. 10.6. Concordia diagram for zircons from Archean quartzite sample 71932 from Mt Narryer, Western Australia (adapted from [92]). Closed squares—data from four angular, inclusion-free zircons; open circles—other zircons from same sample. Zircons from Isua belt are shown for comparison. The very old ages of ≈4150 m.y. for these four zircons indicate silica saturated rocks were probably present in the earth's Earth's crust at that time.

ε_{Nd}+1 to +3 at time of extraction [91]. The mantle source of these rocks was probably already depleted in incompatible elements by that time. More direct evidence (Fig. 10.6) comes from the recent identification of 4.1–4.2 Ae zircons preserved in 3.63 Ae quartzites of the Mount Narryer region, Western Australia [92]. These data indicate that silica saturated rocks were present at about 4.1–4.2 Ae, although it is less clear if it represents the existence of true continental crust or a local occurrence of silicic volcanics [93]. Thus, there is an intriguing possibility of small amounts of continental crust prior to 3.8 Ae.

Is there any way to quantify the rate of crustal growth during the early Archean? Moorbath [94] suggested that only small amounts of continental crust existed during the period 3.8–3.5 Ae (perhaps 5–10%). Crustal recycling models indicate mean crustal growth rates over the period 3.7–3.0 Ae of about 1.2–1.3 km^3 a^{-1} [20]. If we adopt this figure for the period under consideration, it would suggest that no more than 15% of the crust was in place at about 3.2 Ae.

10.4.2 Late Archean crustal growth (3.2–2.5 Ae)

There is considerable evidence that this was the most important period of crustal growth during earth history. The vigorous magmatic activity involved in the development of late Archean granite–greenstone complexes has long

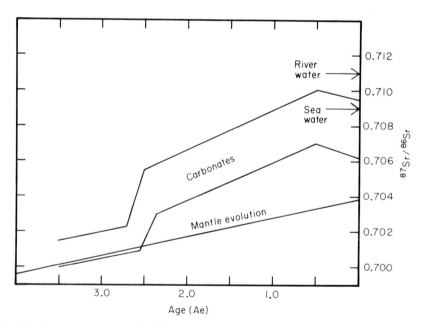

Fig. 10.7. A generalized plot of ^{87}Sr/^{86}Sr variations in sedimentary carbonates during geologic time (adapted from [99, 102]). The carbonate data are considered to represent sea water evolution. Note the departure from the mantle evolution curve at about 2.6–2.5 Ae. Such data are consistent with an increase in area of Rb-rich continental crust. The fact that present-day sea water (and carbonates) have lower ^{87}Sr/^{86}Sr than river waters (representative of continental sources) is a result of upper mantle buffering, due to basalt–sea water interaction [100–102].

been recognized. However, its significance, in terms of crustal evolution, has only recently been appreciated, following detailed isotopic examination.

Both greenstone belt volcanics and most of the surrounding K-rich granitic rocks are characterized by low ($^{87}Sr/^{86}Sr$) ratios (0.701–0.704) (e.g. [95, 96]; also see Section 7.3.2). The fact that the granitic rocks also have these low ratios implies their source material had a very short crustal residence time. The other isotopic data are consistent with this interpretation [e.g. 97, 98]. It is not simple to make quantitative estimates of crustal growth from the available isotopic data, but Moorbath [94] suggested that as much as 50–60% of the continental areas was produced during this time.

Marine carbonates possess low Rb/Sr ratios, and they record the $^{87}Sr/^{86}Sr$ of the marine environment [99]. Archean carbonate $^{87}Sr/^{86}Sr$ ratios are similar or slightly higher than upper mantle values suggesting that extensive continental masses (with higher $^{87}Sr/^{86}Sr$) were not exposed. The ratio was probably buffered by extensive exchange of sea water Sr with oceanic crust Sr. This process would have been enhanced by higher Archean geothermal gradients [100]. There is, however, a marked increase in $^{87}Sr/^{86}Sr$ (≈ 0.0025) at about 2.6–2.5 Ga [101], followed by a more gradual rise during the rest of post-Archean time (Fig. 10.7). The sharp increase in $^{87}Sr/^{87}Sr$ at the Archean–Proterozoic boundary is best explained by a major expansion in the area of Rb-rich continental crust. The sea water Sr-isotopic source changed from predominantly mantle (i.e. oceanic volcanics) during the Archean to predominantly continental during the post-Archean [100, 102], consistent with a major period of crustal growth during the late Archean.

The magnitude of crustal growth in this time was revealed by McCulloch & Wasserburg [97] in their Nd–Sr isotopic study of the Canadian Shield (Table 10.6). They found, as expected, that composite samples representing major areas of the Superior and Slave structural provinces gave T^{Nd}_{CHUR} and T^{Sr}_{CHUR} model ages of 2.4–2.7 Ae. Using depleted mantle parameters results in T^{Nd}_{DM} model ages of 2.7–2.9 Ae. Such ages are probably dating the major Sm–Nd fractionation from the mantle [97]. More surprisingly, the Proterozoic

Table 10.6. Nd and Sr model ages for large-scale composites from the Canadian Shield [97].

Composite	Structural province	$T_{geologic}$ (Ae)	Undepleted mantle		Depleted mantle	
			T^{Nd}_{CHUR}	T^{Sr}_{UR}	T^{Nd}_{DM}	T^{Sr}_{DM}
New Quebec	Superior	2.7–2.5	2.6	2.7	2.8	2.7
North Quebec	Superior (–Churchill–Grenville)	(2.7–2.5)	2.4	2.5	2.7	2.5
Fort Enterprise Gneiss	Slave	2.7–2.5	2.6	2.5	2.9	2.6
Fort Enterprise Granite	Slave	2.7–2.5	2.4	2.5	2.7	2.6
Baffin Island	Churchill	1.9–1.8	2.6	2.5	2.9	2.5
Saskatchewan	Churchill	1.9–1.8	2.6	2.4	2.8	2.4
Quebec	Grenville	(1.2–0.9)	0.8	1.0	1.4	1.1

Depleted mantle parameters assume depletion at 3.8 Ae and present-day mantle $\varepsilon_{Nd}=+10$ and present-day $\varepsilon_{Sr}=-30$ (see Appendix 4).

Churchill province, characterized by K–Ar ages of 1.8–1.9 Ae also give T_{DM}^{Nd} ages of about 2.8–2.9 Ae (T_{DM}^{Sr} are slightly younger). Thus, most of the source material of the Canadian Shield (excluding parts of the Nain and Grenville Provinces) separated from the mantle at about 2.6–2.8 Ae. From the Sr-isotopic evidence, the subsequent fractionation of Rb–Sr into a Rb rich upper crust and Rb-deficient lower crust probably followed quickly (also see [103]).

The isotopic data from the Canadian Shield accordingly argue for a truly major period of crustal growth and differentiation during the latest Archean. Much of the Proterozoic mobile belt terrains are probably comprised of reworked late Archean crustal material [e.g. 97] as has been proposed by geologists working in Proterozoic rocks [104].

On most continents, the period of late Archean crustal growth is concentrated at about 2.8–2.6 Ae with crustal stabilization by about 2.5 Ae. Comparable activity around the Barberton Mountain Land (South Africa) occurred much earlier (see especially Section 8.4). The emplacement of dominantly K-rich granitic rocks at about 3.2–2.9 Ae (Second Magmatic Cycle) is a major magmatic event in this area [82]. These rocks possess low initial $^{87}Sr/^{86}Sr$ ratios and it is not entirely clear if they represent differentiation of the earlier formed crust at 3.6–3.4 Ae (see Section 10.4.1) or are a separate crust formation-differentiation event. Late Archean crustal growth thus spans nearly a billion years but is greatly concentrated at 2.8–2.6 Ae.

10.4.3 *Post-Archean crustal growth (<2.5 Ae)*

In addition to Archean crustal growth, Moorbath [105] also recognizes three periods of post-Archean crustal growth and differentiation at about 1.9–1.6 Ae, 1.2–0.9 Ae and 0.5–0.0 Ae. There has been considerable dispute about the role of crustal reworking versus crustal growth in interpreting much of the Sr and Pb isotopic data [80]. Nd data has shown clearly that formation of granitic rocks from pre-existing continental crust was much more prevalent in the post-Archean [106–110]. The period around 1.8 Ae has been considered as an important time of crust generation, based on low ($^{87}Sr/^{86}Sr$) ratios (0.701–0.703) in granitic rocks of that age [105]. A major period of crust extraction from a depleted mantle source, at 1.8 Ae has been recognized [108] in south-west USA [108] (Fig. 10.8). Crust-forming events at 1.8 Ae are confirmed in this area and in Finland on the basis of Hf isotopic analyses [98]. In central Australia, a comparable event may have occurred somewhat earlier, closer to 2.0 Ae [111].

Abundant radiometric ages are also recorded during the period 1.2–0.7 Ae, and this may also represent an important period of crustal growth. Composite samples from the Grenville Province have $\varepsilon_{Nd}(0) = -7.1$ [97]. This corresponds to a chondritic model age of $T_{CHUR}^{Nd} = 0.8$ Ae. If a more realistic depleted mantle model is used with average crustal Sm/Nd ratios, the age increases significantly to about 1.2–1.4 Ae, depending on the details of the model parameters (Table 10.6). This period also may have been an important time of

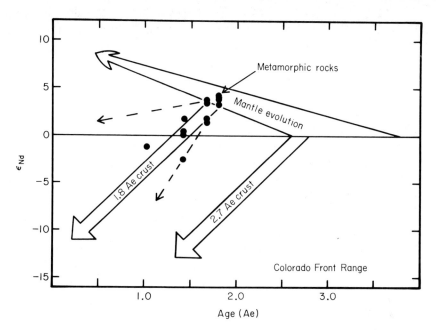

Fig. 10.8. Plot of ε_{Nd} against time, showing initial ε_{Nd} values for 1.8 Ae metamorphic rocks and associated younger suites of granitic rocks from the Colorado Front Range (adapted from [108]). Also plotted is a model mantle evolution curve and inferred evolution for continental crust of 1.8 and 2.7 Ae. The dashed arrows represent the evolution of a metatholeiite sample (high ε_{Nd}) and granitic sample (lowest ε_{Nd}). These data indicate that crustal material was extracted from a depleted mantle source at about 1.8 Ae and that subsequent intrusive events (at 1.7–1.0 Ae) represented intracrustal differentiation of this (1.8 Ae) crustal addition [108].

crust formation in eastern Australia [110, 112]. In addition, Hf-isotopic evidence suggests significant portions of crust may have been generated in Britain and Saudi Arabia at about 0.7 Ae [98].

Can we quantify the rate of crustal growth during the post-Archean? From the discussion in Section 10.3.3, the value of 1.0 km^3 a^{-1} represents the uppermost limit for crustal growth rates during the Phanerozoic. If this rate is extrapolated for all post-Archean time, it would indicate about ⩽36% of the crust formed since 2.5 Ae. A more realistic model is that this rate only applies to the intervals discussed above, where radiometric ages of rock forming events are concentrated (2.0–1.6 Ae, 1.2–0.7 Ae, 0.5–0.0 Ae). For other times, crustal additions would be minimal. In this scenario, ⩽20% of the crust would have formed since 2.5 Ae. These values, though not well constrained, are in agreement with the suggestion, based on freeboard arguments, that up to 70–90% of the continents were in place by about 2.5 Ae.

10.5 Summary: a preferred model of crustal growth

1 Models for the development of the continental crust fall into two basic divisions: recycling of an early formed crust, or steady or episodic growth throughout geological time. Our assessment of the geological, geochemical and

isotopic constraints leads us to adopt the latter alternative. We recognize that the recycling of continental crust through the mantle is a viable mechanism but is not the dominant process. The magnitude of this phenomenon is inadequately understood.

2 The freeboard constraint indicates that continental and oceanic volumes have been relatively constant since the Proterozoic, consistent with models which produce the bulk ($\geq 70\%$) of the continental crust by about 2500 m.y ago.

3 Current sedimentation rates appear to be insufficient to supply enough material to subduction zones to support steady-state models which recycle the continental crust through the mantle.

4 In our view, the following four conclusions are supported overwhelmingly by the geological, geochemical and isotopic data:

(a) While crustal growth may occur continually throughout much of geologic time (since at least 3.8 Ae and possibly before), it is greatly concentrated in episodes of accelerated crustal growth during the periods 3.8–3.5 Ae, 3.2–2.5 Ae (especially 2.8–2.6), 2.0–1.6 Ae, 1.2–0.7 Ae and 0.5–0.0 Ae. There is some suggestion of crustal growth prior to 3.8 Ae [e.g. 92] but this remains to be substantiated.

(b) From the late Archean on, times of crustal growth generally are followed rapidly (≤ 100–200 m.y.?) by intracrustal melting and differentiation into upper crust (LIL-enriched) and lower crust (LIL-depleted).

(c) The period of crustal growth and differentiation during the period 3.2–2.5 Ae (mainly 2.8–2.6 Ae) was especially important. Possibly as much as 60–75% of the continental crust was generated at this time.

(d) The growth of the continental crust was nearly complete (within 70–90% of its present mass) by 2.5 Ae.

In Fig. 10.9 we show our generalized model of crustal growth. The initiation of crustal growth is assumed to be at 3.9 Ae, after the intense

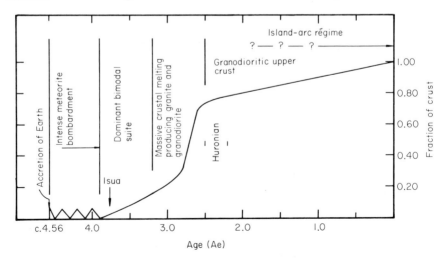

Fig. 10.9. Generalized model for the growth of the continental crust throughout geologic time proposed in this book. Details are discussed in the text.

meteorite bombardment which was endemic in the inner solar system to this time. The model does not preclude earlier crust [e.g. 92], but any present would have been massively altered by meteoritic bombardment (see Chapter 12). The model also assumes linear crustal growth during the periods 3.9–3.5 Ae and 2.5–0.0 Ae. This is almost certainly an oversimplification because of the probable episodic nature of crustal growth. However, given the lack of quantitative estimates for these periods of growth, a linear model is adopted as a first approximation [113]. The crustal growth rate reaches a maximum during the period 2.8–2.6 Ae, the time of major greenstone belt development.

Notes and references

1 Dewey, J.F. & Windley, B.F. (1981) Growth and differentiation of the continental crust. *Phil. Trans. Roy. Soc. Lond.*, **A301**, 189.
2 Reymer, A. & Schubert, G. (1984) Phanerozoic addition rates to the continental crust and crustal growth. *Tectonics*, **3**, in press.
3 A recent statement of this interpretation can be found in Moorbath, S. & Taylor, P.N. (1982) Isotopic evidence for continental growth in the Precambrian. *In:* Kröner, A. (ed.), *Precambrian Plate Tectonics*. Elsevier.
4 Armstrong, R.L. (1968) A model for the evolution of strontium and lead isotopes in a dynamic earth. *Rev. Geophys.*, **6**, 175; Armstrong, R.L. & Hein, S.M. (1973) Computer simulation of Pb and Sr isotope evolution of the earth's crust and upper mantle. *GCA*, **37**, 1.
5 Armstrong, R.L. (1981) Radiogenic isotopes: the case for crustal recycling on a near-steady-state no-continental-growth Earth. *Phil. Trans. Roy. Soc. Lond.*, **A301**, 443.
6 DePaolo, D.J. (1983) The mean life of continents: estimates of continent recycling rates from Nd and Hf isotopic data and implications for mantle structure. *GRL*, **10**, 705.
7 Doe, B.R. & Zartman, R.E. (1979) Plumbotectonics, the Phanerozoic. *In:* Barnes, H.L. (ed.), *Geochmistry of Hydrothermal Ore Deposits*. Wiley, Interscience. 22.
8 Jacobsen, S.B. & Wasserburg, G.J. (1981) Transport model for crust and mantle evolution. *Tectonophysics*, **75**, 163.
9 O'Nions, R.K. & Hamilton, P.J. (1981) Isotope and trace element models of crustal evolution. *Phil. Trans. Roy. Soc. Lond.*, **A301**, 473.
10 Allegre, C.J. (1982) Chemical geodynamics. *Tectonophysics*, **81**, 109.
11 Patchett, J. & Chauvel, C. (1984) The mean life of continents is currently not constrained by Nd and Hf isotopes. *GRL* **11**, 151.
12 Harrison, E.R. (1960) Origin of the Pacific basin: a meteorite impact hypothesis. *Nature*, **188**, 1064; Gilvarry, J.J. (1961) The origin of ocean basins and continents. *Nature*, **190**, 1048.
13 Donn, W.L. et al. (1965) On the early history of the earth. *GSA Bull.*, **76**, 287.
14 Hargraves, R.B. (1976) Precambrian geologic history. *Science*, **193**, 363.
15 Fyfe, W.S. (1978) Evolution of the earth's crust: modern plate tectonics to ancient hot spot tectonics? *Chem. Geol.*, **23**, 89.
16 Walker, C.T. & Dennis, J.G. (1983) Exogenic processes and the origin of the sialic crust. *Geol. Rund.*, **72**, 743.
17 Hurley, P.M. (1968) Absolute abundance and distribution of Rb, K and Sr in the earth. *GCA*, **32**, 273.
18 Hurley, P.M. & Rand, J.R. (1969) Pre-drift continental nuclei. *Science*, **164**, 1229.
19 Veizer, J. (1976) $^{87}Sr/^{86}Sr$ evolution of seawater during geologic history and its significance as an index of crustal evolution. *In:* Windley, B.F. (ed.), *The Early History of the Earth*. Wiley.
20 Veizer, J. & Jansen, S.L. (1979) Basement and sedimentary recycling and continental evolution. *J. Geol.*, **87**, 341.
21 Veizer, J. (1983) Geologic evolution of the Archean–early Proterozoic earth. *In:* Schopf, J.W. (ed.), *Earth's Earliest Biosphere: Its Origin and Evolution* Princeton University Press, 240.
22 Veizer, J. (1984) Recycling on the evolving earth: geochemical record in sediments. *Proc. IGC*, **11**, 325.

23 In the Veizer & Jansen curve [20], the model plots the data points at the end of the time intervals considered rather than the mid-points. This is mathematically correct but geologically (i.e. time) distorting. If plotted at the mid-points of the time intervals considered, the steep period of crustal growth at the Archean–Proterozoic boundary would be about 300 m.y. older (J. Veizer, pers. comm.).
24 McLennan, S.M. & Taylor, S.R. (1982) Geochemical constraints on the growth of the continental crust. *J. Geol.*, **90**, 342.
25 Egyed, L. (1956) Change of earth dimensions as determined from palaeogeographical data. *Geofis. Pura e Appl.*, **33**, 42.
26 Strakhov, N.M. (1948) *Outlines of Historical Geology*. Govt. Printing Office, Moscow.
27 Termier, H. & Termier, G. (1952) *Histoire Geologique de la Biosphere*. Masson.
28 Egyed, L. (1956) Determination of changes in the dimensions of the earth from paleogeographical data. *Nature*, **178**, 534; (1969) The slow expansion hypothesis. *In:* Runcorn, K. (ed.), *Application of Modern Physics to Earth and Planetary Interiors*. Wiley, Interscience. 65.
29 Veizer, J. (1971) Do palaeogeographic data support the expanding earth hypothesis? *Nature*, **229**, 480. The entire question of the incompleteness of the geological record, for essentially episodic geological events (e.g. sedimentation), has received considerable attention recently. The likelihood of obtaining a meaningful average increases as larger time intervals are considered. Sadler, P.M. (1981) Sediment accumulation rates and the completeness of stratigraphic section. *J. Geol.*, **89**, 569; Tipper, J.C. (1983) Rates of sedimentation, and stratigraphical completeness. *Nature*, **302**, 696.
30 Wise, D.U. (1972) Freeboard of continents through time. *GSA Mem.*, **132**, 87; (1974) Continental margins, freeboard and the volumes of continents and oceans through time. *In:* Buck, C.A. & Drake, C.L, (eds), *The Geology of Continental Margins*. Springer-Verlag. 45.
31 The freeboard equations may be expressed [30]:

$$A_E = A_C + A_O \quad (1)$$

$$V_O = A_O \times T_O \quad (2)$$

$$V_C = A_C \times T_C \quad (3)$$

$$T_C = 13.3 + 6.05 \, T_O \quad (4)$$

where A_E is area of earth; A_C is area of continents; A_O is area of oceans; V_C is volume of crust; V_O is volume of oceans; T_C is thickness of crust; T_O is depth of oceans. The final equation is the isostatic link which is dependent on the model of crustal structure used (see [30]). The assumptions are: (1) no earth expansion; (2) constant ocean crust thickness; (3) simple isostatic link between continental crust and oceanic crust (see equation (4)).
32 Hallam, A. (1977) Secular changes in the marine inundation of USSR and North America through the Phanerozoic. *Nature*, **269**, 769.
33 Hallam, A. (1981) *Facies Interpretation and the Stratigraphic Record*. Freeman.
34 Vinogradov, A.P. (ed.) (1967–1969) *Atlas of the Lithological–Palaeogeographic Maps of the USSR*. 4 vols. Ministry of Geol., Moscow.
35 Cook, T.D. & Bally, A.W. (1977) *Stratigraphic Atlas of North and Central America*. Princeton University Press.
36 Vail, P.R. *et al.* (1978) Seismic stratigraphy and global changes of sea level, 4. Global cycles of relative changes in sea level. *Am. Ass. Petrol Geol. Mem.*, **26**, 83. The main calibration datum used for the Vail sea-level curve has been questioned recently. Vail adopted a 350 m drop in sea-level over the past 100 m.y. but recent data suggest it may have been only about 100–150 m (see Kerr, R.A. (1984) *Science*, **223**, 472).
37 Windley, B.F. (1977) Timing of continental growth and emergence. *Nature*, **270**, 426.
38 Knoll, A.H. (1978) Did emerging continents trigger metazoan evolution. *Nature*, **276**, 701.
39 Turcotte, D.L. & Burke, K. (1978) Global sea-level changes and the thermal structure of the earth. *EPSL*, **41**, 341.
40 Burke, K. & Kidd, W.S.F. (1978) Were Archean continental geothermal gradients much steeper than those of today? *Nature*, **272**, 240.
41 Condie, K.C. (1973) Archean magmatism and crustal thickening. *GSA Bull.*, **84**, 2981.
42 Davies, G.F. (1979) Thickness and thermal history of continental crust and root zones. *EPSL*, **44**, 231.
43 Drummond, B.J. *et al.* (1981) Crustal structure in the Pilbara and northern Yilgarn blocks from deep seismic sounding. *Geol. Soc. Austr. Spec. Pub.*, **7**, 33.

44 Glikson, A.Y. & Lambert, I.B. (1976). Vertical zonation and petrogenesis of the early Precambrian crust in Western Australia. *Tectonophysics*, **30**, 55.
45 Embleton, B.J.J. (1978) The palaeomagnetism of 2900 m.y. old rocks from the Australian Pilbara Craton and its relation to Archean–Proterozoic tectonics. *PCR*, **6**, 275.
46 Grambling, J.A. (1981) Pressures and temperatures in Precambrian metamorphic rocks. *EPSL*, **53**, 63.
47 Newton, R.C. (1978) Experimental and thermodynamic evidence for the operation of high pressures in Archaean metamorphism. *In:* Windley, B.F. & Naqvi, S.M. (eds), *Archaean Geochemistry*. Elsevier. 221.
48 Cook, F.A. & Turcotte, D.L. (1981) Parameterized convection and the thermal evolution of the earth. *Tectonophysics*, **75**, 1.
49 Fanale, F.P. (1971) A case for catastrophic degassing of the Earth. *Chem. Geol.*, **8**, 79.
50 Schopf, T.J.M. (1980) *Paleoceanography*. Harvard University Press.
51 Fyfe, W.S. (1976) Hydrosphere and continental crust: growing or shrinking? *Geosci. Can.*, **3**, 82; (1980) Crust formation and destruction. *Geol. Ass. Can. Spec. Pap.*, **20**, 77.
52 McLennan, S.M. & Taylor, S.R. (1983) Continental freeboard, sedimentation rates and growth of continental crust. *Nature*, **306**, 169.
53 It should be noted that the calculations used in DePaolo's work have been seriously criticized (see [11]).
54 Milliman, J.D. & Meade, R.H. (1983) World-wide delivery of river sediment to the oceans. *J. Geol.*, **91**, 1.
55 Kuenen, P.H. (1950) *Marine Geology*. Wiley; Holeman, J.N. (1968) Sediment yield of major rivers of the world. *Water Resources Res.*, **4**, 737.
56 Garrels, R.M. & Mackenzie, F.T. (1971) *Evolution of Sedimentary Rocks*. Norton.
57 Meybeck, M. (1976) Total mineral dissolved transport by world major rivers. *Hydrol. Sci. Bull.*, **21**, 265.
58 Douglas, I. (1967) Man, vegetation and the sediment yields of rivers. *Nature*, **215**, 925.
59 Judson, S. (1968) Erosion of the land, or what's happening to our continents. *Am. Sci.*, **56**, 356.
60 Judson, S. (1971) Estimates of masses of earth materials moved by man (abst.). *GSA Abst.*, **3**, 615.
61 Gregor, B. (1970) Denudation of the continents. *Nature*, **228**, 273.
62 Garrels, R.M. et al. (1976) Controls of atmospheric O_2 and CO_2: past, present and future. *Am. Sci.*, **64**, 306.
63 Meade, R.H. (1982) Sources, sinks, and storage of river sediments in the Atlantic drainage of the United States. *J. Geol.*, **90**, 235.
64 Gregor, B. (1983) The mass-age distribution of Phanerozoic sediments (abst.). *EOS*, **64**, 334.
65 Schwab, F.L. (1976) Modern and ancient sedimentary basins: comparative accumulation rates. *Geology*, **4**, 723.
66 Potter, P.E. (1978) Significance and origin of big rivers. *J. Geol.*, **86**, 13.
67 Harold Reading has termed such material in the Bengal and Indus Fans, 'quasi-continental crust' (see [69]).
68 Uyeda, S. (1983) Comparative subductology. *Episodes*, 1983, 19.
69 Reading, H.G. (1982) Sedimentary basins and global tectonics. *Proc. Geol. Ass.*, **93**, 321.
70 Lisitsyn, A.P. et al. (1982) The relation between element influx from rivers and accumulation in ocean sediments. *Geochem. Int.*, **19**, 102.
71 Dacey, M.F. & Lerman, A. (1983) Sediment growth and aging as Markov chains. *J. Geol.*, **91**, 573.
72 Sclater, J.G. et al. (1981) Oceans and continents: similarities and differences in the mechanisms of heat loss. *JGR*, **86**, 11535.
73 J. Veizer, pers. comm.
74 Watkins, J.C. et al. (1981) Accretion, underplating, subduction and tectonic evolution, Middle America Trench, Southern Mexico: results from DSDP Leg 66. *Oceanol. Acta Spec. Pub.*; (1982) Tectonic synthesis, Leg 66: transect and vicinity. *Init. Repts. DSDP*, **66**, 837; Moore, J.C. (1975) Selective subduction. *Geology*, **3**, 530; Moore, J.C. et al. (1982) Summary of accretionary processes, deep sea drilling project Leg 66: offscraping, underplating, and deformation of the slope apron. *Init. Repts. DSDP*, **66**, 825.

75 Karig, D.E. & Kay, R.W. (1981) Fate of sediments on the descending plate at convergent margins. *Phil. Trans. Roy. Soc.*, **A301**, 233.
76 Moorbath, S. (1977) Age and isotopic evidence for the evolution of continental crust. *Phil. Trans. Roy. Soc.*, **A288**, 401. See Appendix 4 for isotopic notation.
77 Brown, G.C. (1979) The changing pattern of batholith emplacement during earth history. *In:* Atherton, M.P & Tarney, J. (eds), *Origin of Granite Batholiths: Geochemical Evidence.* Shiva. 106.
78 Brown, G.C. & Mussett, A.E. (1981) *The Inaccessible Earth.* George Allen & Unwin.
79 Turcotte, D.L. & Schubert, G. (1982) *Geodynamics.* Wiley.
80 Collerson, K.D. & Fryer, B.J. (1978) The role of fluids in the formation and subsequent development of early continental crust. *CMP*, **67**, 151; Collerson, K.D. et al. (1981) Geochronology and evolution of Late Archaean gneisses in northern Labrador: an example of reworked sialic crust. *Geol. Soc. Austr. Spec. Pub.*, **7**, 205; Collerson, K.D. & McCulloch, M.T. (1982) The origin and evolution of Archaean crust as inferred from Nd, Sr and Pb isotpic studies in Labrador (abst.). *5th Int. Conf. Geochron. Cosmochron. Isotope Geol.*, 61.
81 DePaolo, D.J. & Wasserburg, G.J. (1976) Nd isotopic variations and petrogenetic models. *GRL*, **3**, 249; O'Nions, R.K. et al. (1979) Geochemical and cosmochemical applications of Nd isotope analysis. *Ann. Rev. Earth Planet. Sci.*, **7**, 11; DeLaeter, J.R. et al. (1981) Early Archean gneisses from the Yilgarn Block, Western Australia. *Nature*, **292**, 322; Pettingill, H.S. & Patchett, P.J. (1981) Lu–Hf total-rock age for the Amîtsoq gneisses, West Greenland. *EPSL*, **55**, 150.
82 Anhaeusser, C.R. & Robb, L.J. (1981) Magmatic cycles and the evolution of the Archaean granitic crust in the eastern Transvaal and Swaziland. *Geol. Soc. Austr. Spec. Pub.*, **7**, 457.
83 Bickle, M.J. et al. (1980) Horizontal tectonic interaction of an Archean gneiss belt and greenstones, Pilbara block, Western Australia. *Geology*, **8**, 525.
84 Hickman, A.H. (1981) Crustal evolution of the Pilbara Block, Western Australia. *Geol. Soc. Austr. Spec. Pub.*, **7**, 57.
85 Pidgeon, R.T. (1978) 3450 m.y.-old volcanics in the Archaean layered greenstone succession of the Pilbara Block, Western Australia. *EPSL*, **37**, 421.
86 Hamilton, P.J. et al. (1979) Sm–Nd dating on Onverwacht Group volcanics, southern Africa. *Nature*, **279**, 298.
87 Hamilton, P.J. et al. (1981) Sm–Nd dating of the North Star Basalt, Warrawoona Group, Pilbara Block, Western Australia. *Geol. Soc. Austr. Spec. Pub.*, **7**, 187.
88 Bickle, M.J. et al. (1983) A 3500 Ma plutonic and volcanic calc-alkaline province in the Archaean east Pilbara Block. *CMP*, **84**, 25.
89 Carlson, R.W. et al. (1983) Sm–Nd age and isotopic systematics of the bimodal suite, ancient gneiss complex, Swaziland. *Nature*, **305**, 701.
90 Basu, A.R. (1981) Eastern Indian 3800-million-year-old crust and early mantle differentiation. *Science*, **212**, 1502.
91 Hamilton, P.J. et al. (1978) Sm–Nd isotopic investigations of Isua supracrustals, West Greenland: implications for mantle evolution. *Nature*, **272**, 41; (1983) Sm–Nd studies of Archaean metasediments and metavolcanics from West Greenland and their implications for the Earth's early history. *EPSL*, **62**, 263.
92 Froude, D.O. et al. (1983) Ion microprobe identification of 4,100–4,200 Myr-old terrestrial zircons. *Nature*, **304**, 616.
93 Moorbath, S. (1983) The most ancient rocks? *Nature*, **304**, 585.
94 Moorbath, S. (1977) The oldest rocks and the growth of continents. *Sci. Am.*, **236** (3), 92.
95 Jahn, B.-M. & Sun, S.-S. (1979) Trace element distribution and isotopic composition of Archean greenstones. *PCE*, **11**, 597.
96 Condie, K.C. (1981) Geochemical and isotopic constraints on the origin and source of Archaean granites. *Geol. Soc. Austr. Spec. Pub.*, **7**, 469.
97 McCulloch, M.T. & Wasserburg, G.J. (1978) Sm–Nd and Rb–Sr chronology of continental crust formation. *Science*, **200**, 1003. See Appendix 4 for isotopic notation.
98 Patchett, P.J. (1983) Importance of the Lu–Hf isotopic system in studies of planetary chronology and chemical evolution. *GCA*, **47**, 81.
99 Veizer, J. & Compston, W. (1976) $^{87}Sr/^{86}Sr$ in Precambrian carbonates as an index of crustal evolution. *GCA*, **40**, 905.
100 Veizer, J. et al. (1982) Mantle buffering of the early oceans. *Naturwissenschaften*, **69**, 173.

101 The evolution to higher $^{87}Sr/^{86}Sr$ at the Archean–Proterozoic boundary is much more rapid than originally thought [99] because of recent dating on the Hamersley Group, which indicates an age of about 2.5 Ae rather than the previously determined age of 2.1–2.0 Ae. (Compston, W. et al. (1981) A revised age for the Hamersley Group (abst.). *Geol. Soc. Austr. 5th Ann. Conv. Abst.*, **3**, 40.)

102 Veizer, J. (1983) Trace elements and isotopes in sedimentary carbonates. *Rev. Mineral.*, **11**, 265.

103 Bell, K. et al. (1982) Evidence from Sr isotopes for long-lived heterogeneities in the upper mantle. *Nature*, **298**, 251.

104 Kröner, A. (1982) Archaean to early Proterozoic tectonics and crustal evolution: a review. *Rev. Brasil. Geosc.*, **12**, 15; (1984) Evolution, growth and stabilization of the Precambrian lithosphere. *PCE*, **15** (in press).

105 Moorbath, S. (1976) Age and isotope constraints for the evolution of Archean crust. In: Windley, B.F. (ed.), *The Early History of the Earth*. Wiley. 351.

106 Allegre, C.J. & Ben Othman, D. (1980) Nd–Sr isotopic relationship in granitoid rocks and continental crust development: a chemical approach to orogenesis. *Nature*, **286**, 335.

107 Hamilton, P.J. et al. (1980) Isotopic evidence for the provenance of some Caledonian granites. *Nature*, **287**, 279.

108 DePaolo, D.J. (1981) Neodymium isotopes in the Colorado Front Range and crust–mantle evolution in the Proterozoic. *Nature*, **291**, 193.

109 DePaolo, D.J. (1981) A neodymium and strontium isotope study of the Mesozoic calc-alkaline granitic batholiths of the Sierra Nevada and Peninsular Ranges, California. *JGR*, **86**, 10470.

110 McCulloch, M.T. & Chappell, B.W. (1982) Nd isotopic characteristics of S- and I-type granites. *EPSL*, **58**, 51.

111 Wyborn, L.A.I. & Page, R.W. (1983) The Proterozoic Kalkadoon and Ewen Batholiths, Mount Isa Inlier, Queensland: source, chemistry, age, and metamorphism. *BMR J.*, **8**, 53.

112 Compston, W. & Chappell, B.W. (1979) Sr-isotope evolution of granitoid source rocks. In: McElhinny, M.W. (ed.), *The Earth: Its Origin, Structure and Evolution*. Academic. 377.

113 The model, as presented in Fig. 10.9, infers an average post-Archean crustal growth rate of 0.7 km^3 a^{-1} but given the uncertainties expressed, the rate could be in the range 0.3–0.9 km^3 a^{-1}. It is worth pointing out that because crustal growth is probably an episodic process, measured rates taken over short periods of time would represent upper limits.

11

Crust–Mantle Relationships

11.1 The origin of the Earth

In this chapter, we discuss the general relationship of the continental crust and the mantle. The development of the continental crust appears to be a terrestrial phenomenon, not observed on other planets (see Chapter 12), and opens up a wide range of questions. What was the composition of the primitive terrestrial mantle, before crust formation began? How has it been modified by the extraction of the oceanic and the continental crusts? What effect has recycling of the oceanic crust had on mantle evolution? For that matter what is the composition and nature of the oceanic crust, and does it play any significant role in the formation of the continental crust?

Another series of questions concerns the relationships of the core and the mantle, and of the bulk earth composition to that of the moon and the other inner planets. The general similarity in density between the terrestrial mantle and the moon has encouraged speculation that the moon was derived by fission from the mantle. This is directly testable by comparison of lunar and terrestrial compositions; such significant compositional differences are revealed that the hypothesis appears unlikely [1].

In order to investigate these problems, and to place them in proper perspective, it is necessary to consider the origin of the Earth. About 4.56 billion years ago, a cold disk of dust and gas, the solar nebula, separated from a larger molecular cloud. Such clouds typically contain 100 000 solar masses, with densities of 10 000–1 million molecules per cubic centimetre, at temperatures of 20–100 K. The composition of the solar nebula apparently is preserved in the type I carbonaceous chondrites (referred to as CI where I=Ivuna). The resemblance between the elemental abundances in these meteorites for non-gaseous elements and those observed in the solar photosphere is the principal piece of evidence that they represent the original bulk nebular composition (Fig. 11.1). The CI meteorites are nevertheless mixtures of high and low temperature minerals recording evidence of complex precursor events in the solar nebula.

The most recent element abundance data for the CI meteorites is given in Table 11.1. They are multiplied by a factor of 1.5 to allow for loss of volatiles (H_2O, carbonaceous material). The data set are now becoming reasonably well established. Most of the values come from the recent review by Anders & Ebihara [2]. With minor exceptions, these are similar to those given by Taylor & McLennan [3], Taylor [1], and Landolt-Bornstein [4]. Significant differences between the values used previously [1, 3] and the present set occur for the following elements: Be, V, Nb, Mo, Te, W and Hg. The U and Th values

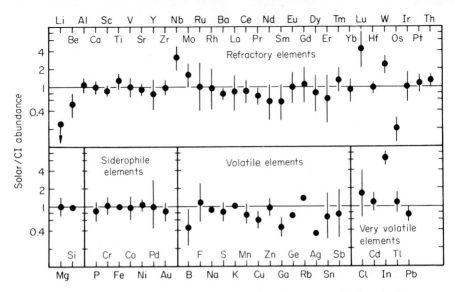

Fig. 11.1. Comparison between the abundances in the solar photosphere with those in CI carbonaceous chondrites. There is little significant difference (apart from Li, consumed in nuclear reactions in the Sun) between these two sets of data. Only Nb, Lu, W, Os, Ga, Ag and In show discrepancies, possibly due to lack of precision in the solar measurements. Recent revisions of the solar data have removed previous anomalies for Zr, Rh, Pd, Eu and Ir. (Adapted from [2].)

Table 11.1. Abundances in CI meteorites (original values ×1.5 to allow for loss of volatiles). [2, except for REE data from 5.]

	%						
SiO_2	34.2						
TiO_2	0.11						
Al_2O_3	2.44						
FeO	35.8						
MgO	23.7						
CaO	1.89						
Na_2O	0.98						
K_2O	0.10						
Σ	99.2						
Li	2.4 ppm	Cu	168 ppm	In	117 ppb	Tm	36 ppb
Be	40 ppb	Zn	462 ppm	Sn	2.52 ppm	Yb	248 ppb
B	1.9 ppm	Ga	15.2 ppm	Sb	233 ppb	Lu	38 ppb
Na	7245 ppm	Ge	48.3 ppm	Te	3.42 ppm	Hf	179 ppb
Mg	14.3%	As	2.87 ppm	Cs	279 ppb	Ta	26 ppb
Al	1.29%	Se	27.3 ppm	Ba	3.41 ppm	W	89 ppb
Si	16.0%	Rb	3.45 ppm	La	367 ppb	Re	55 ppb
K	854 ppm	Sr	11.9 ppm	Ce	957 ppb	Os	1049 ppb
Ca	1.35%	Y	2.25 ppm	Pr	137 ppb	Ir	710 ppb
Sc	8.64 ppm	Zr	5.54 ppm	Nd	711 ppb	Pt	1430 ppb
Ti	654 ppm	Nb	375 ppb	Sm	231 ppb	Au	218 ppb
V	85 ppm	Mo	1.38 ppm	Eu	87 ppb	Hg	585 ppb
Cr	3975 ppm	Ru	1071 ppb	Gd	306 ppb	Tl	215 ppb
Mn	2940 ppm	Rh	201 ppb	Tb	58 ppb	Pb	3.65 ppm
Fe	27.8%	Pd	836 ppb	Dy	381 ppb	Bi	167 ppb
Co	764 ppm	Ag	330 ppb	Ho	85 ppb	Th	42.5 ppb
Ni	1.65%	Cd	1010 ppb	Er	249 ppb	U	12.2 ppb

of Anders & Ebihara [2], which are slightly lower than those used previously, have a better statistical base, and are employed here. The REE data set [5], previously used by Taylor & McLennan [3], are used here for consistency. The values of Anders & Ebihara [2] are a factor of 0.964 lower. Boron presents a special case. We use here the calculated value from Anders & Ebihara [2] rather than the lower values, based on data from the interior portions of chondrites by Curtis *et al.* [6]. The present estimate allows for the preferential concentration of boron, a volatile element, in the matrix rather than chondrules. The extreme ease with which samples can be contaminated by boron rivals that of lead, and makes most boron determinations suspect [6].

The nebula was initially cold, not hot, with temperatures not exceeding 300 K. This important information is inferred principally from the observation that isotopic anomalies, mainly oxygen, are preserved in various classes of meteorites; in a hot nebula, (1500–2000 K) they would be homogenized. The primitive nebula was probably oxidized, that is, elements were present mainly as compounds, although some small metallic nuggets of Pt group elements may have been present. Two processes appear to have operated at a very early stage in the nebula. These were reduction, producing free metal (mainly iron) and volatilization, and they had important implications for the compositions of the inner planets. The production of metal, and the depletion of the volatile elements occurred independently. This is deduced from the observation that the type I enstatite chondrites, in which all the iron has been reduced to metal, contain CI abundances of the volatile elements.

Both processes occurred very early in the nebula, for the meteorite ages, which record such events, are all close to 4.56 Ae. Although the basic mechanisms remain obscure, they probably occurred in the inner solar system within a few AU of the sun. Possible causes include early solar heating, solar flares, and T Tauri type strong solar winds and other phenomena associated with early stages of stellar evolution. Such pre-main sequence and early main sequence effects last for short time-scales of the order of a few million years at most and provide mechanisms such as localized high temperatures and strong solar winds, to remove gas (H, He) and volatile material from the inner solar system.

The Galilean satellites of Jupiter [7] provide some critical information. Since they and the Saturnian satellites are of low density, free metal was probably not present beyond about 5 AU. They also contain substantial quantities of water ice, indicating that temperatures were not high enough to deplete the solar nebula in volatile elements at 5 AU. Evidence of local heating in a Jovian nebula is provided by the monotonic decrease in density from Io to Callisto, as the water ice/rock ratio increases outwards. This change is interpreted as due to mild warming of the Jovian nebula by proto-Jupiter, and provides a small-scale analogue to the solar nebula. Much more severe conditions existed in the inner solar system close to the early sun. From astrophysical observations, early intense solar flare activity and strong solar winds occur within about one million years of the arrival of the Sun on the main sequence. The meteorite age data are consistent with this time-scale.

Accordingly, loss of volatile elements from the inner portions of the solar nebula probably occurred within a few million years of the formation of the sun, although the accretion of the inner planets from planetesimals occurred on much longer time-scales, of the order of 100 m.y.

Although a large amount of the rare gases and considerable amounts of the volatile elements (e.g. 80% of the potassium in the case of the Earth) were removed from the region of the inner planets, enough volatile material remains to provide for the terrestrial budget. Such material was probably trapped, perhaps in CI-type material, in bodies large enough to survive the early intense solar activity, and to remain as condensed bodies in the inner solar system. The terrestrial K/U ratio (10 000) is much lower than the CI value of 60 000 (Fig. 11.2). Likewise, the terrestrial mantle Rb/Sr ratio of 0.031 is an order of magnitude less than the CI value of 0.30, which illustrates the large scale on which these refractory/volatile element fractionations occur. (Lunar ratios are even more extreme, the moon having a K/U ratio of 2500 and an Rb/Sr ratio of 0.01 [1].)

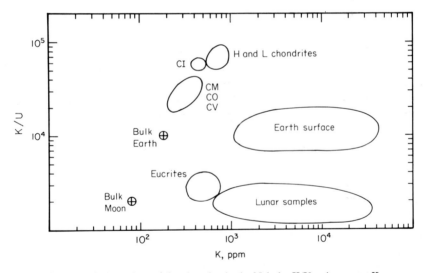

Fig. 11.2. Volatile-refractory element fractionation in the Nebula. K/U ratios versus K abundances, to illustrate the separation of the CI ratio from those of other meteorites, the Earth and the Moon. The data for Venus probably overlap those for the Earth. The data for Mars are uncertain, but the K/U ratio may be higher than that of the Earth if the shergottite, chassignite and nakhlite classes of meteorites are derived from Mars. (Adapted from [1].)

The time-scale for the growth and sweep-up of planetesimals to form the inner planets is of the order of 100 m.y. [8]. This, the planetesimal hypothesis for the formation of the inner planets, is preferred to the alternative giant gaseous proto-planet hypothesis. This latter model may be applicable to the growth of Jupiter and Saturn, which almost certainly grew before the inner planets (accounting for the small size of Mars), before H and He had been removed from the nebula.

The overall composition of the Earth depends, according to the planetesimal hypothesis, on the compositions of the particular suite of planetesimals which were accreted. The differing densities among the inner planets are accounted for by the varying proportions of metal and silicate (and sulphide) phases which were accreted. The metal–silicate–sulphide equilibria were mainly established in low pressure precursor planetesimals, rather than in the Earth itself, following accretion. Only limited reaction would occur during accretional melting. Thus the oxidation state of the mantle, the Fe/Mg ratio, the mantle budget of Ni and other siderophile elements, and the volatile/refractory element ratios (e.g. K/U, Rb/Sr) are all established prior to accretion. There is, of course, no way to deplete the Earth in elements such as K or Rb following accretion.

Melting during accretion and rapid sinking of metal and sulphide phases appears very likely, but this provides only localized opportunity for metal-silicate reactions. If equilibrium was only attained locally during core and mantle separation, then the composition of the core is not directly related to that of the present mantle, since metal-silicate equilibration was established mainly in planetesimals before accretion. The light element in the core is most likely to be sulphur; it is abundant enough in most meteorites to be a viable candidate, even allowing for volatile depletion [9]. Sulphur has the advantage of forming a low melting point Fe–FeS eutectic. Core formation can thus proceed at an early stage of terrestrial accumulation. This is in accord with the evidence of a large core in Mercury, and the evidence of much metal–silicate fractionation in the meteorites, especially in the eucrite parent body. The other principal candidate for the light element in the core is oxygen [10]. The formation of an Fe–FeO mixture in which FeO might have metallic properties would occur at high pressures (one megabar), thus restricting core formation to a stage when the earth was close to its present size. If the Earth melted during accretion, the metallic phases and sulphides would form a core well before this stage, thus rendering the oxygen scenario less likely.

11.2   The primitive terrestrial mantle

In the following section, we discuss the composition of the primitive mantle, which comprises the present day mantle plus crust. Constraints and assumptions about the composition of the mantle, in terms of the planetesimal hypothesis are:

1   Metal and silicate phases were segregated in the planetesimal stage, and accreted as such with only local equilibration occurring during accretion.
2   Depletion of volatile (e.g. K, Rb) relative to refractory elements (e.g. U, Sr) occurred before accretion.
3   The refractory elements were not fractionated among themselves during the planetesimal stage, but were accreted in their CI relative proportions. Thus the Sm/Nd ratio for the bulk Earth will be chondritic.
4   The chalcophile elements were accreted mainly in sulphide phases.

11.2.1 Upper and lower mantle compositions

The major element composition of the upper mantle can be estimated from xenolith data. This approach encounters certain difficulties. The samples are restricted to the upper 200 km. They frequently display residual or depleted chemistry, consistent with the extraction of basaltic melts, so that they may be derived from areas with complex histories. Furthermore, they come from a region which has most probably been the source of much of the continental crust. Although the mantle xenolith data must be approached with caution, there is a fair degree of consensus for the major element composition of the upper mantle (Table 11.2). Such estimates must be consistent with the experimental petrology and high pressure studies. The combination of the processes of melt extraction, redistribution by metasomatism, and crustal contamination make attempts to use the nodule data for trace element

Table 11.2. Major element composition (wt %) of the upper mantle.

	1	2
SiO_2	45	45.1
TiO_2	0.15	0.2
Al_2O_3	3.3	3.3
FeO	8.0	8.0
MnO	0.13	0.15
MgO	39.8	38.1
NiO	0.25	—
Cr_2O_3	0.44	0.4
CaO	2.6	3.1
Na_2O	0.34	0.4
K_2O	0.02	0.03
Σ	100.0	98.8

1. This work.
2. Pyrolite [11].

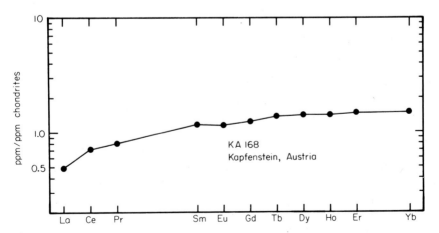

Fig. 11.3. REE abundances in mantle nodule KA 168 from Kapfenstein, Austria. This is one of the best examples of a primitive mantle nodule, but shows LREE depletion relative to chondrites indicating LIL element extraction has occurred [12].

determinations hazardous. Even the best cited examples, such as Ka 168 [12] show light REE depleted patterns, relative to chondrites (Fig. 11.3) indicating removal of LIL elements.

In contrast to the upper mantle, the lower mantle, which comprises about 50% of the Earth, is barely accessible to study. Evidence from seismology and from phase transition studies has raised the question whether the lower mantle is different in composition to the upper mantle. The major discontinuity at 650 km depth, taken here as the boundary between the upper and lower mantle, appears to be very sharp (±4 km) and hence may be a chemical boundary [13]. Studies of mantle convection patterns are about equally divided between two-layer and whole mantle convection. The latter process would presumably homogenize the mantle, so that upper and lower mantle compositions would be the same [14]. The mantle phases established by high-pressure research are given in Fig. 11.4 [15].

Cosmochemical considerations raise major questions concerning the deep mantle composition. For example, the upper mantle Mg/Si weight ratio of 1.14 contrasts with the CI value of 0.90. If the whole mantle has the upper mantle Mg/Si ratio then the bulk Earth is depleted in Si relative to CI abundances. Meteorites show variation in Mg/Si ratios, although generally in the direction

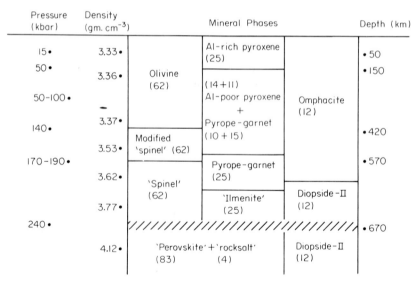

Fig. 11.4. Mineral phases as a function of pressure in the mantle, corresponding to olivine and pyroxene components in the upper mantle (adapted from [15]). The numbers in parentheses indicate the weight percentage of each mineral. Mineral compositions are listed below, with the co-ordination numbers (CN) for the cations with O^{2-} to illustrate the increase in CN with pressure. *Al-rich pyroxene:* $(Mg, Fe)_2(SiAl)_2O_6$ with Mg and Fe in 6 CN and Si in 4 CN. *Al-poor pyroxene:* $(Mg, Fe)SiO_3$ with same CN as Al-rich pyroxene. *Pyrope garnet:* $(Mg, Fe)_3Al_2Si_3O_{12}$ with Mg, Fe in 8 CN, Al in 6 CN and Si in 4 CN. *'Ilmenite':* $(Mg, Fe, Al)(Si, Al)O_3$ with all cations in 6 CN. *'Perovskite':* $(Mg, Fe, Al)(Si, Al)O_3$ with Mg, Fe, Al in 8 CN and Si, Al in 6 CN. *Olivine:* $(Mg, Fe)_2SiO_4$ with Mg and Fe in 6 CN and Si in 4 CN. *'Spinel':* Same as olivine. *'Rock salt':* (ferropericlase) $(Mg, Fe)O$ with Fe, Mg in 6 CN. *Omphacite:* $(Ca, Na)(Mg, Al)Si_2O_6$ with Ca and Na in 8 CN, Mg and Al in 6 CN and Si in 4 CN. *Diopside II:* Same composition as omphacite with Ca, Na and Mg in 8–12 CN and Al and Si in 6 CN.

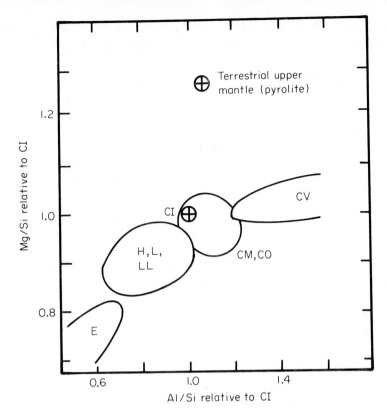

Fig. 11.5. The relationship between Mg/Si and Al/Si in chondritic meteorites. The terrestrial upper mantle is highly enriched in Mg relative to other solar system abundances. We interpret the bulk earth composition to be equivalent to the CI ratios, which implies significant differences between upper and lower mantle compositions. (Adapted from [16].)

of lower, not higher values (Fig. 11.5). If the Earth has CI ratios for Mg and Si, then the lower mantle must be different in composition from the upper mantle. This problem has been addressed most recently by Liu [13] and Anderson [17]. They conclude that the bulk modulus (K_0) data for the lower mantle 'are coincident with perovskite ($MgSiO_3$) (density=4.1 g cm^{-3}; K_0=2.6 mbar). Substantial amounts of (Mg, Fe)O as in the pyrolite models, cannot be tolerated because of the very low K_0 of (Mg, Fe)O.'

The overall evidence favours a difference in composition between the upper and lower mantle. Although no single piece of information is yet definitive, the combined evidence from seismology, from high-pressure studies and from cosmochemistry make a presently heterogeneous mantle a strong possibility. We adopt this view here [18].

11.2.2 Primitive mantle compositions

The bulk mantle composition is derived essentially on the basis of cosmochemistry (Table 11.3). We assume an FeO content of 8.0%, consistent with density and seismic properties. The refractory major elements Si, Ti, Al, Mg

Table 11.3. Element abundances in the primitive mantle (=present mantle plus crust). See text for references.

	1 %	2 %
SiO_2	49.9	49.3
TiO_2	0.16	0.21
Al_2O_3	3.64	3.93
FeO	8.0	7.86
MgO	35.1	34.97
CaO	2.89	3.17
Na_2O	0.34	0.27
K_2O	0.02	0.018
Σ	100.1	99.7

	1	2		1	2		1	2
Li (ppm)	0.83	2.1	Rb (ppm)	0.55	0.39	Eu (ppb)	131	130
Be (ppb)	60	—	Sr (ppm)	17.8	16.2	Gd (ppb)	459	—
B (ppm)	0.6	—	Y (ppm)	3.4	3.26	Tb (ppb)	87	90
Na (ppm)	2500	2040	Zr (ppm)	8.3	13	Dy (ppb)	572	—
Mg (%)	21.2	21.1	Nb (ppm)	0.56	0.97	Ho (ppb)	128	—
Al (%)	1.93	2.08	Mo (ppb)	59	—	Er (ppb)	374	—
Si (%)	23.3	23.0	Ru (ppb)	4.3	—	Tm (ppb)	54	—
K (ppm)	180	151	Rh (ppb)	1.7	—	Yb (ppb)	372	320
Ca (%)	2.07	2.27	Pd (ppb)	3.9	—	Lu (ppb)	57	60
Sc (ppm)	13	15	Ag (ppb)	19	3	Hf (ppm)	0.27	0.33
Ti (ppm)	960	1260	Cd (ppb)	40	20	Ta (ppm)	0.04	0.04
V (ppm)	128	77	In (ppb)	18	6	W (ppb)	16	12
Cr (ppm)	3000	2342	Sn (ppm)	<1	0.60	Re (ppb)	0.25	0.21
Mn (ppm)	1000	1016	Sb (ppb)	25	—	Os (ppb)	3.8	2.90
Fe (%)	6.22	6.11	Te (ppb)	22	—	Ir (ppb)	3.2	2.97
Co (ppm)	100	101	Cs (ppb)	18	20	Pt (ppb)	8.7	—
Ni (ppm)	2000	1961	Ba (ppm)	5.1	5.22	Au (ppb)	1.3	0.50
Cu (ppm)	28	29	La (ppb)	551	570	Tl (ppb)	6	0.01
Zn (ppm)	50	37	Ce (ppb)	1436	1400	Pb (ppb)	120	120
Ga (ppm)	3	4	Pr (ppb)	206	—	Bi (ppb)	10	3.3
Ge (ppm)	1.2	1.13	Nd (ppb)	1067	1020	Th (ppb)	64	76.5
As (ppm)	0.10	0.02	Sm (ppb)	347	320	U (ppb)	18	19.6
Se (ppb)	41	—						

1. This work.
2. Anderson [16].

and Ca are assumed to be present in their CI proportions. In order to arrive at the abundances for the trace elements we work from a variety of isotopic and chemical data, to obtain a self-consistent trace element composition. This approach is similar to that used previously [1, 3]. The first assumption is that the lithophile refractory elements are present in their CI ratios and that no relative fractionation has occurred. For example, the REE patterns are parallel to those observed in chondrites. From the major element abundances for Ca, Al and Ti, the concentration levels for the upper mantle for the refractory elements are 1.50 times the volatile-free CI abundances. This approach enables values for B, Ba, Be, Hf, Nb, REE, Sc, Sr, Ta, Th, U, V, Y, and Zr to be obtained.

The abundance of uranium (18 ppb) so derived provides an estimate of 180 ppm K from the K/U ratio of 10 000. Further constraints on the volatile/refractory ratios come from the Sm/Nd ratio. If this is chondritic, then from the relationship between ε_{Nd} and ε_{Sr}, the mantle Rb/Sr ratio is 0.031. Sr is refractory and its concentration on this basis is 18 ppm. The Rb abundance, from the Rb/Sr ratio, is 0.55 ppm, giving a Ba/Rb ratio of 9.3, close to the value of 11.3 [19]. If the terrestrial K/Rb ratio is 300, then K is 165 ppm. Clearly, estimates of the K content of the mantle between 150–200 ppm are realistic, and are also consistent with values obtained from K–Ar degassing models. K/Rb values for the mantle are probably greater than the CI value of 250. The K value, from K/U, of 180 ppm gives a K/Rb ratio of 327, which would indicate some depletion of Rb relative to less volatile K. Support for this view comes from the crustal abundance of cesium, the most volatile of the alkali elements. The crustal Rb/Cs ratio is 30, compared with a CI value of 12. Cs is not only more volatile, but also more incompatible than Rb or K and accordingly should be highly concentrated in the continental crust. The MORB Rb/Cs ratio of 79 [19] may be too high, being derived from sources already depleted in cesium. Here we use the crustal Rb/Cs ratio, which provides a mantle abundance of 18 ppb, probably an upper limit. Tl is derived from the K/Tl ratio of 3×10^4. An upper limit is placed on the depletion of relatively volatile elements by the lack of Eu and Yb anomalies. These two REE are somewhat less volatile than K, and are not depleted in the bulk Earth. The value for Li comes from Li/Zr=0.1 [20]. The data for the chalcophile and siderophile elements are on a less secure footing, since it is necessary to rely on the nodule data. Values for Ni, Co, Cr and Mn come from this source [1]. Data for Cu, Zn, Ga, Ge, As, and Se, are from [12]. The W value of 12 ppb and the Mo abundance at 59 ppb are from [21]. Values for Ru, Rh, Pd, Re, Os, Ir, Pt, Au are from [22] and data for Ag, Cd, In, Sb, Te, Bi are from [23]. The Pb data are from [24].

Siderophile and chalcophile elements show interesting patterns. The strongly siderophile elements (Au, Ge, Ir, Os, Pd, Pt, Re, Rh, Ru) are present in upper mantle nodules at about one-hundredth of their chondritic abundance. Elements which have greater chalcophile tendencies (Ag, As, Co, Cu, Ga, Mo, Ni, Sb, Sn, W) are present at about one-tenth of their chondritic abundances, a factor of 10 higher. Of these elements, Ni and Co are also lithophile and would be expected to be accreted in olivine as well as in sulphide phases. These abundances observed in upper mantle nodules show that they could not have been in equilibrium with a metal phase in the Earth during core formation. Wide differences among the distribution coefficients would have altered the abundance patterns from their CI pattern. Among other consequences, this information shows that the upper mantle siderophile element signature is not uniquely terrestrial.

A rather different approach to calculating bulk mantle and crust compositions has been adopted by Anderson [16] who has estimated a composition for the primitive mantle (upper+lower+crust). This is listed in Table 11.3 where it may be compared with our estimates of mantle compositions. For a number of elements (Si, Mg, Sc, Mn, Fe, Ni, Co, Cu, Ga, Ge, Y, Ba, REE, Ta, Re, Ir and Pb) the two compilations are very close. Many other trace elements are within 20% in both sets of data. Our values for the refractory elements Ti, Al and Ca are slightly higher.

11.3   Effect of crustal extraction

The continental crust comprises 0.47% of the mass of the mantle. Extraction of the present upper crustal composition from the mantle has made only minimal changes to the major element composition of the upper mantle, well within the tolerances of estimating mantle composition. This is in strong contrast to the Moon, where the lunar highland crust comprises 10% of lunar volume and has concentrated perhaps 50% of the Al content of the bulk moon.

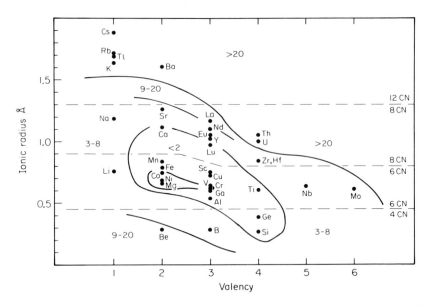

Fig. 11.6. The enrichment of lithophile elements in the continental crust relative to abundances in the primitive mantle (data from Tables 3.5 and 11.3), plotted against ionic radius and valency. Ionic radii data from Shannon [25] are plotted for 4-, 6-, 8- and 12-fold co-ordination with oxygen, as indicated. Elements are enriched in the crust to the degree that their radius or valency differs from that of the principal mantle cations (Mg, Fe) in 6-fold co-ordination. This enrichment is a consequence of crystal–liquid fractionation.

The trace elements tell a different story. Figure 11.6 illustrates that the degree of enrichment of an element in the crust depends on the difference between its ionic radius and valency and those of the cations forming the major mantle minerals (Mg, Fe). This indicates that crystal–liquid fractionation is the principal factor responsible for the derivation of the crust from the mantle. The very large percentage of elements such as Cs, Rb, K, Ba, U, Th, La and Ce in the continental crust indicates the processing of very large fractions of the mantle to produce the crust.

11.3.1 Extent of terrestrial differentiation

The extraction of the LIL-enriched crust from the mantle results in a complementary LIL-depleted reservoir in the mantle. It is commonly held that the depleted N-type mid-ocean ridge basalts represent melts from this depleted reservoir. Has this depletion affected the mantle in a homogeneous fashion or is only a portion of the mantle involved in the crust forming processes? If only a fraction of the mantle is involved, then several questions arise. How much mantle is involved? Does the depleted portion occur as blobs throughout the mantle or does it form discrete layers?

On the basis of chemical and petrological arguments, Ringwood [11] suggested that about 70% of the mantle had been differentiated to form the crust. This estimate was later revised to 30–60% [10]. More recently, considerable effort has gone into modelling isotopic characteristics of mantle and crust to address these questions. O'Nions et al. [26] have suggested that about one-half of the mantle was involved in crust formation while Jacobsen & Wasserburg [27] suggested that only about one-third of the mantle was involved. The difference in these estimates can be related to variations in the assumed present-day concentrations of key elements (e.g. Rb, Sr, Sm, Nd) in the various reservoirs.

For the crustal and mantle elemental abundances adopted here, an approximate upper limit of mantle involvement can be calculated from simple mass balance considerations. Rubidium is a highly incompatible element which is strongly enriched in the crust and strongly depleted in MORB. Cesium is even more incompatible but dependable data are not available. The procedure involves calculating the effects of extracting the crustal complement of Rb from varying volumes of primitive mantle and forming a ratio to a relatively compatible element (in this case Yb). By comparing the calculated Rb/Yb ratio to the known MORB Rb/Yb ratio, we can estimate the degree of mantle involvement. These calculations are shown in Table 11.4 and indicate that one-third to one-half of the mantle has undergone crust extraction. A similar answer is reached if barium is used as the incompatible element in place of rubidium. Such a procedure is only approximate since even 10% differences in crustal or mantle abundances or crustal mass estimates can significantly affect the result. More mafic compositions for the bulk crust result in lower estimates of mantle involvement.

If the base of the upper mantle is taken to be the major seismic

Table 11.4. Rubidium and ytterbium mass balance in the mantle and crust.

	Primitive mantle	Continental crust	Predicted depleted mantle for various degrees of involvement				MORB
			100%	50%	33%	25%	
Rb (ppm)	0.55	32	0.37	0.19	0.003	—	2.2
Yb (ppm)	0.37	2.2	0.36	0.35	0.34	0.33	5.1
Rb/Yb	1.49	14.5	1.03	0.55	0.01	—	0.43

discontinuity at about 650 km [13], then the upper mantle volume corresponds almost exactly to one-third of the mantle. If an equivalent amount of mantle was involved in crust formation, chemically layered mantle models (upper depleted zone and lower primitive zone) become attractive. This also bears on the question of whole mantle convection versus separation into two (upper and lower mantle) convecting systems. Although this is a fascinating area of inquiry, the chemical data do not really specify which portions of the mantle are involved.

11.4 Crustal recycling: evidence from island-arc volcanics

Island-arc volcanic rocks show a number of important geochemical and isotopic characteristics [28], which include:
1 enrichments in various incompatible elements, notably K, Sr, Ba and Pb;
2 radiogenic Sr and Pb isotopic characteristics and ε_{Sr}–ε_{Nd} correlations in some arcs;
3 low $^{143}Nd/^{144}Nd$ and $^{176}Hf/^{177}Hf$ ratios;
4 high concentrations of ^{10}Be.

A common explanation of these has been the involvement of sedimentary material in the origin of island-arc volcanics. Recognition of a sedimentary component in island-arc volcanics is important for several reasons:
1 it presents direct evidence for the involvement of the down-going slab;
2 it demonstrates that sediment can be subducted to mantle depths (80–150 km for the origin of island-arc volcanics);
3 it provides a minimum estimate of continental crustal recycling. The actual value of recycling could be higher if some sedimentary material remixed with the mantle and remains behind.

11.4.1 Geochemical evidence

Evaluation of the Th/U ratios limited the involvement of oceanic sediment in island-arcs to 2% or less [29]. From multicomponent melting-mixing models for the genesis of arc magmas, using constraints from K, Rb, Pb, Ba, Sr, REE abundances and Nd and Sr isotopic ratios, it was proposed that at least 10% of the pelagic sediment column was involved in arc magmatism [30]. Arculus & Johnson [31] criticized this approach because it failed to consider the lack of Th and U enrichment in arc lavas. They also pointed out that Sr, Ba and Pb enrichments also existed in volcanics remote from island arcs, and suggested that contamination by lower continental crustal rocks, rather than sediments, best explained the chemistry of arc magmas. McLennan & Taylor [32] constrained the role of sediment by modelling the negative Eu-anomaly which crustal rocks would impart to island-arc volcanic sources. The upper limits of sediment involvement were:
1 10% for MORB source;
2 1% for primitive or single-stage depleted mantle;
3 <0.3% for two-stage, highly depleted mantle.

11.4.2 Pb, Sr, Nd and Hf isotopes

The isotopic character of arc volcanics is consistent with some sedimentary component. The arc volcanic data are distinctly radiogenic with respect to $^{206}Pb/^{204}Pb$ and $^{207}Pb/^{204}Pb$ compared to MORB (Fig. 11.7). They form arrays which generally are consistent with about 1–2% sediment in the source [33]. It is not necessary to invoke a sediment component because E-type MORB has Pb-isotopic compositions which encompass the island-arc data [28]. Accordingly, arc volcanics could be derived from a variety of E-type MORB sources.

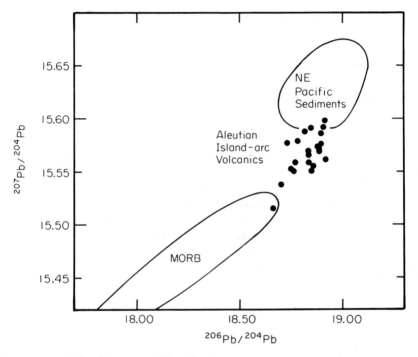

Fig. 11.7. Plot of $^{207}Pb/^{204}Pb$ against $^{206}Pb/^{204}Pb$ for island-arc volcanics for the Aleutian Islands. Also shown are fields for MORB and sediments from the north-east Pacific (data from [33]). The island-arc data appear to form a mixing array between MORB and the sediments. Such data have been used to infer a 1–2% sediment component in the source of island-arc volcanic rocks.

Another isotopic argument cited for sediment contribution comes from the Sr–Nd isotopic data (Fig. 11.8). Arc volcanics generally have more radiogenic $^{87}Sr/^{86}Sr$ ratios than the mantle array. They also tend to have lower $^{143}Nd/^{144}Nd$ than typical MORB. Such data are used to argue for a slab component derived from sea water and/or upper continental crust (i.e. sediments). Such data may equally well be explained by contamination from the lower continental crust [31], although this explanation cannot apply to island-arcs far removed from the continents, (e.g. Marianas Arc). In some arcs there is evidence for a correlation between $^{87}Sr/^{86}Sr$ and ε_{Nd} which has also been used to argue for sediment involvement.

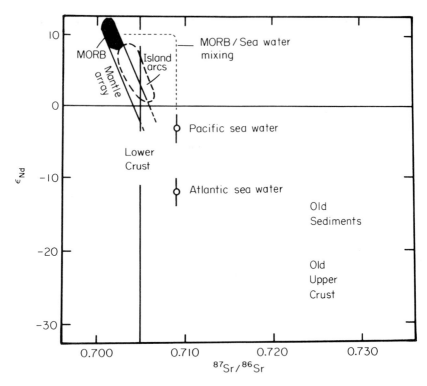

Fig. 11.8. ε_{Nd}–$^{87}Sr/^{86}Sr$ diagram showing the fields for MORB and island-arc volcanics relative to the mantle array. The values for Atlantic and Pacific sea water are also shown.

Strong evidence for sediment subduction is known from the Banda island-arc lavas of Indonesia [34]. These data show a clear mixing line between the mantle array and upper continental crust. Such mixing is also confirmed by oxygen isotope data [35] and Pb-isotopic compositions [34]. An extremely large amount of sediment (>50%) is needed to explain some of the Sr–Nd isotopic compositions. These levels of crustal contamination are unique to the Banda island-arc and may be related to the proximity of the Australian continent [34].

Hf-isotopic data from island-arc volcanics are consistent with a small ($\approx$1–2%) sediment component [36]. The Lu–Hf isotopic system has the potential to provide the best evidence concerning crust–mantle interaction at island-arcs and in the development of ocean island basalts (OIB) [37]. During sedimentary processes, zircon is concentrated in coarser-grained sediments (see Section 2.3.3) resulting in fractionation of Lu/Hf ratios between continental sands (low Lu/Hf) and deep-sea muds (high Lu/Hf). This contrasts with the lack of fractionation of Sm/Nd ratios during sedimentation. The Hf–Nd isotopic variations of mixing mantle and various types of sediment can thus be predicted and compared to island-arc volcanic and OIB data. Results [37] are consistent with <2% of nearly equal proportions of turbidite sand and pelagic sediment being combined with mantle sources to produce the Hf–Nd isotopic array, although the geological feasibility of such a sediment mixture is not obvious [36].

11.4.3 ^{10}Be evidence

Recent analytical developments have allowed the precise determination of ^{10}Be abundances at very low levels ($<10^6$ atoms g^{-1}) [38]. ^{10}Be is produced in the upper atmosphere by cosmic rays, primarily by spallation reactions on nitrogen and oxygen. It has a half-life of about 1.5 m.y. ^{10}Be is a potentially important tracer of subducted sedimentary material in island-arc volcanics because young sedimentary material has a high ^{10}Be content which is derived from the upper atmosphere. The short half-life ensures rapid disappearance so it does not monitor long-term recycling. A summary of some of the available ^{10}Be data are presented in Fig. 11.8 and Table 11.5 [38, 39]. Sediments are enriched in ^{10}Be compared to volcanics by large factors. Island-arc volcanics have variable ^{10}Be concentrations but are up to 10–100 times enriched over the levels in other volcanics. This has been used as evidence for a sediment component, suggesting that as much as 2–3% of the ^{10}Be which could potentially reach the base of an island-arc by sediment subduction is included in the arc volcanics [39].

Table 11.5. ^{10}Be data in volcanic rocks and sediments.

	n	^{10}Be (10^6 atoms per g^{-1})
Island-arc volcanics		
Nicaragua	2	6.5
Guatemala	2	3.1
Aleutians	9	4.3
Japan	1	≈0.1
Taiwan	1	0.3
Other volcanics		
Pribilof Islands	2	≈0.2 (≈0.1, 0.3)
Hawaii	1	≈0.1
Columbia Plateau	1	1.0
Marine sediments	16	5600±1800
Non-marine sediments	10	362±67

Volcanic data from [39] and sedimentary data from [38].

The interpretation of the island-arc data is not entirely clear-cut for several reasons. The data base is very small, being restricted to five arc systems, with two of the arcs (Japan, Taiwan) showing evidence of zero or trivial sediment contribution. One of the other volcanic rocks which has been measured (Columbia Plateau basalt) has ^{10}Be abundances approaching the island-arc values. ^{10}Be can be produced through ^{7}Li$(\alpha, p)^{10}$Be and ^{10}Be begins to accumulate in volcanic rocks after eruption. Finally, because of slow sedimentary rates for pelagic sediments, ^{10}Be is concentrated in the upper

50–100 m of the pelagic column and accordingly only this sediment is traced by the ^{10}Be method.

Despite these limitations, the ^{10}Be data provide the least equivocal evidence for the role of sediment contribution to arc magmatism. A synthesis of all the geochemical and isotopic data appears to limit the sediment component in island-arc volcanics to about 3%. This contribution appears to be sporadic, and to be essentially unrelated to the fundamental processes responsible for the generation of island-arc volcanism.

11.4.4   Th/U ratios

The Th/U ratios in island-arc volcanic rocks are typically low, in the range 1.5–3.0. These values are similar to or less than those observed in MORB, and are very much lower than those typical of igneous rocks in the upper crust (3.8) or of sedimentary rocks, in which the ratio, except in first cycle volcanogenic sediments, is greater than 4. Uranium is preferentially incorporated from sea water into altered oceanic crustal basalts. Pristine mantle sources have ratios of 3.8 equivalent to that of the bulk earth.

Information about the Th/U ratios in the source materials of the island-arc rocks may be obtained from a study of the ^{230}Th–^{238}U disequilibrium system, in which the activity ratio ^{230}Th/^{232}Th represents the current Th/U ratio in the source of the magmas [40]. A study of island-arc volcanics from the Marianas and the Aleutians shows that for each arc the Th/U ratio in the source is relatively constant. No effect is discerned in the Aleutian arc in passing from oceanic to continental crust along the arc. In the Marianas arc, the ^{230}Th/^{232}Th data indicate that the Th/U ratio of the source is similar to MORB and 'imply no significant incorporation of subducted sediment or altered MORB' [40, p. 270]. For the Aleutians data, a small contribution of altered MORB, high in U, appears to be required. In summary, there appears to be constant but different low Th/U ratios in the sources of these two volcanic arc systems. Enrichment of U relative to Th has occurred 'probably due to preferential enrichment of the magma in U by fluids derived from the subducted slab' [40, p. 270]. In the context of the present discussion, these data confirm the conclusions from other isotopic and geochemical studies that the involvement of recycled continental crustal sediments in the production of island-arc magmas is sporadic and limited to a few per cent.

11.5   The oceanic crust

Although this book is concerned with the composition and evolution of the continental crust, the nature and composition of the oceanic crust is relevant to the discussion. At first sight, the oceanic crust appears rather simple, but in fact many problems remain to be solved. Away from the ridges, the crust is comprised of three main parts. Layer 1, the sedimentary cover, may be up to 1 km thick. Layer 2, variable but generally about 2.5 km thick is composed of basalt and lenses of sediment [41]. Layer 3 is very uniform, both seismically

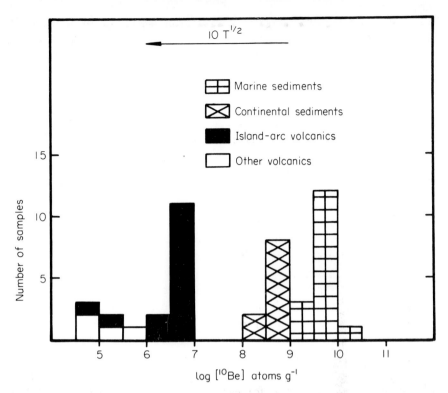

Fig. 11.9. The abundance of ^{10}Be in sedimentary and volcanic rocks. Note that island-arc volcanics contain more ^{10}Be than volcanic rocks from other sources, and that continental sediments are enriched relative to marine sediments. (Data from Table 11.5.)

and in thickness (4.5 km). It is thought to be composed principally of MORB-type basalt.

Estimates of the composition of the oceanic crust are dominated by this MORB component. This simple picture is complicated by several factors. Firstly the number of guyots or seamounts superimposed on the ocean floor is difficult to estimate. If their composition is largely alkali basalt, then our estimates of the bulk composition of the oceanic crust may seriously underestimate the amount of the larger ion lithophile elements in the oceanic crust available for subduction. The general opinion that insufficient LIL elements are available from a subducting oceanic crust to supply the amounts found in calc-alkaline volcanics may thus not be a valid constraint on that hypothesis [28].

A second problem concerns the interaction of water with the basalts of the oceanic crust. This process has implications for heat flow, for ocean water chemistry and is a major geochemical source and sink of elements. Thus large amounts of uranium are adsorbed on altered oceanic basalt. A number of studies have addressed these problems. Of particular interest here is the fate of the rare earth elements. Studies by Staudigel & Hart [42] indicate that alterations of basaltic glass to palagonite does not alter REE patterns, in contrast to studies for example by Ludden & Thompson [43] (see Section

2.4.4). The REE compositions of downflowing sea water is rapidly equilibrated with the REE in the basalts. 'Thus a significant change of REE patterns or REE isotopic systematics of the bulk crust during alteration appears unlikely' [42]. The average depth of sea water penetration into the oceanic crust is quite variable, but probably much less than 5 km as assumed by Gregory & Taylor [44]. Ito & Clayton [45] found that abundant sea water penetration, giving water/rock mass ratios greater than one, is restricted to the upper part of the plutonic layer.

Ophiolite complexes have been studied extensively as possible exposed examples of the oceanic crust [46]. Commonly they are thinner (4 km) than the 5–7 km thicknesses of typical oceanic crust. Their chemistry may also be distinct. Thus Sm–Nd studies of the Troodos massif suggest an island-arc rather than ocean floor environment [47]. The differences in thickness may be accounted for by tectonic thinning during emplacement.

Table 11.6 gives our assessment for the composition of the basaltic layer (layer 3) of the oceanic crust [1] with the following additions and revisions. Data for Ge, Se, Pd, Ag, Cd, In, Sn, Sb, Te, Os, Ir, Au, Tl and Bi are from Wolf & Anders [48] and Hertogen *et al.* [49]. Values for Rb, Cs, Ba are from Hoffman & White [19]. Data for Ru, Rh, Re and Pt come from Chou *et al.* [22]. The abundances for U and Th are from [50] based on K/U=12 500. The lead values are from Tatsumoto [51]. The composition of the oceanic crust as a

Table 11.6. Composition of the oceanic crust. (See text for sources.)

	%		Norm
SiO_2	49.5	Or	0.89
TiO_2	1.5	Ab	23.7
Al_2O_3	16.0	An	30.7
FeO	10.5	Di	21.0
MgO	7.7	Hy	5.90
CaO	11.3	Ol	14.6
Na_2O	2.8	Il	2.6
K_2O	0.15		
Σ	99.5		

Li	10 ppm	Cu	86 ppm	In	72 ppb	Tm	0.54 ppm
Be	0.5 ppm	Zn	85 ppm	Sn	1.4 ppm	Yb	5.1 ppm
B	4 ppm	Ga	17 ppm	Sb	17 ppb	Lu	0.56 ppm
Na	2.08%	Ge	1.5 ppm	Te	3 ppb	Hf	2.5 ppm
Mg	4.64%	As	1.0 ppm	Cs	30 ppb	Ta	0.3 ppm
Al	8.47%	Se	160 ppb	Ba	25 ppm	W	0.5 ppm
Si	23.1%	Rb	2.2 ppm	La	3.7 ppm	Re	0.9 ppb
K	1250 ppm	Sr	130 ppm	Ce	11.5 ppm	Os	<0.004 ppb
Ca	8.08%	Y	32 ppm	Pr	1.8 ppm	Ir	0.02 ppb
Sc	38 ppm	Zr	80 ppm	Nd	10.0 ppm	Pt	2.3 ppb
Ti	0.90%	Nb	2.2 ppm	Sm	3.3 ppm	Au	0.23 ppb
V	250 ppm	Mo	1.0 ppm	Eu	1.3 ppm	Hg	20 ppb
Cr	270 ppm	Ru	1.0 ppb	Gd	4.6 ppm	Tl	12 ppb
Mn	1000 ppm	Rh	0.2 ppb	Tb	0.87 ppm	Pb	0.8 ppm
Fe	8.16%	Pd	<0.2 ppb	Dy	5.7 ppm	Bi	7 ppb
Co	47 ppm	Ag	26 ppb	Ho	1.3 ppm	Th	0.22 ppm
Ni	135 ppm	Cd	130 ppb	Er	3.7 ppm	U	0.10 ppm

whole may be estimated by combining appropriate amounts of ocean floor basalt, pelagic clay (Table 2.3) and ocean island basalts. The mixture is left to individual taste.

Over half the total heat loss from the earth is through the Cenozoic oceanic crust, which covers only 31% of the earth's surface [52, 53]. Typical heat flow values are 250 mW m^{-2} for oceanic crust less than 4 m.y. old. This value declines sharply to 46 mW m^{-2} for oceanic crust 120–140 m.y. old and an equilibrium heat flow of 38 mW m^{-2} is reached when the oceanic lithosphere has cooled for 200 m.y.

11.6   Summary

1   The Earth accreted from a series of planetesimals in which the volatile/refractory element ratios and metal/silicate/sulphide equilibrium were already established.

2   The composition of the primitive terrestrial mantle (=present mantle plus crust) is estimated mainly from cosmochemical data. In comparison with primitive solar nebula abundances it is depleted in volatile, chalcophile and siderophile elements.

3   The upper and lower mantles are different in composition, with the lower mantle having lower Mg/Si ratios than the upper mantle.

4   The extraction of the continental crust from the mantle has affected 30–50% of the mantle.

5   Recycling of continental crustal material by subduction in island-arc environments and eruption of calc-alkaline lavas appears to be sporadic and restricted to less than a few per cent (<3%).

6   Table 11.6 provides our estimate for the composition of the basaltic fraction of the oceanic crust.

Notes and references

1 Taylor, S.R. (1982) *Planetary Science: A Lunar Perspective*. LPI Houston, 481 pp. This source also provides detailed references for much of the following discussion in this section.
2 Anders, E. & Ebihara, M. (1982) Solar system abundances of the elements. *GCA*, **46**, 2363.
3 Taylor, S.R. & McLennan, S.M. (1981) The composition and evolution of the continental crust: rare earth element evidence from sedimentary rocks. *Phil. Trans. Roy. Soc.*, **A301**, 381.
4 Landolt-Bornstein (1981) Group VI, Vol. 2. Abundances of elements in the solar system, 257.
5 Evensen, M.N. *et al.* (1978) Rare-earth abundances in chondritic meteorites. *GCA*, **42**, 1203.
6 The matrix of chondrites is more likely to preserve the primitive solar nebula abundance levels for volatile elements than the chondrules, formed by high temperature processes. Curtis, D. *et al.* (1980) *GCA*, **44**, 1945.
7 Morrison, D. (ed.) (1982) *Satellites of Jupiter*. University of Arizona Press.
8 Wetherill, G.W. (1980) Formation of the terrestrial planets. *Ann. Rev. Astron. Astrophys.*, **18**, 77.
9 Ahrens, T.J. (1979) Equations of state of iron sulfide and constraints on the sulfur content of the Earth. *JGR*, **84**, 985.
10 Ringwood, A.E. (1979) *The Origin of the Earth and Moon*. Springer-Verlag.
11 Ringwood, A.E. (1975) *Composition and Petrology of the Earth's Mantle*. McGraw Hill.
12 Jagoutz, E. *et al.* (1979) The abundances of major, minor and trace elements in the earth's mantle as derived from primitive ultramafic nodules. *PLC*, **10**, 2031.

13 Liu, L. (1979) On the 650 km seismic discontinuity. *EPSL*, **42**, 202; (1982) Chemical inhomogeneity of the mantle. *GRL*, **9**, 124.
14 Spohn, T. & Schubert, G. (1982) Models of mantle convection and the removal of heat from the Earth's interior. *JGR*, **87**, 4682.
15 Liu, L. & Bassett, W.A. (1985) *Elements, Oxides and Silicates: High Pressure Phases*. Oxford University Press.
16 Kerridge, J.F. (1979) Fractionation of refractory lithophile elements among chondritic meteorites. *PLC*, **10**, 989.
17 Anderson, D.L. (1983) Chemical composition of the mantle. *PLC*, **14**, *JGR*, **88**, B41.
18 This is in contrast to the view expressed by Shakespeare (1604) who considered that the whole mantle was composed of olivine. 'If heaven would make me another world of one entire and perfect chrysolite, I'd not have sold her for it.' Shakespeare, W. (1604) *The Tragedy of Othello, The Moor of Venice*, Act V, Scene II, lines 171–174. Chrysolite is an archaic term for olivine.
19 Hoffmann, A.W. & White, W.M. (1983). Ba, Rb and Cs in the Earth's mantle. *Z. Naturforsch.*, **38a**, 256.
20 Dreibus, G. *et al.* (1977) The bulk composition of the moon and the eucrite parent body. *PLC*, **8**, 211.
21 Newsom, H. & Drake, M.J. (1984) The depletion of siderophile elements in the Earth's mantle: new evidence from molybdenum and tungsten. *LPS*, **XV**, 607.
22 Chou, C-L. *et al.* (1983) Siderophile trace elements in the Earth's oceanic crust and upper mantle. *PLC*, **13**, *JGR*, **88**, A507.
23 Morgan, J.W. *et al.* (1980) Composition of the Earth's upper mantle. II. Volatile trace elements. *PLC*, **11**, 213.
24 Sun, S-S. (1982) Chemical composition and origin of the earth's primitive mantle. *GCA*, **46**, 179.
25 Shannon, R.D. (1976) Revised effective ionic radii and systematic studies of interatomic distances in halides and chalcogenides. *Acta Cryst.*, **A32**, 751.
26 O'Nions, R.K. *et al.* (1979) Geochemical modelling of mantle differentiation and crustal growth. *JGR*, **84**, 6091.
27 Jacobsen, S.B. & Wasserburg, G.J. (1979) The mean age of mantle and crustal reservoirs. *JGR*, **84**, 7411; (1981) Transport models for crust and mantle evolution. *Tectonophysics*, **75**, 163.
28 Recent reviews on this topic have been given by several authors. Arculus, R.J. (1981) Island-arc magmatism in relation to the evolution of the crust and mantle. *Tectonophysics*, **75**, 113; Gill, J.B. (1981) *Orogenic Andesites and Plate Tectonics*. Springer-Verlag; Thorpe, R.S. (ed.) *Orogenic Andesites and Associated Rocks*. Wiley; Perfit, M.R. *et al.* (1980) Chemical characteristics of island-arc basalts: implications for mantle sources. *Chem. Geol.*, **30**, 227.
29 Church, S.E. (1973) Limits of sediment involvement in the genesis of orogenic volcanic rocks. *CMP*, **39**, 17.
30 Kay, R.W. (1980) Volcanic arc magmas: implications of a melting-mixing model for element recycling in the crust—upper mantle system. *J. Geol.*, **88**, 497.
31 Arculus, R.J. & Johnson, R.W. (1981) Criticism of generalized models for the magmatic evolution of arc-trench systems. *EPSL*, **39**, 118.
32 McLennan, S.M. & Taylor, S.R. (1981) Role of subducted sediments in island-arc magmatism: Constraints from REE patterns. *EPSL*, **54**, 423.
33 Meijer, A.(1976) Pb and Sr isotopic data bearing on the origin of volcanic rocks from the Mariana island arc system. *GSA Bull.*, **87**, 1358; Kay, R.W. *et al.* (1978) Pb and Sr isotopes in volcanic rocks from the Aleutian Islands and Pribilof Islands, Alaska. *GCA*, **42**, 203; Sun, S-S. (1980) Lead isotopic study of young volcanic rocks from mid-ocean ridges, ocean islands and island arcs. *Phil. Trans. Roy. Soc.*, **A297**, 409.
34 Whitford, D.J. & Jezek, P. (1979) Origin of late-Cenozoic lavas from the Banda arc, Indonesia. *CMP*, **68**, 141; Whitford, D.J. *et al.* (1981) Neodymium isotopic composition of Quaternary island arc lavas from Indonesia. *GCA*, **45**, 989; McCulloch, M.T., pers. comm.
35 Margaritz, M. *et al.* (1978) Oxygen isotopes and the origin of high $^{87}Sr/^{86}Sr$ andesites. *EPSL*, **40**, 220.
36 White, W.M. & Patchett, J. (1984) Hf–Nd–Sr isotopes and incompatible element abundances in island arcs: implications for magma origins and crust–mantle evolution. *EPSL*, **67**, 167.
37 Patchett, P.J. *et al.* (1984) Hafnium/rare-earth element fractionation in the sedimentary system and crustal recycling into the earth's mantle. *EPSL*, **69**, 365.

38 Brown, L. *et al.* (1981) Beryllium-10 in continental sediments. *EPSL*, **55**, 370.
39 Brown, L. *et al.* (1982) ^{10}Be in island-arc volcanoes and implications for subduction. *Nature*, **299**, 718.
40 Newman, S. *et al.* (1984) ^{230}Th–^{238}U disequilibrium in island arcs: evidence from the Aleutians and the Marianas. *Nature*, **308**, 268.
41 Anderson, R.N. *et al.* (1982) DSDP hole 504B, the first reference section over 1 km through Layer 2 of the oceanic crust. *Nature*, **300**, 589.
42 Staudigel, H. & Hart, S.R. (1983) Alteration of basaltic glass: Mechanisms and significance for the oceanic seawater budget. *GCA*, **47**, 337.
43 Ludden, J.N. & Thompson, G. (1979) An evaluation of the behaviour of the rare earth elements during the weathering of sea-floor basalt. *EPSL*, **43**, 85.
44 Gregory, R.T. & Taylor, H.P. (1981) An oxygen isotope profile in a section of Cretaceous oceanic crust. *JGR*, **86**, 2737.
45 Ito, E. & Clayton, R.N. (1983) Submarine metamorphism of gabbros from the Mid-Cayman Rise: An oxygen isotope study. *GCA*, **47**, 535.
46 Moores, E.M. (1982) Origin and emplacement of ophiolites. *RGSP*, **20**, 735.
47 McCulloch, M.T. & Cameron, W. (1983) Nd–Sr isotopic studies of primitive lavas from the Troodos ophiolite, Cyprus: Evidence for a subduction-related setting. *Geology*, **11**, 727.
48 Wolf, R. & Anders, E. (1980) Moon and Earth: compositional differences inferred from siderophiles, volatiles and alkalis in basalts. *GCA*, **44**, 2111.
49 Hertogen, J. *et al.* (1980) Trace elements in ocean ridge basalt glasses: implications for fractionations during mantle evolution and petrogenesis. *GCA*, **44**, 2125.
50 Jochum, K.P. *et al.* (1983) K, U and Th in mid-ocean ridge basalt glasses and heat production, K/U and K/Rb in the mantle. *Nature*, **306**, 431.
51 Tatsumoto, M. (1978) Isotopic composition of lead in oceanic basalt and its implication to mantle evolution. *EPSL*, **38**, 63.
52 Pollack, H.N. (1980) The heat flow from the Earth: a review. *In:* Davies, P.A. & Runcorn, S.K. (eds), *Mechanisms of Continental Drift and Plate Tectonics*. Academic Press. 183.
53 Sclater, J.G. *et al.* (1980) The heat flow through oceanic and continental crust and the heat loss from the Earth. *RGSP*, **18**, 269.

12

Early Planetary Crusts

12.1 The 800 million year gap

The observable geological record ceases at about 3.8 Ae on the Earth, although zircon grains with Pb ages of 4.1 Ae imply the existence of older crustal rocks [1]. Speculation concerning events before that time has been widespread, and a wide range of possible compositions and processes have been suggested. A long-standing favourite has been an early world-encircling sialic crust. Other candidates have included basaltic crusts, while anorthositic crusts, from the lunar analogy, enjoyed a brief popularity [2]. Additions from outer space have also been popular and have come in two basic models. The first invokes direct plastering of a silica-rich material [3], while the second involved giant basin-forming impacts which dug the ocean basins. The continents were formed by the ejecta piled up around the rims of the basins [4]. Many of these models have not survived scientific testing, so that it is not profitable to discuss them at length. In this chapter we examine the record from other planetary bodies in an attempt to gain some insights into events in the period from the isolation of the solar nebula at 4.56 Ae to the beginning of the geological record on the earth at about 3.8 Ae. Much of the discussion in the following sections is based on the review of planetary science by Taylor [5], which should be consulted for detailed references to the lunar and planetary literature. From this perspective we conclude with some speculations on the course of events during the first 600–800 m.y. years of Earth history.

12.2 Primary and secondary crusts

It is important at the outset to make a fundamental distinction between planetary crusts of primary and secondary origin. Most of the terrestrial planets have acquired a surface composition which is distinct from that of their interior, or from their bulk composition. When the lunar samples were examined, the crustal composition of the moon was found to be so highly differentiated that models involving heterogeneous accretion were invoked, plastering on a layer of refractory material as the last episode of formation. Such models were in direct contrast to earlier views that the Moon might be a primitive object [6]. The surfaces of Mars, Mercury and Venus likewise turned out to be different from reasonable estimates of bulk planetary compositions. The Galilean satellites of Jupiter present four differing crustal types for our inspection, from one of the most ancient surfaces in the Solar System on Callisto, to surfaces so fresh that few craters are seen (Io and Europa).

It has now become clear that surficial crusts on planets may arise in two ways, either as a consequence of early melting and differentiation, or by

Fig. 12.1. The contrast between primary and secondary crusts is well shown in this view of farside lunar highlands and maria. The large circular crater, filled with mare basalt, is Thomson (112 km diameter) in the north-east sector of Mare Ingenii (370 km diameter, 34°S 164°E). The large crater in the right foreground is Zelinskiy (54 km diameter). The stratigraphic sequence, from oldest to youngest, is (1) formation of highland crust, (2) excavation of Ingenii basin, (3) formation of Thomson crater, (4) formation of Zelinskiy, (5) flooding of Ingenii basin and Thomson crater with mare basalt, (6) production of small craters on mare surface including a probable chain of secondary craters. There is no evidence of expansion on the ancient mare basalt surfaces which are over 3 Ae in age. (NASA AS-15-87-11724.)

derivation from the planetary mantles by partial melting long after accretion, this time being measured in billions of years. The highland crust of the Moon represents a well understood example of the first type, while the basaltic lunar maria (Fig. 12.1) and continental crust of the Earth are familiar examples of the second process [7].

12.3 The lunar crust

The geophysical data indicate an average thickness of 73 km, ranging from a nearside value of 64 km to a farside averaging 80 km. This thicker farside crust remains the simplest explanation for the centre of figure–centre of mass offset in the Moon of about 2 km and is also consistent with the very limited amount of mare basalt on the farside (Fig. 12.2). The question of the variation in composition of the crust with depth has been much debated, but the overall density and moment of inertia constraints call for a relatively low density crust. The principal effect of this is reflected by the Al_2O_3 content. If the overall crustal composition has a lower alumina content (e.g. 20% Al_2O_3) then

Fig. 12.2. The lunar highland crust. A farside view showing heavily cratered terrain. The scarcity of mare basalt, in contrast to the nearside, is attributed to the greater thickness of the feldspathic highland crust on the farside. Mare Crisium is the circular mare on the NW limb. The other patches of basalt are Mare Smythii and Marginis. (NASA Apollo 16 metric frame 3023.)

it must be thicker; conversely if it has a higher content (e.g. 30% Al_2O_3) then it must be thinner to accommodate the geophysical constraints.

The origin of this low density crust is now generally accepted to be by crystallization and flotation in a 'magma ocean'. Geochemical balance calculations indicate that about 40% of the total lunar budget of Al_2O_3 has been concentrated in the highland crust, a little over 10% of lunar volume. Similar amounts of the trace REE, europium, have also been concentrated. The geochemical evidence thus demands the operation of crystal-liquid fractionation for large volumes of the Moon. Whether this occurred in a magma ocean or in a series of smaller regions is not immediately apparent from the chemical data, although the evidence suggests whole Moon involvement, and a magma ocean is assumed for the remainder of the discussion.

How deep was the magma ocean, and how much of the Moon was involved in the production of the lunar highland crust? A minimum volume of 40% may be calculated, assuming total extraction of Al_2O_3 and its flotation as a plagioclase-rich crust. Estimates have called from depths as low as 200 km, up to total Moon melting. The geochemical estimates are consistent with whole-Moon melting, since appropriate amounts of Al_2O_3, for example, have to be retained in the interior. Mare basalts, derived by partial melting from such regions, contain typically about 8–10% Al_2O_3.

The composition of the lunar highland crust can be understood in terms of three principal components. The first of these is composed mainly of plagioclase feldspar (anorthite) with some trapped interstitial minerals. It begins to crystallize in the magma ocean when sufficient olivine and orthopyroxene have crystallized to build up the Ca and Al contents to precipitate plagioclase. This floats in the anhydrous lunar magma (no water, even at ppb levels, has been identified as indigenous to the Moon) and comprises the bulk of the lunar highland crust.

The crystallization of the magma ocean continues with the formation of zoned cumulate source regions depleted in europium. The subsequent melting and extrusion of mare basaltic lavas from these regions bears the imprint of this europium depletion. The final crystallization of most of the magma ocean results in a residuum, enriched in incompatible elements (e.g. K, REE, P, Zr, Nb, U, Th, etc.). Being of low density, it pervades the crust, with which it is mixed intimately by the continuing massive meteoritic bombardment. This component, known as KREEP, has uniform Sm–Nd isotopic systematics, and is additional evidence favouring derivation from a Moon-wide melting event, rather than from a series of localized intrusions. A third component, also intimately intermingled with the others, is the so-called Mg-suite. This comprises gabbroic rocks which have apparently crystallized separately from the main magma ocean, and may represent a series of intrusive episodes. The nearest terrestrial analogue to the lunar highland crust is the Stillwater layered intrusion, which has analogues both of the lunar anorthosites and the Mg-suite.

The crystallization of the magma ocean and the formation of the crust were complex in detail, although the general outlines are clear. The establishment of the crust was greatly complicated by the raining meteorite bombardment [8]. Forty-three major basins have been identified and most estimates of the flux rate propose many more which have now been totally obliterated. This question of whether saturation was achieved is still debatable, but a consensus appears to be emerging that the oldest observable lunar highland craters and basins represent a saturated surface. This conclusion is reinforced by the identification of many multi-ring basins on Mercury [9] so that large basin collisions were not unique to the Moon. The depth of excavation of the large basins is most probably in the range 30–60 km. The projectiles responsible for the multi-ring basins are also about this diameter.

How long did the magma ocean take to crystallize? Many dates obtained for lunar highland rocks have been reset by the large basin forming collisions. There is a tendency for the crustal ages to peak at about 3.82 Ae, the date of the Imbrium collision. This event, followed shortly thereafter by the formation of the Orientale basin, marks the steep decline in the flux of large objects striking the Moon. Greater ages are observed among the highland samples, but those older than about 4.2 Ae are subject to increasing uncertainty in view of the massive bombardment history. It is possible that the crust was not able to become established as a unit impervious to destruction until about 4.2 Ae [8]. The time of crystallization given by the lead isotopic systematics indicate an age of 4.47 Ga, which marks the closure of the isotope systems in the source regions of the mare basalts. These are formed as complementary differentiates during the crystallization of the lunar highland crust. The Rb–Sr, Sm–Nd and Pb isotopic systematics indicate closure at about 4.4 Ae. The mare basalts, characterized by depletion in Eu, were derived by partial melting from these regions at later periods from about 4.2 Ae, down to ages somewhat younger than 3.0 Ae. It thus appears from the isotopic systematics that the differentiation which produced the lunar crust was effectively completed by about 4.4 Ae. The lunar highland crust accordingly results from an initial melting and planetary differentiation.

Following the formation of that crust, the slow accumulation of radiogenic heat initiated partial melting in the deep interior (200–400 km), and the mare basalt lavas were erupted from about 4.2 Ae to 2.5 Ae. Their volume is small, less than around 0.1% of lunar volume, so that no massive heating of the lunar interior is required. Indeed, more than 20 differing types of basalt have been erupted, indicating the local variations and heterogeneous nature of the lunar interior. This activity ceased by about 2.5 Ae.

This final stage of lunar history resembles the outpouring of oceanic basalts on the Earth, and represents a secondary stage of crustal formation. The production of the continental crust on the Earth is derived by further stages of partial melting of mantle derived material, or, in the case of the Archean crust, from mixtures of basaltic material and felsic material derived by partial melting of basalts. Accordingly, the formation of the lunar crust bears no genetic relationship to that of the continental crust of the Earth, and it is pointless to pursue analogies based on any supposed resemblance.

Absence of an early crust on the Earth, in contrast to the Moon, is due to differences in the crystallization history of the terrestrial mantle compared with that of the lunar magma ocean. The lunar crust, although greatly complicated by the details of its crystallization history during the raining meteoritic bombardment, results essentially from a single stage of melting. The upper continental crust of the Earth is the product of at least three successive partial melting events.

12.4 The Mercurian crust

The accretion of Mercury must have been closely followed by differentiation [5]. Two principal observations support this conclusion: (a) the high density of Mercury (5.44 g cm^{-3}), and (b) the presence of silicate material and a lunar-like topography at the surface. These observations lead to the conclusion that the planet has a high iron content, segregated into a core about 0.75–0.80 of Mercurian radius, overlain by a silicate crust. The heavy cratering of the crust was early, by analogy with the Moon, and the crust must have been thick enough and cold enough to preserve the record of this bombardment well before 4.0 Ae.

No chemical data are available for the Mercurian crust. The reflectance spectra indicate a silicate surface much resembling that of the Apollo 16 soils, possibly indicating a composition containing less than about 5% FeO.

The resemblance between the reflectance spectra of the lunar highlands and of Mercury may mean that the crust of that planet is of anorthositic gabbro composition. The remote-sensing data are integrated over the whole planetary surface, and resolution of individual areas, for example the smooth plains, is not possible. No spectra typical of mare basalts have been observed and the absence of contrast in albedo among the various crustal units on Mercury does not encourage the view that the plains units on Mercury are analogous to lunar maria. The preferred explanation is that they are derived from impact-produced ejecta from large basin-forming collisions, as is the case for the Cayley Plains on the Moon. However, it should be recalled that the Moon has a distinctly lower bulk density and hence a different composition and interior structure, so that the mantle of Mercury may not resemble that of the Moon, and lavas erupted from it may differ from lunar basalts in albedo.

The heavily cratered terrain bears a close resemblance to the lunar highlands crust. Nineteen multi-ring basins have been identified including the well-known Caloris basin on the photographed portion of Mercury. The total number of basins with diameters larger than 200 km probably exceeds 50, comparable with the number on the Moon. This observation suggests that basin production was continuous and widespread in the solar system rather than due to a late spike or cataclysm. By analogy with the Moon, the heavily cratered terrain is dated as older than about 4.0 Ae. Crater counting techniques give best estimates of 3.8 Ae for the Caloris basin and 3.9–4.1 Ae for the heavily cratered south polar uplands. These ages provide evidence that the massive cratering events persisted for several hundred million years after accretion, as occurred also on the Moon. The best assessment for the crust of Mercury is that it is derived, like the highland crust of the Moon, by crystallization following massive planetary melting. The compositional evidence suggests an anorthositic composition. This implies that Mercury is depleted in volatiles, a reasonable assumption, since only a dry anorthositic crust would float during initial melting.

Lobate scarps are unique to Mercury. They vary in length from 20 to 500 km and in height from a few hundred metres to about 3 km. They are reverse thrust faults, formed due to compressive stresses, and appear to have a rather uniform distribution over the planet. The decrease in estimates of the surface area, associated with these scarps, corresponds to a decrease in Mercurian radius by about 1–2 km. The lobate scarps appear to have formed relatively early in Mercurian history. They occur mainly on the intercrater plains and on the older parts of the heavily cratered terrain, cutting some craters and in turn having craters superimposed on them.

The origin of the lobate scarps is most generally ascribed to contraction of the 600 km-thick silicate mantle around a cooling and shrinking iron core. From the analogy of the heavily cratered terrain with that of the lunar highland surface, the scarps must have formed prior to 4.0 Ae and hence preserve a record that the radius of Mercury is essentially unchanged for that immense period of time. The wider significance is that the lack of evidence for any expansion on Mercury (and the Moon) places serious limits on models both for Earth expansion and variation of G with time (see Section 12.8) [10].

12.5 The Martian crust

Several distinct geological provinces occur on Mars [11], but crustal development may not have proceeded very far. We are slightly better informed about the surface composition than for Mercury. Taken at its face value, the surface composition is basaltic. The limited data for K/U ratios from the Russian gamma-ray experiments indicates values of about 3000, similar to the lunar values, and hence implying of a general depletion of volatile elements. However, if the differentiated meteorites with crystallization ages of 1.3 billion years (shergottites, nakhlites and chassignites) are derived from Mars [12], then their K/U ratios indicate values in excess of 10^4. Until this point is resolved, speculation about the bulk composition of Mars resembles the old debates over the existence of Martian canals, since the presently available data are inadequate to resolve the question.

Crustal development appears not to have proceeded past the stage of basaltic volcanism. Mars is dominated by two differing crustal terrains with a major boundary inclined at about 28 degrees to the equator. To the north are plains and volcanic ridges. To the south lies the oldest exposed surface on Mars, containing a large number of craters, typically in the size range 20–100 km diameter, with generally smooth floors. Intercrater plains are common.

The cratered terrain occupies one-half of the planetary surface, and represents the oldest visible crust (>4.0 Ae) (Fig. 12.3). Extensive plains units lie to the north. These are generally interpreted to be of volcanic origin. The chief evidence for this is the presence of scarps, which are probably flow fronts. Although these plains are often compared to the lunar maria, there are many distinctions. The Martian plains are light, not dark, and they are comparatively high-standing (perhaps equivalent to the terrestrial continental flood basalts). Volcanic constructional features are frequent, whereas they are rare on the lunar maria.

The plains are younger, on the basis of crater counting techniques, than the heavily cratered terrain. Their ages range from 3.6 Ae down to about 3 Ae. Younger regions such as the Tharsis Plateau, have ages from about 1.6 down to 0.3 Ae. All these ages are subject to assumptions about the cratering flux and have large uncertainties. The ages given here are the conventional estimates although some interpretations of the cratering record give much older ages, compressing most of Martian history into an early epoch. Volcanic activity appears to have been of long duration on Mars. The Tyrrhena Patera volcanoes, in Hesperium Planum, are dated at 3.1 Ae, the Elysium volcanics at 1.9 Ae, the three large shield volcanoes of Ascraeus Mons, Pavonis Mons and Arsia Mons at about 1.7 Ae, whereas Olympus Mons is dated at about 300 m.y. (range 100–700 m.y.). Olympus Mons is 550 km across and rises 21 km above its base. The central caldera is 70 km in diameter.

Other large volcanoes include Elysium Mons, 170 km in diameter and 14 km high, with a 12 km diameter caldera. Other rather lower volcanic constructs (e.g. Alba Patera) are of enormous extent, 1500–2000 km in diameter, although only about 3 km high. Extremely rapid effusion of lava seems to be required to account for this extraordinary volcano.

The most striking elevated feature on Mars is the Tharsis Plateau. Early

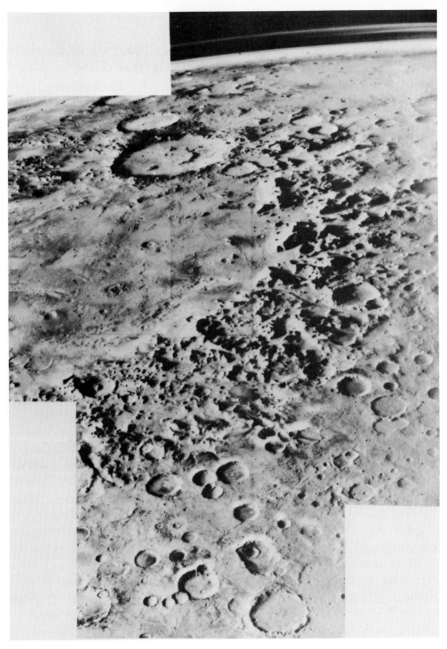

Fig. 12.3. View north-east across heavily cratered ancient crust on Mars. The large circular feature is the Argyre basin about 750 km in diameter, centred at 50°S 43°W. The large crater on the farside rim of Argyre is Galle, 210 km in diameter. Note the high cloud haze. (NASA SP 76 27774.)

estimates of 10 km elevation have been reduced to 3–5 km elevation by later earth-based radar measurements. Superimposed on the Tharsis plateau are the volcanic mountains of Arsia, Ascraeus, Olympus and Pavonis Mons. The plateau is perhaps primarily of volcanic construction, although apparently some more ancient terrain is included, so that possibly internal dynamic processes are also operating.

We know much more about Mars than Mercury, but many puzzles still remain. In the context of this book, we can summarize the crustal evolution of Mars as follows:

1 There is an ancient heavily cratered crust for which no direct compositional evidence is available.

2 Extensive basaltic volcanism has constructed plains, plateaux and circular volcanic mountains. The compositional evidence from the Viking landers, 4000 km apart, supports an overall basaltic composition for the crust. The most probable scenario for Mars is that the crust is dominated by basalt produced by partial melting from the mantle. The heavily cratered terrain has not provided a separate decipherable signature observable in the Viking data so we do not know if an early crust of differing composition was present.

There is no evidence that the early cratered terrain is anorthositic, similar to the Moon. This must be considered unlikely since Mars has at least a small volatile content, so that plagioclase would not float in a wet melt. There is no real evidence of any sialic crust. Crustal processes on Mars do not appear to have progressed beyond the primitive stage of basaltic lava effusion. The ancient cratered terrain may be the best analogy for a primitive crust on the Earth and attempts to secure its composition should be given high priority. On the flimsy evidence available, it is likely to be basaltic.

12.6 Venus

Two large 'continent-sized' areas occur on Venus [5, 13]. The largest, comparable in size to Australia, is Aphrodite Terra, elevated 2–5 km above the rolling plains which make up so much of the surface. The other, Ishtar Terra, is about the size of Antarctica, and contains the Maxwell Mountains, which rise to 11 km. Other smaller high relief areas (e.g. Beta Regio) are interpreted as shield volcanoes. Otherwise, the surface has little relief, about 60% being within 500 m of the average lowland elevation (see Fig. 1.1). The composition of the surface at the Venera 13 and 14 sites is basaltic, the latter analysis resembling MORB, whereas the Venera 13 data resembles terrestrial alkali basalts [14]. This latter observation is critical to the interpretation of the Russian gamma-ray data from the Venera 8, 9 and 10 landers. High values for K, U and Th, resembling the abundances in granitic rocks, were obtained by Venera 8, and occasionally are cited as evidence for the possible presence of granite (and hence continental crust) on Venus. However, it is more probable that the high values reported are due to extensive eruptions of incompatible element enriched alkali basalts with 4% potassium, as measured from the Venera 13 XRF experiment. The large high-standing massifs, on this hypothesis, result from basaltic volcanic activity like Olympus Mons on Mars. The high resolution orbital data from Venera 15 and 16 greatly strengthens the interpretation of the surface as basaltic. From the observed meteorite impact craters, the general age of the surface appears to be about one billion years.

Lightning sources correlate with Beta Regio and Aphrodite [15] and the presence of large excess concentrations of SO_2 in the Venusian atmosphere [16] accords with the notion that volcanoes are currently active. Consequently, it appears that typical continental crust may well be absent on Venus as it probably is on Mercury and Mars. An alternative scenario is that the continent-sized masses are horsts, and owe their elevation to internal dynamics rather than to chemical differences. If such is the case, they are probably also basaltic in composition. The topography of Venus shows a very close correspondence with gravity, indicating that the surface features are supported against subsurface creep. The general impression from the Russian chemical data, and the Venera 15 and 16 orbital radar topography [17], is of a planetary surface dominated by basaltic volcanism, intermediate in planetary evolution between Mars and the Earth. Venus is thus not a twin planet to the Earth.

In summary, the planets in the inner solar system, except the Earth, do not appear to have produced crustal material comparable to that of terrestrial continents.

12.7 The Galilean and Saturnian satellites

All these bodies are of low density and most exhibit crusts of water ice [18]. Most information is available for the four Galilean satellites of Jupiter. The outermost, Callisto, has the most heavily cratered crust yet observed in the solar system (Fig. 12.4). The crust is probably composed of water ice, as shown by the high albedo of fresh crater ejecta. Callisto (radius=2410 km) is nearly as large as Mercury (r=2439 km) but has a much lower density (1.83 g cm^{-3}) compared with Mercury (5.435 g cm^{-3}). Some planetary melting and transport of water to the exterior has probably occurred. The initial K, U and Th content in the silicate fraction, if resembling CI values, is adequate to ensure melting and outward transport of water. These icy crusts thus form a separate planetary crustal type, due to early melting of a low-melting point phase.

Ganymede, the third of the four Galilean satellites of Jupiter, with a radius of 2638 km, is a little larger than Mercury and is the largest satellite in the solar system. It has a density of 1.99 g cm^{-3}, consistent with a silicate–ice mixture. Areas of an older cratered crusts are separated by younger grooved terrain. The most reasonable interpretation is that the older crust has been split, and new material injected from below. Expansion on such a planet can occur by the following mechanism. A mixture of silicate and ice accretes and the ice deep within the planet undergoes polymorphic change to higher density forms with densities ranging from 1.16 to 1.66 g cm^{-3}. The initial planetary crust is heavily cratered at times earlier than 4000 m.y. Heating on a longer time-scale due to the radioactive elements K, U and Th present in the silicates causes melting in the ice in the interior. The water migrates outward. The change from high density ice to lower density water provides an expansion of the planet of up to 5–7%. The water refreezes near the surface in the lower density Ice I polymorph, after disrupting the older frozen crust. Thus we find some evidence for minor expansion in Ganymede. This is explicable by the

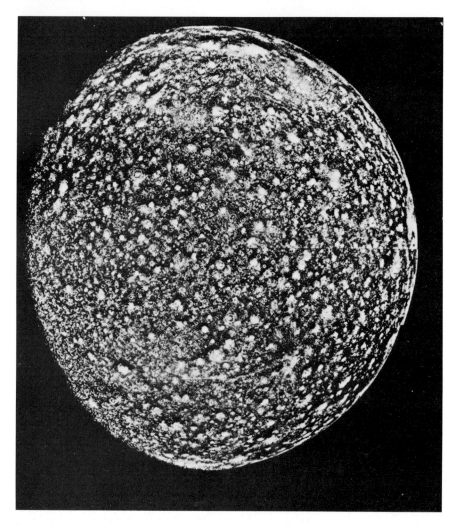

Fig. 12.4. Callisto (r=2140 km), the outermost Galilean satellite of Jupiter, viewed from a distance of 390 000 km (close to Earth–Moon distance), showing a crater saturated surface. The Asgard basin is near the upper right limb. (NASA.)

known properties of ice and it is not necessary to invoke any more mysterious mechanism to account for the expansion [18].

Europa (r=1536 km) is a little smaller than the Moon. It has a freshly resurfaced icy crust, with no impact craters. Tidal heating effects from Jupiter are probably remelting the icy crust and erasing the cratering record.

There is no sign of ice on Io (r=1816 km) which is a little larger than the Moon (r=1738 km). The crust is being covered rather rapidly by volcanic plume ejecta and sulphur lava flows. Current activity on Io resurfaced an area of 10^4 km^2 around the Pele Vent in the four months between the successive Voyager missions [19]. Sodium ions are being removed and form a torus around Jupiter. Ancient high mountains (9 km high) exist on Io and may be

Fig. 12.5. The icy Saturnian satellite Tethys (r=525 km) showing craters down to about 5 km in diameter, viewed from a distance of 282 000 km. The icy surface records a complex history, with a heavily cratered region (top right) separated from a more lightly cratered region (lower right). Note the prevalence of central peak craters. The large crater (upper right) is close to the large trench system which extends nearly three-quarters of the circumference of the satellite. This trench system may have formed from expansion of Tethys as a warmer interior froze, rifting the outer solid crust. (NASA P.24065 BW.)

remnants of an early crust. The current volcanism, due to partial melting within the planet, is analogous to basaltic volcanism on other planets, and represents an extreme case of planetary volcanism.

The satellites of Saturn display icy crusts. A typical example, Tethys [20] (r=535 km), is shown in Fig. 12.5. Like Ganymede, this satellite shows some evidence of crustal fracturing and slight expansion due to ice crystallization.

Enceladus, a similar small satellite of Saturn (r=250 km) also has planetary wide trenches. Their individual cratering histories are of much interest, but cannot be discussed in the present context.

In summary, the outer satellites apparently produce icy crusts rather rapidly due to radiogenic heating, but planetary differentiation does not proceed beyond this point, except in the special case of Io, due to a superimposed heat source.

12.8   Planetary constraints on expanding Earth hypotheses

Hypotheses which attempt to explain the present distribution of the continental crust by appealing to massive changes in the radius of the Earth continue to appear in the literature [e.g. 21]. Commonly they assume fragmentation of a world encircling sialic shell, and involve an increase in the radius of the Earth by a factor of two in times ranging from the Permian to the Proterozoic. Tests from palaeomagnetic data do not support such changes in the palaeoradius of the Earth [10].

The progress of science depends on the erection of testable hypotheses, but such ideas are inherently difficult to examine by the study of structural features on the surface of the Earth. This is mobile on time-scales of 100–200 m.y. What evidence exists elsewhere in the solar system? Unless the Earth possesses unique properties, it is reasonable to suppose that other planetary bodies should show evidence of massive changes in radius if the hypothesis is correct [22].

The Moon (Section 12.3) provides us with a dated surface of great antiquity (Fig. 12.1). Only very mild tectonic forces have affected the basaltic mare surfaces, producing wrinkle ridges and rilles. If these are due to planetary wide, rather than to local, stresses then they involve radius changes due to thermal expansion and contraction of about 1 km (lunar radius=1738 km). There is no sign of distortion on the older highland crust, much battered by the meteorite bombardment. The 'lunar grid', occasionally cited as due to internal forces, results from the patterns produced by overlapping ejecta blankets from the large impact basins. In effect, the Moon presents us with a frozen fossil surface, unchanged in major outline for nearly 4 Ae.

The surface of Mercury provides additional evidence. The lobate scarps (Section 12.4) indicate contraction, not expansion, at times before about 4 Ae. They are probably due to shrinking of a 600 km thick silicate shell around a cooling iron core. The contraction in Mercurian radius (r=2439 km) amounts to 1–2 km. No apparent change can be detected on the surface since about 4 Ae.

Mars has a much more complex geological history, with basaltic volcanism extending to rather recent times (Section 12.5). If the Tharsis plateau or bulge, which is the centre of the principal radial fracture system on Mars, is formed by internal dynamic forces, then a maximum planetary increase in radius of 19 km has occurred (Mars radius=3390 km). An alternative possibility is that the Tharsis plateau is of volcanic constructional origin.

The evidence for a few per cent expansion on Ganymede (Section 12.7) and probably also on the small icy satellites of Saturn such as Enceladus and Tethys, is attributed to the thermal properties of ice and has no larger cosmic significance.

Accordingly, this test of the expansion hypotheses is negative. Further tests from astronomy are equally discouraging [23]; one example must suffice, since it has geologically testable consequences. The luminosity of main sequence stars such as the sun is proportional to the third power of the mass. If the expansion hypotheses apply generally, then the sun is reduced to half its radius and one-eighth of its mass in the Permian or Proterozoic according to the particular model used. The resultant reduction in solar luminosity, by a factor of well in excess of 10^{-2}, results in the virtual extinction of solar radiation reaching the Earth. No such effects are discernible in the geological record. The existence of sedimentary rocks, and hence liquid water, is apparent back to 3.8 Ae. The occurrence of photosynthesizing algae is also documented well back into the Precambrian [24].

12.9 Meteorite flux and the cratering record

A vital question concerned with the formation of early planetary crusts before 4 Ae is the meteoritic flux. Intense bombardment will modify early crustal development, although it is unlikely that it would prevent the formation of a crust, or completely destroy one already formed. Nevertheless the impact of large planetesimals, creating basins thousands of kilometres in diameter, together with the multitude of smaller craters, will brecciate and modify any early crust. On the Earth the present flux will form a >20 km diameter crater about every 30 m.y. The present terrestrial and lunar production rate for craters greater than 20 km in diameter is about 0.35×10^{-14} km^2 a^{-1} [25] and is accounted for by the Earth-crossing population of Apollo asteroids. These rates persist back to between 3 and 4 Ae for the inner planets, but then show a steep rise indicating a massive increase in the flux rate. Figure 12.6 shows the cratering record for the Earth throughout time, based mostly on the lunar record. Similar curves can be constructed for Mars and Mercury. In the latter case, the existence of many multi-ring basins is strong evidence for a cratering history similar to that of the Moon. This indicates that impacts by basin-forming projectiles (typically >100 km diameter) were endemic in the early solar system, rendering less likely the concept of a late spike or cataclysm in the cratering record. Although there has been much dispute, the broadest view of the lunar and other planetary cratering history is consistent with a steep decline in the cratering flux, without a major late 'spike' in the record, although some small spikes may be superimposed on the main curve.

It is not clear whether the craters on old planetary surfaces represent a continuing production rate, or whether the heavily cratered surfaces represent saturation. In the former case, it will be possible to establish the cratering flux and history, and the relative ages of various surfaces. In the latter case, only minimum estimates can be given. From the analogies with the dated lunar highlands surface, and the similar surfaces on Mercury and Mars, it is

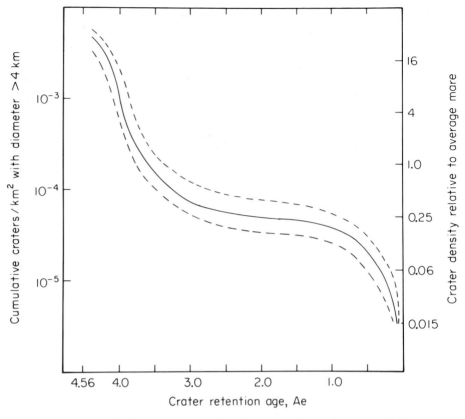

Fig. 12.6 The relationship between crater density and crater retention age for craters with diameters greater than 4 km for the Earth. The curve is based principally on the observed cratering history observed on the Moon. Note the steep increase in cratering between 3.5 and 4.5 Ae. (Adapted from [27].)

assumed that an early, pre-4 Ae bombardment struck the Earth. Calculations of the number of impacts and the resulting crater dimensions are subject to large uncertainties. If we extrapolate from the lunar highland record, making due allowance for size and gravitational differences, then one estimate predicts that at least 200 basins with diameters of several hundred kilometres formed on the Earth before 3.9 Ae [26]. In addition to these large multi-ring basins, a host of smaller craters will form. The effect of these can be studied from lunar examples (Fig. 12.2).

Figure 12.7 shows a view of the Orientale basin, 900 km in diameter. It was the last multi-ring basin to form on the Moon, at about 3.8 Ae. If the calculations noted above are correct, then perhaps 200 of these basins existed on the early crust of the Earth. It is difficult to evaluate the effects of this bombardment, but the survival of recognizable rock units is remote. Intense and repeated periods of brecciation and the production of extensive impact melts will take place. The magnitude of these effects, and the influence on the thermal regime of the upper mantle are all difficult to assess because of the large extrapolations required. Because of this, the petrological consequences of

Fig. 12.7. Mare Orientale, a classic example of a multi-ring impact basin. The diameter of the outer mountain ring (Montes Cordillera) is 920 km. The second ring (Montes Rook) has a diameter of 620 km, followed by the Inner Rook ring (480 km) and the Inner Basin Scarp (320 km diameter) which encloses the central dark area of mare basalt. Perhaps 200 such basins were formed on the Earth before 3.8 Ae, in addition to the many smaller craters. (NASA Orbiter IV 194 M.)

the massive bombardment for the early history of the crust are dimly understood, and a host of questions remain to be answered.

12.10 The early Earth

Based on the discussion given in previous sections, we can begin to construct a scenario. We assume that the accretion of the Earth was complete by 4.4 Ae, and that separation into mantle and core had occurred [28]. Large-scale early melting of the mantle was likely. The surface was still being pulverized by meteorite impact which continued down to about 4 Ae. Large-scale differentiation within the mantle, due to initial melting, may have occurred. In contrast to the situation on the Moon, no unambiguous evidence of this event seems to have survived. Initial melting and crystallization of the mantle, at depths below a few hundred kilometres, is unlikely to generate a less dense melt which will rise to form an initial crust [29]. No evidence exists for a primary sialic crust. An anorthositic crust if formed would not have floated in a wet terrestrial magma. There is no sign of the presence of primitive ^{87}Sr which would have been isolated in such anorthosites. The stability limit for plagioclase on the Earth is about 40 km, so that any convection below that depth will also hinder the formation of an anorthositic crust. Accordingly, there is no evidence for a lunar-style thick anorthositic crust. However, small amounts of anorthosite may well form as differentiates in a basaltic crust, and there appears to be some isotopic evidence from eclogites in support of this [30].

The most likely early crust on the Earth was basaltic or komatiitic, generated by partial melting within the solid mantle. As radiogenic heat builds up in the mantle, small amounts of basalt will appear at the surface, and will probably be subducted by primitive sea-floor spreading. We can estimate how extensive this crust was by examining the Sm–Nd evidence for the existence of depleted mantle in the oldest basalts. Production of more silicic crusts, as noted earlier, is a much less efficient process, although it had begun on a local scale by 3.8 Ae, as is shown by the rocks at Isua, and possible by 4.1 Ae, as suggested by the zircons in the Mt. Narryer quartzite [1]. It is, of course, difficult to decide whether the early crust of which the zircon grains are a remnant was acidic or basic. Zircon is much more characteristic of acidic than basic rocks, and the size of the grains is suggestive of an acid parent [31].

This survey of planetary crusts shows that continental crust of 'granitic' composition is apparently restricted to the Earth. The underlying cause for this unique behaviour is probably due to the presence of liquid water at the surface of the Earth. Water is essential for the production of granites which so restricts their occurrence in large quantities to the Earth [32]. It is interesting to note that water, essential for the evolution of life from the beginning, also plays a major role in the production of platforms above sea-level, on which life has evolved to its present diversity.

12.11 Summary

1 Crusts may form on planets either crystallizing directly from early planetary wide melting (primary crusts, e.g. lunar highlands) or by later partial melting of the solid mantle (secondary crusts, e.g. lunar maria, terrestrial crusts).

2 The lunar highland crust formed at about 4.4 Ae, principally by flotation of plagioclase from an anhydrous magma ocean, probably involving whole moon melting.

3 An early crust, analogous to the lunar highlands, formed on Mercury. The presence of lobate scarps indicates slight planetary contraction at about 4.0 Ae.

4 Extensive basaltic volcanism on Mars has extended to fairly recent times, but the southern hemisphere retains an ancient heavily cratered crust.

5 The crust on Venus appears to be dominated by basaltic volcanic activity including alkali-rich varieties. Volcanism is probably still active. There is no definitive evidence of an acidic crust.

6 The crusts on the Jovian and Saturnian satellites illustrate the evolution of rock-ice and icy bodies. Fracturing of early crusts due to mild thermal explansion is well shown on Ganymede, and probably accounts for the trench features on Enceladus and Tethys.

7 The surface features of the moon, Mercury and Mars provide us with no evidence to support hypotheses involving massive changes in Earth radius during geological time.

8 The present meteorite flux on the Earth is principally due to the

Earth-crossing Apollo asteroids. From analogy with the moon, an early (pre-3.8 Ae) heavy bombardment is inferred. The presence of many large multi-ring basins on Mercury indicates that the record on the moon is not unique. Probably at least 200 multi-ring basins with diameters of several hundred kilometres formed on the Earth before 3.8 Ae.

9 Speculations about the history of the Earth prior to 3.8 Ae suggest that although the Earth melted during accretion, no sialic or anorthositic crust developed. The most probable early crust was basaltic or komatiitic.

Notes and references

1 Froude, D.O. et al. (1983) Ion microprobe identification of 4100–4200 Myr-old terrestrial zircons. *Nature*, **304**, 616.
2 Shaw, D.M. (1976) Development of the early continental crust. In Windley, B.F. (ed.), *The Early History of the Earth*. Wiley. 33.
3 Donn, W.L. et al. (1965) On the early history of the Earth. *GSA Bull.*, **76**, 287.
4 Gilvarry, J.J. (1961) The origin of ocean basins and continents. *Nature*, **190**, 1048.
5 Taylor, S.R. (1982) *Planetary Science: A Lunar Perspective*. Lunar and Planetary Institute, Houston, Texas. 481 pp.
6 It is curious to record, in this book which is principally concerned with examining the terrestrial sedimentary record, that water-laid sedimentary rocks were proposed just before the Apollo 11 mission as candidates for the material filling the lunar maria (Gilvarry, J.J., (1968) Observational evidence for sedimentary rocks on the moon. *Nature*, **218**, 336).
7 Taylor, S.R. (1982) Lunar and terrestrial crusts: a contrast in origin and evolution. *PEPI*, **29**, 233.
8 Hartmann, W.K. (1980) Dropping stones in magma oceans: Effects of early lunar cratering. *In:* Papike, J.J. & Merrill, R.B. (eds), *Proc. Conf. Lunar Highlands Crust*. Pergamon Press. 155.
9 Spudis, P.D. & Strobell, M.E. (1984) New identification of ancient multi-ring basins on Mercury and implications for geologic evolution. *LPS*, **XV**, 814.
10 McElhinny, M.W. et al. (1978) Limits to the expansion of Earth, Moon, Mars and Mercury and to changes in the gravitational constant. *Nature*, **271**, 316.
11 Carr, M.H. (1981) *The Surface of Mars*. Yale University Press.
12 Ashwal, L.D. et al. (1982) SNC meteorites: Evidence against an asteroidal origin. *PLC*, **13**, *JGR*, **87**, A393; Shih, C-Y. et al. (1982) Chronology and petrogenesis of young achondrites, Shergotty, Zagami and ALHA 77005; late magmatism on a geologically active planet. *GCA*, **46**, 2323. For a contrary view, see Treimon, A.H. et al. (1984) The SNC/Mars connection: Geochemical inconsistencies. *LPS*, **XV**, 864.
13 Hunten, D.M. et al. (1983) *Venus*. University of Arizona Press. 1143 pp.
14 Surkov, Yu A. et al. (1983) Determination of the elemental composition of rocks on Venus by Venera 13 and 14. *PLC*, **13**, *JGR*, **88**, A481.
15 Scarf, F.L. & Russell, C.T. (1983) Lightning measurements from the Pioneer Venus data. *GRL*, **10**, 1192.
16 Esposito, L.W. (1984) Sulfur dioxide: Episodic injection shows evidence for active Venus volcanism. *Science*, **223**, 1072.
17 Barsukov, V.L., pers. comm.
18 Squires, S.W. (1980) Volume changes in Ganymede and Callisto and the origin of the grooved terrain. *GRL*, **7**, 593; Parmentier, E.M. et al. (1982) The tectonics of Ganymede. *Nature*, **295**, 290; Poirier, J.P. (1982) Rheology of ices: a key to the tectonics of the ice moons of Jupiter and Saturn. *Nature*, **299**, 683.
19 Morrison, D. (ed.) (1982) *Satellites of Jupiter*. University of Arizona Press.
20 Moore, J.M. & Ahern, J.L. (1983) The geology of Tethys. *PLC*, **13**, *JGR*, **88**, A577.
21 Carey, S.W. (1981) *The Expanding Earth: A Symposium*. University of Tasmania.
22 Taylor, S.R. (1981) Limits to Earth expansion from the surface features of the Moon, Mercury, Mars and Ganymede. *Ibid.*, 343.

23 Tryon, E.P. (1981) Cosmology and the expanding Earth hypothesis. *Ibid.*, 349. Recent ranging measurements to the Viking Landers on Mars indicate that the rate of change of G is effectively zero ($<0.2 \times 10^{-11}$ a^{-1}) thus removing the possibility that Earth expansion might be caused by variations in the gravitational constant. Hellings, R.W. *et al.* (1983) Experimental test of the variability of G using Viking Lander Ranging data. *Phys. Rev. Lett.*, **51**, 1609.
24 Cloud, P. (1978) *Cosmos, Earth and Man.* Yale University Press; (1980) Early biogeochemical systems. *In: Biogeochemistry of Ancient and Modern Environments.* Australian Academy of Science, Canberra. 7.
25. Grieve, R.A.F. & Dence, M.R. (1979) The terrestrial cratering record. II. The crater production rate. *Icarus*, **38**, 230.
26 Grieve, R.A.F. & Parmentier, E.M. (1984) Considerations of large scale impact and the early Earth. *In: Workshop on the Early Earth.* LPI. 23.
27 *Basaltic Volcanism on the Terrestrial Planets.* Pergamon Press (1981). 1081.
28 The early Earth is discussed in many papers. A useful reference is Smith, J.V. (1981) The first 800 million years of Earth's history. *Phil. Trans. Roy. Soc.*, **A301**, 401.
29 Ohtani, E. (1984) Generation of komatiite magma and gravitational differentiation in the deep upper mantle. *EPSL*, **67**, 261. Herzberg, C.T. (1984) Chemical stratification in the silicate Earth. *EPSL*, **67**, 249. Anderson, D.L. (1984) The Earth as a planet: paradigms and paradoxes. *Science*, **223**, 347.
30 Jagoutz, E. *et al.* (1984) Anorthositic oceanic crust in the Archean Earth. *LPS.*, **XV**, 395.
31 Possibly some clues may be gained from the trace element content of the zircons, although most elements which enter the zircon lattice are incompatible and so equally diagnostic of late basaltic differentiates as of acidic magmas. Scandium, however, forms an exception. It enters the zircon lattice, but is strongly fractionated into basic magmas. Accordingly, one would expect zircons derived from basic rocks to have higher Sc values than the values typically observed in zircons from granites.
32 Campbell, I.H. & Taylor, S.R. (1983) No water, no granites—no oceans, no continents. *GRL*, **10**, 1061.

Appendices

1   **Reference abbreviations**

AAPG : American Association of Petroleum Geologists
AGU: American Geophysical Union
Ap. J: Astrophysical Journal
AJS: American Journal of Science

BMRJ: Bureau of Mineral Resources (Australia) Journal

Chem. Geol.: Chemical Geology
CJES: Canadian Journal of Earth Sciences
CMP: Contributions to Mineralogy and Petrology

DSDP: Deep Sea Drilling Project

Econ. Geol.: Economic Geology
EPSL: Earth and Planetary Science Letters

GCA: Geochimica et Cosmochimica Acta
Geol. Mag.: Geological Magazine
Geol. Soc. Austr.: Geological Society of Australia
GRL: Geophysical Research Letters
GSA Bull.: Geological Society of America Bulletin
GSC: Geological Survey of Canada

IAEA: International Atomic Energy Agency
IGC: International Geological Congress

J. Geol.: Journal of Geology
J. geol. Soc. Lond.: Journal of the Geology Society of London
JGR: Journal of Geophysical Research
J. Petrol.: Journal of Petrology
J. R. Astr. Soc.: Journal of the Royal Astronomical Society
JSP: Journal of Sedimentary Petrology

LPI: Lunar and Planetary Institute, Houston, Texas
LPS XV: Lunar and Planetary Science XV LPI, Houston

N. Jb. Miner.: Neues Jahrbuch für Mineralogie
NZJGG: New Zealand Journal of Geology and Geophysics

PCE: Physics and Chemistry of the Earth
PCR: Precambrian Research
PEPI: Physics of the Earth and Planetary Interiors
Phil. Trans. Roy. Soc.: Philosophical Transactions of the Royal Society, London
PLC: Proceedings of the Lunar and Planetary Science Conference

Rev. Brasil Geosc.: Revista Brasileira de Geosciencias
RGSP: Reviews of Geophysics and Space Physics

SEPM: Society of Economic Palaeontologists and Mineralogists

Trans. R. Soc. Can.: Transactions of the Royal Society of Canada

USGS: United States Geological Survey

2 Chondritic rare earth element normalizing factors

	Chondritic rare earth element normalizing factors*	Calibration data for USGS standard rock BCR-1
La	0.367	24.2
Ce	0.957	53.7
Pr	0.137	6.50
Nd	0.711	28.5
Sm	0.231	6.70
Eu	0.087	1.95
Gd	0.306	6.55
Tb	0.058	1.08
Dy	0.381	6.39
Ho	0.0851	1.33
Er	0.249	3.70
Tm	0.0356	0.51
Yb	0.248	3.48
Lu	0.0381	0.55
Y	2.1	34.0

* Values from Evensen, M.N. et al. (1978) GCA, **42**, 1203, for type I carbonaceous chondrites (volatile-free: 1.5×original data).

The subscript 'N' in the text indicates chondrite normalized values.

The REE data used here have a better cosmochemical rationale than the 'chondrite' data sometimes employed. The CI values represent our best approach to primitive solar nebular values. REE data normalized to these values thus display the amount of fractionation relative to primitive abundances. The CI data expressed on a volatile-free basis have absolute values only a little higher than those of 'ordinary chondrites' such as Leedey, commonly used for normalizing REE data.

The europium anomaly

The ratio Eu/Eu* is a measure of the depletion or enrichment of europium relative to the neighbouring REE, samarium and gadolinium. Eu is the measured elemental abundance in the sample; Eu* is the theoretical concentration for no Eu anomaly, and is calculated by assuming a smooth REE pattern in the region Sm–Eu–Gd (i.e. $Eu/Eu^* = Eu_N/[(Sm_N)(Gd_N)]^{1/2}$). Values of Eu/Eu* less than 0.95 indicate depletion, and values of greater than 1.05, enrichment of europium relative to the neighouring REE.

Analytical calibration

The analytical data for the REE reported from our laboratory have been obtained by spark source mass spectrometry using the methods of Taylor & Gorton (see Taylor, S.R. & Gorton, M.P. (1977) *GCA*, **41**, 1375). The method is calibrated against the set of values reported in column 2 for USGS Columbia River Basalt Standard BCR-1. The data listed represent our assessment of the 'true' REE abundances in BCR-1.

3 Ionic radii for cations in Angstrom units

	Radius Å	CN		Radius Å	CN		Radius Å	CN
Cs^+	1.88	12	Ho^{3+}	1.015	8	Sn^{4+}	0.68	6
Rb^+	1.72	12	Er^{3+}	1.004	8	Nb^{5+}	0.64	6
Tl^+	1.70	12	Tm^{3+}	0.994	8	Ti^{4+}	0.61	6
K^+	1.64	12	Yb^{3+}	0.985	8	Mo^{6+}	0.59	6
Ba^{2+}	1.61	12	Lu^{3+}	0.977	8	W^{6+}	0.60	6
			Y^{3+}	1.019	8			
Pb^{2+}	1.29	8				Cr^{3+}	0.62	6
Sr^{2+}	1.26	8	Ge^{4+}	0.39	4	V^{3+}	0.64	6
Eu^{2+}	1.25	8	Si^{4+}	0.26	4	Fe^{3+}	0.65	6
Na^+	1.18	8	P^{5+}	0.17	4	Sc^{3+}	0.75	6
Ca^{2+}	1.12	8	Be^{2+}	0.27	4	Ti^{4+}	0.61	6
			B^{3+}	0.27	6	Ni^{2+}	0.69	6
La^{3+}	1.160	8				Co^{2+}	0.75	6
Ce^{3+}	1.143	8	Th^{4+}	1.05	8	Cu^{2+}	0.73	6
Pr^{3+}	1.126	8	U^{4+}	1.00	8	Fe^{2+}	0.78	6
Nd^{3+}	1.109	8	Ce^{4+}	0.97	8	Mn^{2+}	0.83	6
Sm^{3+}	1.079	8	U^{6+}	0.86	8	Zn^{2+}	0.74	6
Eu^{3+}	1.066	8	Zr^{4+}	0.84	8	Mg^{2+}	0.72	6
Gd^{3+}	1.053	8	Hf^{4+}	0.83	8	Li^+	0.76	6
Tb^{3+}	1.040	8				Ga^{3+}	0.62	6
Dy^{3+}	1.027	8	Nb^{3+}	0.72	6	Al^{3+}	0.54	6

Data from Shannon, R.D. (1976) *Acta Cryst.*, **32**, 751.
CN = co-ordination number.

For $O^{2-} = 1.26$ Å, 4-fold co-ordination is favoured for cation radii less than 0.52 Å, 6-fold for the range 0.52–0.92 Å, 8-fold for the range 0.92–1.26 Å and 12-fold for the cation radii greater than 1.26 Å.

4   Isotopic notation

It is common for initial ^{143}Nd/^{144}Nd ratios to be reported as deviations in parts per ten thousand (ε_{Nd}) from an undepleted (chondritic) reference mantle reservoir (CHUR):

$$\varepsilon_{Nd}(T) = \left[\frac{(^{143}Nd/^{144}Nd)^T_{INITIAL}}{(^{143}Nd/^{144}Nd)^T_{CHUR}} - 1\right] \times 10^4$$

where:

$$(^{143}Nd/^{144}Nd)^T_{CHUR} = (^{143}Nd/^{144}Nd)^0_{CHUR} - (^{147}Sm/^{144}Nd)^0_{CHUR}(e^{\lambda_{Sm}T} - 1)$$

and present-day values are given as:

$$(^{143}Nd/^{144}Nd)^0_{CHUR} = 0.511836$$

$$(^{147}Sm/^{144}Nd)^0_{CHUR} = 0.1967$$

$$\lambda_{Sm} = 6.54 \times 10^{-12} \, a^{-1}$$

An analogous notation may be used for Sr-isotopes (ε_{Sr}), measuring deviations from an undepleted (*not* chrondritic!) reference mantle reservoir (UR). Present-day values are given as:

$$(^{87}Sr/^{86}Sr)^0_{UR} = 0.7047$$

$$(^{87}Rb/^{86}Sr)^0_{UR} = 0.085$$

$$\lambda_{Rb} = 1.42 \times 10^{-11} \, a^{-1}$$

Model ages may be calculated for estimates for the time of derivation from the undepleted mantle reservoirs:

$$T^{Nd}_{CHUR} = \frac{1}{\lambda_{Sm}} \ln\left[1 + \frac{(^{143}Nd/^{144}Nd)_{MEASURED} - (^{143}Nd/^{144}Nd)^0_{CHUR}}{(^{147}Sm/^{144}Nd)_{MEASURED} - (^{147}Sm/^{144}Nd)^0_{CHUR}}\right]$$

$$T^{Sr}_{UR} = \frac{1}{\lambda_{Rb}} \ln\left[1 + \frac{(^{86}Sr/^{86}Sr)_{MEASURED} - (^{87}Sr/^{86}Sr)^0_{UR}}{(^{87}Rb/^{86}Sr)_{MEASURED} - (^{87}Rb/^{86}Sr)^0_{UR}}\right]$$

Alternatively, model ages may be calculated for estimates of the time of derivation from a depleted mantle reservoir (DM). Such a model assumes that the mantle has progressively evolved to the present-day average value ($\varepsilon_{Nd} = +10$; $\varepsilon_{Sr} = -30$) over some time (τ). For this book, we have used depleted mantle parameters with $\tau = 3.8$ Ae and the model ages may be given as:

$$T^{Nd}_{DM} = \frac{1}{\lambda_{Sm}} \ln\left[1 + \frac{(^{143}Nd/^{144}Nd)_{MEASURED} - (^{143}Nd/^{144}Nd)^0_{DM}}{(^{147}Sm/^{144}Nd)_{MEASURED} - (^{147}Sm/^{144}Nd)^0_{DM}}\right]$$

where $(^{143}Nd/^{144}Nd)^0_{DM} = 0.51235$ and $(^{147}Sm/^{144}Nd)^0_{DM} = 0.217$ and

$$T_{DM}^{Sr} = \frac{1}{\lambda_{Rb}} \ln\left[1 + \frac{(^{87}Sr/^{86}Sr)_{MEASURED} - (^{87}Sr/^{86}Sr)_{DM}^0}{(^{87}Rb/^{86}Sr)_{MEASURED} - (^{87}Rb/^{86}Sr)_{DM}^0}\right]$$

where $(^{87}Sr/^{86}Sr)_{DM}^0 = 0.7026$ and $(^{87}Rb/^{86}Sr)_{DM}^0 = 0.046$.

The present-day Nd and Sr isotopic ratios adopted for depleted and undepleted mantle reservoirs may vary slightly among various laboratories but the overall effect on calculating model ages is not great.

Author Index

Addy, S.K. 141
Ahern, J.L. 295
Ahrens, T.J. 275
Alcock, F.J. 191, 207
Allegre, C.J. 116, 231, 251, 255
Allen, A.R. 80, 94
Allen, J.R.L. 141
Amin, M.A. 141
Anders, E. 256, 258, 274, 275, 277
Anderson, D.L. 263, 264, 265, 276
Anderson, R.N. 277
Anhaeusser, C.R. 144, 155, 186, 187, 208, 231, 254
Archibald, N.J. 188
Arculus, R.J. 71, 83, 95, 268, 276
Armstrong, N.V. 208
Armstrong, R.L. 234, 243, 251
Arndt, N.T. 186, 189
Arth, J.G. 142, 187, 188, 190, 230
Arrhenius, G.O.S. 53
Ashwal, L.D. 232, 295
Aston, S.R. 16, 53
Ayres, L.D. 142, 186, 231

Bacon, F. xi, xiv
Bagnold, R.A. 53
Balashov, Y.A. 54
Bally, A.W. 252
Barager, W.R.A. 186, 187, 188
Barker, F. 186, 188, 230, 231
Barsukov, V.L. 295
Barth, T.F.W. 22, 53
Bassett, W. 276
Bastron, H. 55
Basu, A. 54
Basu, A.R. 254
Bates, R.L. xiv
Bavinton, O.A. 115, 189, 190
Bell, K. 255
Bence, A.E. 55
Ben Othman, D. 80, 94, 231, 255
Bernard-Griffiths, J. 94
Bettenay, L.F. 231
Bhatia, M.R. 141
Bickle, M.J. 186, 189, 230, 254
Birch, F. 55, 58, 71
Blake, T.S. 203, 208
Blatt, H. 140, 142
Bowen, N.L. 71, 93
Bowes, D.R. 52
Boak, J.L. 190

Boles, J.R. 141
Boltwood, B. 54
Brenninkmeyer, B.M. 53
Bridgwater, D. 190
Broecker, W. 16, 22, 53
Brown, G.C. 94, 230, 231, 243, 254
Brown, L. 276, 277
Brunfelt, A.O. 54
Burke, K. 252
Button, A. 208

Calvert, S.E. 24, 53
Cameron, E.N. 115, 190
Campbell, I.H. 8, 189, 218, 231, 296
Card, K.D. 94, 207
Carey, S.W. 295
Carlson, R.W. 187, 254
Carr, M.H. 295
Cerny, P. 231
Chapman, D.S. 58, 71
Chappell, B.W. 95, 141, 232, 255
Chaudhuri, S. 55
Chauvel, C. 251
Chester, R. 16, 53
Chesworth, W. 52
Chou, C-L. 274, 276
Christensen, N.I. 8
Church, S.E. 276
Clarke, F. 26, 29, 54
Clayton, R.N. 274, 277
Cloud, P. 207, 296
Cobbing, E.J. 232
Cogley, J.G. 4, 8
Collerson, K.D. 188, 189, 218, 231, 254
Compston, W. 187, 188, 208, 254, 255
Condie, K.C. 8, 140, 141, 142, 155, 186, 187, 188, 189, 190, 231, 232, 252, 254
Constable, S. 95
Cook, F.A. 239, 253
Cook, T.D. 252
Crocket, J.H. 16, 53
Crook, K.A.W. 122, 140, 188
Cullers, R.L. 54, 55
Curtis, C.D. 53
Curtis, D. 258, 275
Cuvier, M. 140

Dacey, M.F. 115, 253
Daly, R.A. 96, 114
Danchin, R.V. 165, 189

Davies, G.F. 252
DeLaeter, J.R. 254
Demaiffe, D. 232
Dennis, G.T. 251
De Paolo, D.J. 116, 221, 232, 251, 254, 255
Dewey, J.F. 243, 251
De Wit, M.J. 186
Dickinson, B.B. 189
Dickinson, W.R. 122, 140, 141
Dietvorst, P. 94
Dimroth, E. 188
Doe, B.R. 251
Donaldson, J.A. 142
Donn, W. L. 251, 295
Dostal, J. 94
Dott, R.H. 140
Douglas, I. 253
Drake, M.J. 276
Dreibus, G. 276
Drummond, B.J. 94, 252
Drury, S.A. 94, 190
Duchesne, J.C. 232
Duddy, I. R. 55
Duncan, P.M. 95
Dupuy, C. 82, 95
Dymek, R.F. 190
Dypvik, H. 54

Eade, K.E. 11, 47, 48, 52
Easton, R.M. 207
Ebihara, M. 256, 258, 275
Edwards, M.B. 55
Egyed, L. 235, 252
Elderfield, H. 55
Embleton, B.J.J. 253
Emslie, R.F. 232
Engel, A.E.J. 98, 115
England, P.C. 116
Eriksson, K.A. 142, 152, 186, 187, 189
Ermanovics, I.F. 188
Esposito, L.W. 295
Evensen, M.N. 275, 298
Ewart, A. 71

Fahrig, W.F. 11, 47, 48, 52
Fairbairn, H.W. 94, 207
Fanale, F.P. 253
Finlayson, D.M. 8, 94
Flood, R.H. 232
Floyd, P.A. 41, 55
Fountain, D.M. 8, 75, 94
Fowler, C.M.R. 189
Frarey, M.J. 206, 207
Frey, F.A. 55
Friedman, G.M. 53
Frost, C.D. 116
Froude, D.O. 254, 295
Fryer, B.J. 189, 190, 231, 254
Fyfe, W.S. 75, 94, 188, 211, 230, 231, 234, 243, 251, 253

Garrels, R.M. 8, 52, 97, 114, 115, 190, 253
Garrett, D.E. 114

Gee, R.D. 188
Gehl, M.A. 115
Gemuts, I. 190
Gibb, R.A. 52
Gibbs, A.K. 208
Gibbs, R.J. 53
Gill, J. 60, 71, 142, 187, 276
Gill, R.C.O. 188
Gilluly, J. 231
Gilvarry, J.J. 251, 295
Glikson, A.Y. 186, 187, 189, 218, 231, 253
Glover, J.E. 186
Goldberg, E.D. 53
Goldschmidt, V.M. 11, 29, 52, 54, 64, 71
Goldstein, S.L. 116
Goodwin, A.M. 186, 190
Gorman, B.E. 186
Gorton, M.P. 299
Graham, S.A. 141
Grambling, J.A. 189
Greaves, M.J. 55
Green, D.H. 188
Gregor, B. 105, 115, 253
Gregory, R.T. 274, 277
Grieve, R.A.F. 296
Griffiths, J.C. 141
Gromet, P. 55
Grout, F.F. 142
Groves, D.I. 186
Grubb, P.L.C. 208

Hajash, A. 55
Hallam, A. 237, 252
Hallberg, J.A. 187, 230
Hambrey, M.J. 52
Hamilton, P.J. 79, 94, 116, 187, 189, 231, 251, 254
Hamilton, W. xi, xiv
Hanson, G.N. 142, 187, 190, 230, 231, 232
Hargraves, R.B. 188, 210, 230, 251
Harland, W.B. 52, 207
Harris, N.B.W. 189
Harrison, E.R. 251
Hart, S.R. 41, 273, 277
Hartmann, W.K. 295
Haskin, L.A. 30, 54, 101, 115, 141, 190
Haskin, M.A. 115
Hawkesworth, C.J. 95, 116, 187
Hawkins, J.W. 140
Head, J. 8
Hegner, E. 208
Heier, K.S. 54, 76, 80, 94
Hein, S.M. 251
Heinrichs, H. 16, 48, 52, 68, 72
Heller, F. 55
Hellman, P.L. 41, 55
Henderson, J.B. 142, 186, 188, 190
Henderson, P. 41, 55
Hennessy, J. 231
Hellings, R.W. 296
Helz, G.R. 54
Hertogen, J. 274, 277
Herz, N. 232
Hickman, A.H. 186, 187, 254

303

Hildreth, W. 208
Hirst, D.M. 53
Hoffman, A.W. 274, 276
Hoffman, P.K. 207
Holeman, J.N. 253
Holland, H.D. 53
Holland, J.B. 71
Holland, J.G. 52, 64, 71
Hower, J. 53
Huckenholz, H.G. 141
Hunter, D.R. 187, 208, 230, 295
Hurley, P.M. 8, 234, 243, 251
Hussain, S.M. 142
Hyde, R.S. 142

Ito, E. 274, 277

Jackson, G.D. 142
Jackson, I. 95
Jackson, J.A. xiv
Jacobsen, S.B. 95, 251, 267, 276
Jagoutz, E. 275, 296
Jahn, B.-M. 94, 186, 187, 230, 254
Jakeš, P. 190
James, H.L. 207
Jameson, R. 140
Janardhan, A.S. 94
Jansen, S.L. 52, 106, 111, 115, 208, 234, 242, 243, 251
Jarvis, G.T. 189, 218, 231
Jarvis, J.C. 54
Jaupart, C. 55
Jenner, G.A. 115, 142, 187, 189
Jezek, P. 276
Jochum, K.P. 277
Johnson, R.W. 268, 276
Jones, A.G. 95
Jordan, T.H. 8
Judson, S. 253

Karig, D.E. 242, 243, 254
Kay, R.W. 94, 242, 243, 254, 276
Kay, S.M. 94
Kerrich, R. 190
Kerridge, J.F. 276
Kidd, W.S.F. 252
Knoll, A.H. 252
Kossinna, A. 8
Krauskopf, K.B. 26, 29, 54
Kronberg, B.I. 52
Kroner, A. 186, 188, 255
Kuenen, P.H. 253
Kuo, H.Y. 16, 53

La Berge, G.L. 208
Laird, M.G. 141
Lambert, I.B. 253
Lambert, R.St.J. 52, 64
Langmuir, C.H. 190
Land, L.S. 142
Lee, D.E. 55

Le Maitre, R.W. 71
Lerman, A. 115, 253
Leyreloup, A. 95
Li Y.-H. 16, 22, 53
Lisitsyn, A.P. 23, 53, 253
Liu, L. 263, 275, 276
Liu, T.S. 55
Logan, W.E. 191, 206
Lowe, D.R. 141, 142, 151, 186
Ludden, J.N. 273, 277
Lyell, C. 43, 55

MacDonald, J.A. 208
MacGeehan, P.J. 146, 187
Mackenzie, F.T. 8, 52, 97, 114, 115, 253
MacLean, W.H. 146, 187
Magaritz, M. 276
Mahood, G. 208
Manghani, M.H. 95
Martin, J.M. 53, 54
Mason, B. 188
Mawe, J. 140
Maynard, J.B. 138, 140
McCulloch, M.T. 87, 95, 116, 187, 218, 220, 231, 232, 247, 254, 255, 276, 277
McElhinny, M.W. 232, 295
McGlynn, J.C. 188
McGregor, V.R. 71, 188, 190, 231
McLennan, S.M. xiv, 16, 23, 46, 47, 48, 53, 54, 55, 64, 68, 72, 98, 104, 115, 140, 142, 172, 188, 189, 190, 198, 201, 207, 208, 234, 243, 252, 253, 256, 258, 268, 275, 276
McWilliams, M.O. 232
Medawar, P. xiv
Meijer, A. 276
Menzies, M. 55
Meybeck, M. 55, 253
Michard, A. 41, 55
Migdisov, A.A. 52, 100, 114
Miliken, K.L. 142
Miller, R.G. 116
Milliman, J.D. 53, 253
Minami, E. 30, 54, 115
Monod, J. 8
Moorbath, S. 231, 243, 246, 247, 248, 251, 254, 255
Moore, J.C. 253
Moore, J.M. 295
Moores, E.M. 277
Morgan, J.W. 276
Morgan, P. 8, 49, 55, 116
Morrison, D. 275, 295
Morse, R.A. 232
Muecke, G.K. 78, 79, 94
Mueller, P.A. 190
Mussett, A.E. 230, 254

Nance, W.B. 30, 54, 55, 115, 141, 189, 190
Nanz, R.H. 97, 114
Naqvi, S.M. 142, 186, 190
Nathan, S. 53, 115, 141
Needham, R.S. 208

Nekut, A. 95
Nesbitt, H.W. 51, 52, 55, 80, 94
Nesbitt, R.W. 55, 187, 190
Neumann, H. 232
Newman, S. 277
Newsom, H. 276
Newton, R.C. 94, 189, 253
Nicolaysen, L.O. 55
Nieuwland, D.A. 188
Nisbet, E.G. 186, 189
Nutman, A.P. 188, 190

O'Day, P. 208
Ohtani, E. 296
Ojakangas, R.W. 142, 187
Olhoeft, G.R. 90, 95
Oliver, J. 8
Ondrick, C.W. 141
O'Neill, J.R. 232
O'Nions, R.K. 111, 116, 187, 190, 230, 231, 251, 254, 267, 276
Oversby, V.M. 189

Packham, G.H. 140
Padgham, W.A. 188
Padovani, E. 95
Page, R.W. 208, 255
Pakiser, L.C. 64, 70, 71, 86, 95
Pankhurst, R.J. 190, 230, 231
Parekh, P.P. 54
Park, R.G. 188
Parkhomenko, E. 95
Parmentier, E.M. 295
Patchett, J. 251, 254, 276
Peng, T.-H. 16, 22, 53
Percival, J.A. 94, 95
Perfit, M.R. 276
Perkins, D. 189
Peterman, Z.E. 188, 230
Pettijohn, F.J. 53, 140, 142, 187
Pettingill, H.S. 254
Philpotts, J.A. 55
Phinney, W.C. 232
Pidgeon, R.T. 189, 200, 208, 254
Piper, J.D.A. 232
Pitcher, W. 232
Poirer, J.P. 295
Poldervaart, A. 3, 8, 52
Pollack, H.N. 5, 8, 55, 71, 72, 112, 116, 230, 277
Pollard, W.F. 8
Potter, P.E. 253
Price, N.P. 53
Pride, C. 78, 79, 94

Rand, J.R. 8, 234, 243, 251
Rasmussen, S.E. 208
Reading, H.G. 140, 253
Reimer, T.O. 115, 142
Reymer, A. 234, 243, 251
Ricci, C.A. 141
Richardson, S.W. 116

Ringwood, A.E. 267, 275
Robb, L.J. 208, 231, 254
Robinson, R. 64, 70, 71, 86, 95
Rogers, N.W. 95
Ronov, A.B. 23, 52, 54, 64, 71, 97, 98, 100, 114, 115
Roscoe, S.M. 206, 207
Rousseau, D. 116
Rubey, W.W. 71
Russell, C.T. 295
Rutherford, E. 54

Sabatini, G. 141
Sadler, D.M. 252
Saggerson, E.P. 189
Sakai, T. 94
Salisbury, M.H. 75, 94
Sanders, J.E. 53
Scarf, F.L. 295
Schmid, R. 80, 94
Schnetzler, C.C. 55
Schopf, T.J.M. 253
Schubert, G. 234, 243, 251, 254, 276
Schwab, F.L. 98, 115, 140, 253
Sclater, J.G. 230, 253, 277
Shakespeare, W. xiv, 276
Shannon, R.D. 266, 276, 299
Shaw, D.M. 11, 23, 47, 48, 51, 52, 53, 295
Shaw, H. 55
Shaw, S.E. 232
Shih, C.-Y. 295
Sighinolfi, G.P. 94
Sigleo, A.C. 54
Silver, L.T. 55
Simmons, E.C. 232
Sims, P.K. 207
Smith, I.E.M. 187
Smith, J.V. 188, 296
Smith, R.B. 93
Smithson, S.B. 8, 64, 70, 71, 72, 86, 95
Snansieng, S. 141
Sorby, H.C. xiv
Soller, D.R. 8
Spears, D.A. 141
Spencer, L.J. 55
Spohn, T. 276
Spooner, C.M. 94
Spudis, P. 295
Squires, S.W. 295
Staudigel, H. 41, 273, 277
Sterry Hunt, T. 206
Stockwell, C.H. 191, 207
Strakhov, N.M. 252
Strobell, M.E. 295
Suczek, C.A. 141
Sun, S.S. 55, 186, 187, 190, 254, 276
Surkov, Yu.A. 295
Sutton, J. 188

Tankard, A.J. 186
Tarling, D.H. 186
Tarney, J. 54, 64, 69, 70, 71, 72, 78, 79, 85, 94, 188, 189

305

Tatsumoto, M. 274, 277
Taylor, H.P. 231, 274, 277
Taylor, P.N. 190, 231, 251
Taylor, S.R. xiv, 8, 16, 23, 30, 46, 48, 52, 53, 54, 55, 56, 59, 64, 68, 71, 72, 104, 111, 115, 116, 141, 155, 187, 188, 189, 190, 198, 208, 230, 232, 234, 243, 252, 253, 256, 258, 268, 275, 276, 278, 295, 296, 299
Termier, G. 252
Termier, H. 252
Theron, A. 190
Thompson, G. 273, 277
Thoni, M. 116
Thorpe, R.S. 60, 71, 276
Tipper, J.C. 252
Touret, J. 94
Treimon, A.H. 295
Trendall, A.F. 189, 200, 206, 207, 208
Tryon, P. 296
Turcotte, D.L. 239, 252, 254
Turner, C.C. 142
Turner, D.R. 52, 53
Turner, F.J. 94
Tuttle, O.F. 71, 93

Ullman, W. 16, 22
Upton, B.C.J. 95
Uyeda, S. 253

Vail, P.R. 252
Valley, J.W. 232
Valloni, R. 140
Van Schmus, R. 207
Van Schmus, W.R. 207
Veizer, J. 52, 97, 98, 106, 107, 111, 114, 115, 208, 234, 242, 243, 251, 252, 253, 254, 255
Viljoen, M.J. 231

Viljoen, M.P. 231
Vinogradov, A.P. 64, 71, 252
Vitorello, I. 55, 72, 116
Von Richthofen, F. 43, 55

Walker, C.T. 251
Walker, R.G. 142, 187
Wasserburg, G.J. 55, 95, 116, 220, 231, 247, 251, 254, 267, 276, 277
Watkins, J.C. 253
Watson, J.V. 113, 116, 188, 189
Weaver, B.L. 64, 69, 70, 71, 72, 78, 79, 85, 94
Wedepohl, K.H. 8, 50, 51, 55
Wetherill, G.W. 275
Whetten, J.T. 140
White, A.J.R. 95, 232
White, W.M. 274, 276
Whitfield, M. 15, 52, 53
Whitford, D.J. 71, 276
Wiebe, R.A. 232
Wildeman, T.R. 115, 141, 142, 189, 190
Wilson, I.H. 230
Wilson, J.T. xiv, 71
Windley, B.F. 54, 94, 141, 186, 188, 189, 243, 251, 252
Winkler, H.G.F. 77, 94
Wise, D.U. 236, 237, 252
Wolf, R. 274, 277
Wood, B.J. 80, 94
Wood, D.A. 55
Wright, P.L. 53
Wyborn, L.A.I. 255
Wyllie, P.J. 71, 230

Yaroshevsky, A.A. 64, 71
Young, G.M. 51, 52, 188, 193, 207

Zartman, R.E. 251

Subject Index

Accretion tectonics 6, 7
Accretion, planetary 260
Akilia metasediments,
  composition 160–1
  geochemistry 171–3
  REE in 173
Alkali feldspar, REE in 35, 36
Allanite, REE in 37, 38
Amîtsoq gneiss 171, 244,
  Sr isotopes 244
Andesites, Archean 147,
  composition 60
  REE in 61, 148, 211
Andesite model 62,
  and bulk crust 68
  Cr in 62, 63, 216
  Ni in 62, 63, 216
Anorthositic crust,
  Earth 293
  Moon 279 et seq.
Anorthosites 224–8,
  Archean 225–6
  Proterozoic 226–8
Anoxic sediments, composition 23, 24
Apatite, REE in 37, 38
Archean anorthosites 225, 226,
  composition 226
Archean biomodal suite 65, 66,
  REE in 66
Archean crust,
  bulk composition 182–4
  differentiation 184
  thickness 156
Archean felsic rocks, composition 181
Archean greywackes,
  comparison with Phanerozoic 135–8
  composition 131–5
  crustal samples 139
  early versus late 138
  petrography 130
  REE in 134, 135
Archean high grade terrains 153 et seq.,
  composition 160–1
Archean mafic volcanics, composition 181
Archean–Proterozoic boundary 213–16,
  definition 191–2
Archean volcanics,
  Nd isotopes 150
  Sr isotopes 150
Archean upper crust, composition 180–3

Barberton Mountain Land 151–3, 164, 165
Basaltic andesite,
  composition 60
  REE in 61
BCR-1, REE in 298
Be isotopes,
  and crustal recycling 271–3
  in island-arc rocks 271–3
  in sediments 271–3
Bimodal igneous suite 65, 66, 176, 180, 181, 209, 210,
  REE in 66
Biotite, REE in 35, 36
Bournac xenoliths 81, 82, 84
Bulk crust,
  Archean 182–4
  composition 64, 65, 66–8

Calc-alkaline volcanism 59, 62, 63, 68, 209, 216
Callisto 287, 288
Canadian Shield,
  composition 10
  Nd model ages 247
  Sr model ages 247
Cannibalism, sedimentary 107
Carbonaceous chondrites,
  composition 256, 257
  REE in 298
Carbonates,
  composition 15
  Sr isotopic evolution 246
Chondrites,
  normalizing factors 298
  REE in 298
Chemical index of weathering 13, 14
CIA 13, 14
Clay minerals, REE in 37
Clay, pelagic, composition 15
COCORP 6, 73, 79
Colorado Front Range granites 249
Conrad discontinuity 74
Continents,
  area 3, 4
  elevation 3
  thickness 3, 4, 49, 156, 237, 238
Core,
  composition 259
  formation 259

307

Cr,
  abundance 216
  Andesite model 62, 63
  Archean sediments 175
Cratering rate, terrestrial 291–3
Cratonic terrains, Archean 154
Crust,
  bulk composition 64, 65, 66–8
  density 4, 9
  extraction from mantle 266
  growth model 244–50
  impact origin 278
  sampling by sediments 9, 11, 24, 26
  structure 5–7
  variation in composition 96 et seq.
Crustal area, variation with time 238, 239
Crustal growth,
  Early Archean 244–6
  Late Archean 246–8
  Post-Archean 248–50
Crustal recycling,
  Be isotopes 271, 272
  geochemical evidence 268
  Hf isotopes 270
  isotopic evidence 269–71
  Nd isotopes 270
  Sr isotopes 270
  Th/U 272

Dacite, composition 60
Deep sea muds,
  composition 120
  REE in 120
Deep sea sands 117 et seq.
Diagenesis 23,
  effect on REE 41
Differentiation,
  Archean crust 184
  terrestrial 266, 267

Early Archean granites 218
Early Archean greenstones,
  comparison with late 179, 180
  composition 158–60
Early crustal growth 244–6
Earth,
  core composition 260
  core formation 260
  origin of 256 et seq.
  primitive mantle 260 et seq.
Electrical conductivity, crustal 88 et seq.
Element,
  mobility 23
  residence time 12, 15, 17, 22, 23
Epidote, REE in 37, 38
Erosion 16 et seq.
Europa 288
European shale composition, REE in 30, 31
Eu/Eu*, secular variations 101
Europium,
  anomaly 42, 213, 299
  geochemistry 213

Expanding Earth hypothesis, planetary constraints 290, 291

Felsic igneous rocks,
  Archean 149, 181
  composition 181
  REE in 149
Fig Tree Group 134, 135
  composition 158, 159
  REE in 163
  stratigraphy 164, 165
Freeboard, continental 235 et seq.

Galilean satellites 287–90,
  Callisto 287, 288,
  Europa 288
  Ganymede 287, 288
  Io 288, 289
Ganymede 287, 288
  expansion 287, 288
  and expanding earth hypothesis 291
Garnet,
  REE in 37, 38
  in mantle 65, 218
Geotherms,
  continental 5, 214
  oceanic 5
Glacial erosion 11
Godthåb region 171–3
grain size fractions,
  composition 21
  REE in 39
Granite,
  composition 60, 218–23
  Early Archean 218
  formation of 214, 217
  I-type 222–4
  isotopic character 224, 225
  Late Archean 218, 219
  Phanerozoic 221
  Proterozoic 220
  S-type 222–4
Granodiorite, composition 60
Granulites 75 et seq.,
  and crustal thickness 238
  Sm–Nd isotopes 86–8
  Rb–Sr isotopes 86–8
Granulite facies 75 et seq.,
  element mobility in 76–7
  Rb 76–7
  U 76–7
Greenstone belts 144 et seq.,
  basement of 152
  composition 158–61
  comparison 179, 180
  Early Archean 158–60
  Early Proterozoic 205, 206
  Late Archean 165–71
  models 155 et seq.
  provenance of sediments 157
  sedimentary rocks 151 et seq.
  volcanic rocks 145, 146
Grenville province 248

Greywacke,
   Archean 129 *et seq.*
   classification 122, 123
   comparisons 135–8
   petrography 123, 124, 130
   Phanerozoic 121 *et seq.*
   REE in 127–9, 135
Growth, crustal,
   Early Archean 244–6
   estimates 234
   Late Archean 246–8
   Phanerozoic 243
   Post-Archean 248–50
Guyana greenstone belts 205, 206

Hamersley basin 199–202,
   composition 201, 202
   REE in 202
Heat flow 5, 49,
   Archean terrains 58, 59, 183, 184
   crustal component 59
   geological age 113
   mantle component 59
   Mesozoic 58
   Proterozoic 58
   provinces 112
   tectonothermal component 59
   variation with age 58, 113
   Weaver–Tarney model 79
Heat production, upper crust 50
Hf isotopes, crustal recycling 270
High-grade Archean terrains 153 *et seq.*,
   composition 160, 161
Hornblende, REE in 35, 36
Huronian Supergroup 192–6,
   composition 192, 193
   REE in 195
   stratigraphy 192, 193

Ice, polymorphic change 287
Impact models for continental growth 234
Io 288, 289
Ionic radii 299, 300
Island-arc volcanics (see andesites),
   Be isotopes 271–3
   REE in 61
Isotopic constraints on crustal recycling 244 *et seq.*
Isotopic notation: Nd, Sr 300, 301
Isua metasediments 171–3,
   composition 160, 161
   REE in 173
Isukasia region 171–3
I-type granites 221–4,
   characteristics 224
   composition 223
   Nd–Sr isotopes 224
   REE in 223
Ivrea–Verbano zone 80

Jupiter, Galilean satellites 258, 287–90

Kalgoorlie metasediments 134, 135, 166,
   composition 159, 160
Kambalda metasediments 166, 167, 179,
   composition 159, 160
Kapuskasing structural zone 75
Knife Lake Group 134, 135, 171
Komatiites, REE in 147
KREEP 281
K/U ratios,
   in solar system 259

La/Sc, secular variation 104
La/Th correlation 47
La–Th–Sc correlation 174
Late Archean crustal growth 246–8
Late Archean granites 218–20
   REE in 219
Late Archean greenstone belts,
   comparison with early 179–80
   composition 159, 160
Lead isotopes,
   crustal recycling 269
Lesotho xenoliths 81, 83, 84, 87
Lewisian gneiss complex 77 *et seq.*,
   amphibolite facies 78
   composition 78
   granulite facies 78, 79
   Sm–Nd 86–7
   Rb–Sr 86–7
Loess 42–5,
   provenance 44
   REE in 44
   and crustal composition 43
Low grade terrains,
   Archean 143
Lower crust,
   composition 90–2
   REE in 91, 92
Lower mantle 261–3
Low velocity zone 7
LREE/HREE, secular variation 102
Lunar highland crust 279 *et seq.*,
   age 281, 282
   composition 280, 281
   and expanding Earth 291

Mafic volcanics,
   Archean composition 181
Magma ocean, lunar 280
Magnetite, conductivity 90
Major elements, secular trends 96 *et seq.*
Malene metasediments 171–3,
   composition 160, 161
   REE in 173
Mantle,
   composition 261, 264, 265
   crustal extraction 266
   lower 261–3
   Mg/Si 262, 263
   mineralogy 262
   primitive 260 *et seq.*
   REE in 261
   upper 261–3

Marda complex 211
Mare basalt 279, 282
Marine carbonates,
  composition 15
  Sr isotopic evolution 246
Marine sediments,
  Be isotopes 271, 273
Massif anorthosites 226–8,
  composition 226
  isotopic signature 227, 228
  origin 227, 228
  REE in 228
Mars,
  Argyre basin 285
  composition 284
  cratered terrain 284
  crust 284 et seq.
  and expanding Earth 291
  Tharsis plateau 284, 285
  volcanoes 284, 285
Mercury,
  core 282
  craters 283
  crust 282 et seq.
  and expanding Earth 290
  lobate scarps 283
Metamorphic facies 76
Metamorphism, effect on REE 41
Meteorite flux, terrestrial 291–3
Mineralogy,
  mantle 262
  upper crust 49–51
Minerals, REE in 35–7
Mixing models 176–9
Mobility, element 23,
  REE 38–42, 273, 274
Moho 3, 7, 75
Molecular clouds 256
Monazite, REE in 37, 38
Moodies Group,
  composition 158, 159
  REE in 163
  stratigraphy 164, 165
Moon 279 et seq.,
  and expanding Earth 291
  Orientale basin 292, 293
Mt. Narryer quartzite 245, 246, 294
Multi-ring basins 281,
  early Earth 293, 294
  Mercury 283
  Orientale 292, 293
Muscovite, REE in 35, 36

NASC, REE in 30, 31
Nd isotopes,
  Archean volcanics 150
  crustal recycling 270
Nd model ages,
  Canadian Shield 247
  loess 110, 111
  sediments 110, 111
  stratigraphic age 111

Ni,
  abundance 216
  Andesite model 62, 63
  Archean sediments 175
  Normalizing factors,
  chondritic REE 298
North Atlantic craton 153

Ocean water,
  chemistry 22
  residence times 15–17
Oceanic crust 272 et seq.,
  composition 274
  pelagic sediments 15, 16
  REE in basalt 273, 274
Olivine, REE in 35, 36
Orientale basin 293
Orthopyroxene, REE in 35, 36
Oxic sediments 23, 24

PAAS, REE in 30, 31
Partial melting, in crust 77
Pelagic carbonate, composition 15
Pelagic clay, composition 15
Pelagic sedimentation, global rate 241
Phanerozoic crustal growth 243
Phanerozoic granites 221–4,
  characteristics 222
  composition 223
  Nd–Sr isotopes 224
  REE in 223
Phanerozoic greywackes,
  comparison with Archean 135–8
  composition 125, 126
  petrography 124
  REE in 127–9
Pilbara Block 162, 163
  composition 158, 159
  REE in 163
  thickness 237, 238
Pine Creek Geosyncline 196–8,
  REE in 198
Plagioclase, REE in 35, 36
Planetesimals, sweep-up time 259
Plate tectonics,
  development of present regime 229
Pongola Supergroup 204,
  REE in 204
Post-Archean shales, REE in 32 et seq.
Primary crusts 278, 279
Primitive mantle, composition 261, 264, 265
Proterozoic anorthosites 226–8,
  composition 226
  isotopic signature 227, 228
  origin 227, 228
  REE in 228
Proterozoic granites 220, 221,
  REE in 220
Proterozoic, lower boundary 191, 192, 213–16
Provenance, of Archean sediments 157 et seq.
Pyrolite 261, 263

Ratio–ratio plots 177, 178
Recycling, crustal 239 et seq., 268 et seq.,
    Be isotopes 271–3
    geochemical evidence 268,
    Hf isotopes 270
    isotopic constraints 244 et seq., 269–71
Recycling, sedimentary 105 et seq., 203,
    Th/Sc constraint 109
REE (see table of contents),
    chondritic values 298
    in minerals 35–8
    mobility 23, 38–42, 273, 274
    secular variations 100 et seq.
Residence times, element 12, 15, 17, 22, 23
Resistivity, crustal 88 et seq.
Reynolds number 18
Rhyolite, composition 60
Rivers
    sediment composition 27, 33
    sediment load 240, 241
    water composition 15

Sampling, crustal by sediments 9, 11, 24, 26, 42–5
Sands, REE in 33, 34
Saturnian satellites 287–90
    Tethys 289, 290
Sauviat-sur-Vige 80, 81, 84
Scourian granulites 78, 79, 84
Sea-level changes, Phanerozoic 236, 237
Sea water,
    composition 15, 16
    effect on REE 273, 274
    partition coefficients 14–17
    residence times 14–17
    Sr isotopes 246
Secondary crusts 278, 279
Secular trends,
    crustal composition 96 et seq.
    Eu/Eu* 101
    La/Sc 104
    Th/Sc 103
    Th 104
    Th/U 100, 107
    U 105
Sediment,
    Be isotopes 271, 273
    global yield 240
    mass 3, 106
    subduction 241 et seq.
Sedimentation, human influence 240
Seismic velocities 5, 7
    lower crust 84
Settling velocity 19
Solar nebula 256, 257
    temperature 257
South Pass greenstone belt 134, 135, 170
    REE in 170
Southern Africa,
    Barberton Mountain Land 164, 165
    greywackes 130, 132, 134, 135
    Late Archean 204, 205
    Pongola Supergroup 204, 205
    Witwatersrand Supergroup 205

Sphene, REE in 37, 38
Steady state recycling 242
Stokes Law 18, 19
S-type granites 221–4,
    characteristics 222
    composition 223
    Nd–Sr isotopes 224
    REE in 223
Suspect terrains 7

Tectonic setting,
    modern sediments 118
    Phanerozoic greywackes 122
Tektites, as crustal samples 45
Terrestrial differentiation, extent 266, 267
Tethys 289, 290
Th, secular variation 104
Th/Sc,
    secular variation 103
    and recycling 109
Th/U,
    crustal recycling 272
    secular variations 100, 107
Thickness, crustal 3, 4, 49, 156, 237, 238
    Archean 156
    upper crust 49
    variation with age 237–9
Tholeiitic basalts,
    REE in Archean 147
Tonalites,
    composition 60
    source 211, 214
    REE in 210
Trondhjemites,
    composition 60
    Eu enrichment 79
    source 211, 214
    REE in 210
Tourmaline, REE in 37, 38
T Tauri solar winds 258
Turbidites 117

Upper crust, present,
    composition 46 et seq.
    thickness 49
Upper crust, Archean 180–3,
    composition 180 et seq.
Upper mantle 261–3

Venus 286, 287
    Aphrodite terra 286
    Ishtar terra 286
    surface composition 286

Water, in deep crust 89, 90
Water–rock partition coefficients 14, 15
Weathering 13 et seq.
    chemical index 13, 14
    CIA 13, 14
    effect on REE 39, 40

311

Weaver–Tarney composition 69, 79, 85
West Greenland,
  granites 212
  metasedimentary rocks 160, 161, 171–3
Whim Creek Group, REE in 170, 171
Witwatersrand Supergroup 205,
  REE in 205
Wopmay orogen 196

Xenoliths,
  Bournac 81, 82, 84
  Hoggar 81, 83
  Lesotho 81, 83, 84, 87, 88
  SE Australia 83

Yellowknife Supergroup 134, 135, 167, 168,
  composition 159, 160
  REE in 169
Yilgarn Block 165–7,
  composition 159, 160
  REE in 167
  thickness 237, 238

Zircon 294,
  REE in 36–8